Control Theory and Systems Biology

Control Theory and Systems Biology

edited by Pablo A. Iglesias and Brian P. Ingalls

The MIT Press
Cambridge, Massachusetts
London, England

MIT Press books may be purchased at special quantity discounts for business or sales promotional use. For information, please e-mail special_sales@mitpress.mit.edu or write to Special Sales Department, The MIT Press, 55 Hayward Street, Cambridge, MA 02142.

This book was set in Times New Roman on 3B2 by Asco Typesetters, Hong Kong. Printed and bound in the United States of America.

Library of Congress Cataloging-in-Publication Data

Control theory and systems biology / edited by Pablo A. Iglesias and Brian P. Ingalls.
p. cm.
Includes bibliographical references and index.
ISBN 978-0-262-01334-5 (hardcover : alk. paper) 1. Biological control systems. 2. Control theory.
I. Iglesias, Pablo A., 1964– II. Ingalls, Brian P., 1974–
[DNLM: 1. Biochemical Phenomena. 2. Cell Physiological Processes. 3. Feedback, Biochemical—physiology. 4. Models, Biological. 5. Systems Biology. QU 34 C764 2010]
QH508.C67 2010
571.7—dc22 2009007075

10 9 8 7 6 5 4 3 2 1

Contents

Preface

Issues of regulation and control are central to the study of biological and biochemical systems. The maintenance of cellular behaviors and the appropriate response to environmental signals can only be achieved by systems that are robust to certain perturbations and sensitive to others. Because these behaviors demand the use of feedback, it is natural to expect that the tools of feedback control theory, which were developed to analyze and design self-regulating systems, would be useful in the study of these biological mechanisms. Indeed, the application of control-theoretic tools to biochemical systems is currently an area of burgeoning interest.

The application of control theory to biological systems has a long history, dating back more than 60 years. The mathematician Norbert Wiener, through his pioneering work on cybernetics, saw a set of problems, common to both "the machine" and "living tissue," centered around questions of communication and control (Wiener, 1948). Subsequent work dealt with attempts to understand human physiology. For example, in their influential 1954 paper, Fred Grodins and colleagues discussed the response of the respiratory system to CO_2 inhalation as a feedback regulator, using electrical circuit analogs. Representative references include Bayliss (1966), Grodins (1963), Kalmus (1966), and Milhorn (1966).

The current spate of research activity differs from previous efforts in a number of respects. First, the focus is on cell physiology. Our understanding of the molecular mechanisms by which a cell operates has advanced dramatically over the last quarter century; we now have a solid characterization of the chemical basis for the cell's ability to regulate its behavior. Second, in recent years, biologists have developed experimental techniques that have opened up new opportunities for investigating cellular behavior. Traditional molecular biology techniques are restricted to measurements of only a few variables at a time. Using new "high-throughput" technologies, researchers can now take hundreds or even thousands of measurements simultaneously, allowing them to investigate system-wide behavior. Moreover, many of these techniques provide quantitative measurements of time-series data—crucial to an understanding of dynamics. These developments have enabled the development of accurate mathematical models of cellular mechanisms based on physicochemical laws and falsifiable by experimentation. It is these experimental developments that have ushered in systems biology as a "new" field devoted to systems-inspired analysis of cellular biology (Ideker et al., 2001; Kitano, 2001, 2002).

This type of interdisciplinary work requires knowledge of the results, tools and techniques of another discipline, as well as an understanding of the culture of an unfamiliar research community.

Our primary objective is to present control-theoretic research in an accessible manner. Although readers are presumed to have some familiarity with calculus and differential equations, they need have no specific knowledge of control theory. We present the necessary concepts from dynamical systems and control theory in chapter 1, which introduces ordinary differential equation–based modeling and should allow theoretically inclined life scientists to follow the rest of the volume. Because it covers standard material, readers from the mathematical sciences may wish to skip this introduction.

The next two chapters introduce alternative modeling frameworks. In chapter 2, Mustafa Khammash introduces stochastic modeling, used to address the probabilistic nature of processes. In chapter 3, Pablo Iglesias introduces partial differential equation–based modeling, used to address dynamic processes where spatial localization is significant.

Chapters 3 through 14 sample a wide variety of applications of control theory to molecular systems biology. In chapter 4, Simone Frey, Olaf Wolkenhauer, and Thomas Millat show how measures of dynamic behavior, central to control engineering, can be used to study cell signaling systems. The next chapters discuss modularity, a key concept in control, with Hana El-Samad demonstrating an analysis of the bacterial heat-shock response as a modular controller in chapter 5, and Domitilla Del Vecchio and Eduardo Sontag presenting a treatment of "retroactivity to interconnections," a feature that threatens our ability to apply modular thinking to cell biology, in chapter 6. David Angeli and Eduardo Sontag adopt a modular approach to dynamics in chapter 7 by applying graph-theoretic and stoichiometric ideas to interpret complex biochemical systems as compositions of dynamically simple building blocks. Further exploring network stoichiometry in chapter 8, Brian Ingalls discusses its implications for sensitivity analysis.

That feedback loops can be used to reduce sensitivity is a foundational concept in control engineering. Harold Black invented the negative feedback amplifier in 1927 as a means of decreasing the sensitivity of the telephone network to nonlinearities in the underlying components. In the 1980s, there was renewed interest in techniques for analyzing the sensitivity of systems and for designing systems that are insensitive or *robust* to uncertainties. Robustness, most often framed in the context of *homeostasis* (Cannon, 1932), is also a central concept in biology. Much of the recent interest in the robustness of biological systems by biologists and control engineers alike was spurred by the theoretical (Barkai and Leibler, 1997) and experimental (Alon et al., 1999) work of Stan Leibler and colleagues. Not surprisingly, the use of control-theoretic tools to study robustness and sensitivity is the topic of several chapters in this volume. In chapter 9, Jason Shoemaker, Peter Chang, Eric Kwei, Stephanie Taylor, and Frank Doyle discuss sensitivity analysis and present a number of techniques, some from robust control theory, for addressing robustness of cellular networks. Reporting on a number of biological case studies in chapter 10, Camilla Trané and Elling Jacobsen further expand on the use of robust control techniques. And in chapter 11, Jongrae Kim and Declan Bates take a robust control approach to the robustness of oscillatory behavior, using oscillations in the social amoeba *Dictyostelium* as an illustrative example.

The final three chapters concern themselves with network identification. In chapter 12, David Thorsley and Eric Klavins take up the analysis of stochastic models, demonstrating a powerful technique for model comparison and model calibration. In chapter 13, Jorge Gonçalves and Sean Warnick then discuss network reconstruction—the derivation of a system "wiring diagram" from input-output data. Finally, Dirk Fey, Rolf Findeisen, and Eric Bullinger consider the closely related task of calibrating model parameters from experimental data, presenting a novel observer-based design, in chapter 14.

We hope these chapters, which range from surveys of established material to presentation of current developments will provide readers with a useful overview of recent control-theoretic studies in systems biology, an appreciation of the biological insights such studies can provide, and perhaps even the inspiration to join us in the continuing task of unraveling the complexities of biological regulation and control.

We want to thank those without whose help this project would not have reached fruition. First and foremost, the contributors for their excellent chapters and for dealing with numerous requests for revisions in a timely manner. Second, Bob Prior at the MIT Press for his phenomenal encouragement. Bob has been a supporter of the systems biology community for more than a decade. It was at his urging that we undertook this work. And last, but most important, we thank our families—Elizabeth, Vicente, Miguel, Angie, Logan, Alexa, Sophia, and Ruby—for the patience they have shown. To them, we dedicate this book.

Control Theory and Systems Biology

1 A Primer on Control Engineering

Brian P. Ingalls and Pablo A. Iglesias

The field of control engineering grew out of the need to analyze and design regulatory mechanisms. To meet this need, a vast array of mathematical tools has been developed. Because of the conceptual similarities between engineering and biological regulatory mechanisms, it is not surprising that these tools are now being used to analyze biochemical and genetic networks. *Control Theory and Systems Biology* brings together some examples of this work. This chapter introduces background material that will be helpful in reading the chapters to follow.

1.1 System Modeling

The current volume addresses the dynamic behavior of biochemical and genetic networks. This time-varying behavior is often left implicit in the cartoon models of biological networks that are standard in the biological literature. Though such models are static, they typically attempt to describe interactions that evolve over time. For example, the activation of a transcription factor does not lead to an immediate increase in the target protein, but instead produces this effect through a sequence of events, each of which unfolds on a particular time scale. Our descriptions of these interactions indicate how the *rate* of a process, such as expression, degradation, phosphorylation, or cleavage, depends on the availability of an affector species, such as a transcription factor, metabolite, or enzyme.

1.2 Differential Equation Description of Biochemical Reaction Networks

The most successful way to describe quantitatively dynamic processes is with models based on differential equations. The simplest such models, consisting of ordinary differential equations (ODEs), can be applied to biochemical networks only under two key assumptions, however.

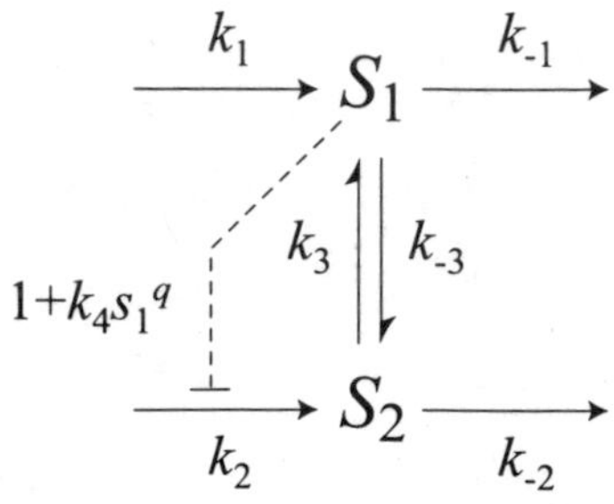

Figure 1.1
Hypothetical biochemical pathway. Molecules of two interacting species, S_1 and S_2, are produced with rates k_1 and k_2 nMh^{-1} and are degraded at rates k_{-1} and k_{-2} h^{-1}. Interconversion between the two species occurs at rates k_3 and k_{-3} nMh^{-1}. Finally, species S_1 inhibits the formation of S_2 (*dashed line*).

The first assumption, called the *continuum hypothesis*, allows us to measure species abundance as a continuously changing concentration rather than a discrete number of molecules. This is usually considered a valid assumption provided the number of molecules is not less than about 1,000 (corresponding to a concentration of about 10 nM in a cell of volume of 0.1 picoliter). If the number drops well below 1,000, it is advisable to use a formalism that allows a discrete measure of the molecule number and that captures the randomness that is significant on this scale. Such stochastic methods are the subject of chapter 2.

The second required assumption is that the reactants find one another immediately and equally, the so-called *well-mixed assumption*. This is valid provided that the time scale of the process under investigation is longer than the time scale of diffusion of its components. This second assumption can be relaxed through the use of partial differential equations, which are introduced in chapter 3.

Under these assumptions, each of the interactions in a cartoon model such as the one illustrated in figure 1.1 can be characterized by an appropriate mathematical formalism (for example, mass-action or Michaelis-Menten kinetics) and these terms can be combined into a description of the rate of change of the abundance (that is, concentration) of the various species in the model.

In general, given a network with n interacting components, we denote their respective concentrations by s_i, for $i = 1, \ldots, n$. We organize these into a vector:

$$\boldsymbol{s}(t) = \begin{bmatrix} s_1(t) \\ \vdots \\ s_n(t) \end{bmatrix}.$$

(In this volume, the shorthand notation $\boldsymbol{s} \in \mathbf{R}^n$ will sometimes be used to indicate that $\boldsymbol{s}$ is a vector with n components.) Given a vector-valued function $\boldsymbol{f} = [f_1 \ldots f_n]^T$

whose components describe the rate of change of the concentration of each species the system dynamics can be described by a differential equation of the form

$$\dot{\boldsymbol{s}}(t) = \frac{\mathrm{d}}{\mathrm{d}t}\boldsymbol{s}(t) = \boldsymbol{f}(\boldsymbol{s}(t)). \tag{1.1}$$

(Both the Leibniz ($\frac{\mathrm{d}}{\mathrm{d}t}$) and Newtonian "overdot" ($\dot{\boldsymbol{s}}$) notation for time derivative will be used.) The vector $\boldsymbol{s}(t)$ of concentrations is referred to as the *state* of the system and can be interpreted as the system's memory: together with the differential equation (1.1), knowledge of the state at any given time t_0 allows us to determine the behavior of the system for all future time $t \geq t_0$.

Example Consider the biochemical system shown in figure 1.1. Denoting the concentrations of S_1 and S_2 by s_1 and s_2, respectively, the system is described by the following two differential equations:

$$\frac{\mathrm{d}}{\mathrm{d}t}s_1(t) = k_1 + k_3 s_2(t) - (k_{-1} + k_{-3})s_1(t), \tag{1.2a}$$

$$\frac{\mathrm{d}}{\mathrm{d}t}s_2(t) = \frac{k_2}{1 + k_4 s_1^q(t)} + k_{-3}s_1(t) - (k_{-2} + k_3)s_2(t). \tag{1.2b}$$

We can represent this system with

$$\boldsymbol{s} = \begin{bmatrix} s_1 \\ s_2 \end{bmatrix}, \quad \text{and} \quad \boldsymbol{f}(\boldsymbol{s}) = \begin{bmatrix} f_1(s_1, s_2) \\ f_2(s_1, s_2) \end{bmatrix} = \begin{bmatrix} k_1 + k_3 s_2 - (k_{-1} + k_{-3})s_1 \\ \dfrac{k_2}{1 + k_4 s_1^q} + k_{-3}s_1 - (k_{-2} + k_3)s_2 \end{bmatrix}. \quad \blacksquare$$

In nearly all cases of interest, the function $\boldsymbol{f}$ does not depend linearly on the state $\boldsymbol{s}$, in which case the vector equation (1.1) is known as a set of nonlinear differential equations. Although, in general, it is not possible to obtain explicit solutions to such nonlinear equations, the evolution of the system from particular initial states can be simulated numerically using generic packages such as Matlab or Mathematica, or programs tailored to biological systems (reviewed by Alves et al. 2006). Alternatively, features of the system behavior can be analyzed directly from the differential equations, as will be illustrated below.

The points $\bar{\boldsymbol{s}}$ satisfying

$$\boldsymbol{f}(\bar{\boldsymbol{s}}) = \boldsymbol{0}$$

are *fixed points* of the system and are referred to as *steady states* or *(mathematical) equilibria.* Note that the term "steady state" is also used to describe persistent dynamic behaviors such as limit cycle oscillations, discussed below.

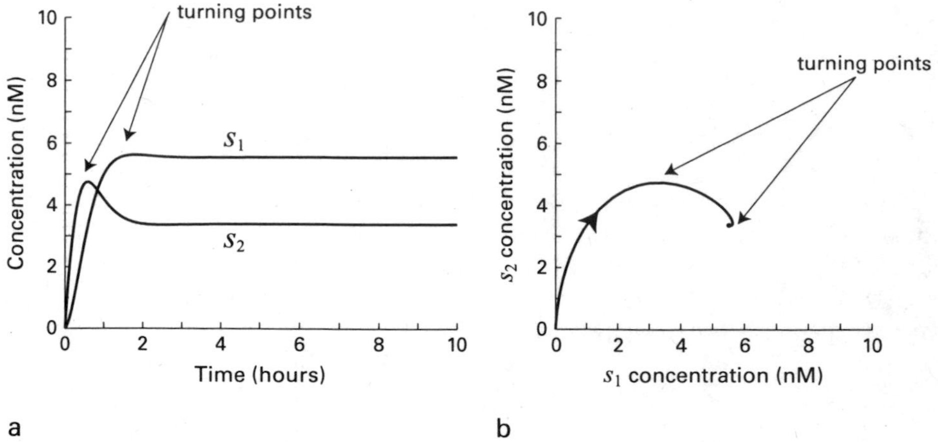

Figure 1.2
Evolution of the system described by equation (1.2). (a) The time-varying nature of the concentrations is emphasized by plotting each species concentration as a function of time, t. (b) The relationship between the concentrations of the two species is emphasized by plotting them in the s_1-s_2 plane. The curve starts at the initial condition and continues until it reaches an equilibrium. The arrowhead indicates progression with time. The turning points are instances where the phase plot changes direction. Parameters used are $k_1 = 1$ nM h^{-1}, $k_{-1} = 1$ h^{-1}, $k_2 = 20$ nM h^{-1}, $k_{-2} = 1$ h^{-1}, $k_3 = 2$ h^{-1}, $k_{-3} = 0.4$ h^{-1}, $k_4 = 0.05$ nM^{-1}, and $q = 2$.

1.2.1 Phase-Plane Analysis

Figure 1.2a shows how system behavior is typically visualized by plotting the concentrations of the two species (s_1 and s_2) as functions of time, which is consistent with an experimental time course. An alternative approach is to plot the time-varying behavior on the s_1-s_2 plane, which is referred to as the *phase plane*. These plots, called *phase portraits* cannot be generated for systems with more than two (or three) species. Nevertheless, the enhanced understanding they provide in these cases leads to valuable insights into more complex networks.

As an example, figure 1.2b shows precisely the same behavior as in figure 1.2a: the system starts from initial condition $(s_1, s_2) = (0, 0)$ and converges to the steady state $(\bar{s}_1, \bar{s}_2) \approx (5.53, 3.37)$; the plot emphasizes the time-varying relationship between the two variables, but de-emphasizes the relationship with the time variable t itself. Indeed, although each point on the curve corresponds to the value $(s_1(t), s_2(t))$ at a particular time instant t, the only time points that can be easily identified are at $t = 0$ (where the curve starts) and the longtime behavior $t \to \infty$ (where the curve ends).

In figure 1.3, the time courses (or *trajectories*) corresponding to a number of different initial conditions are displayed simultaneously. The trajectories begin at different

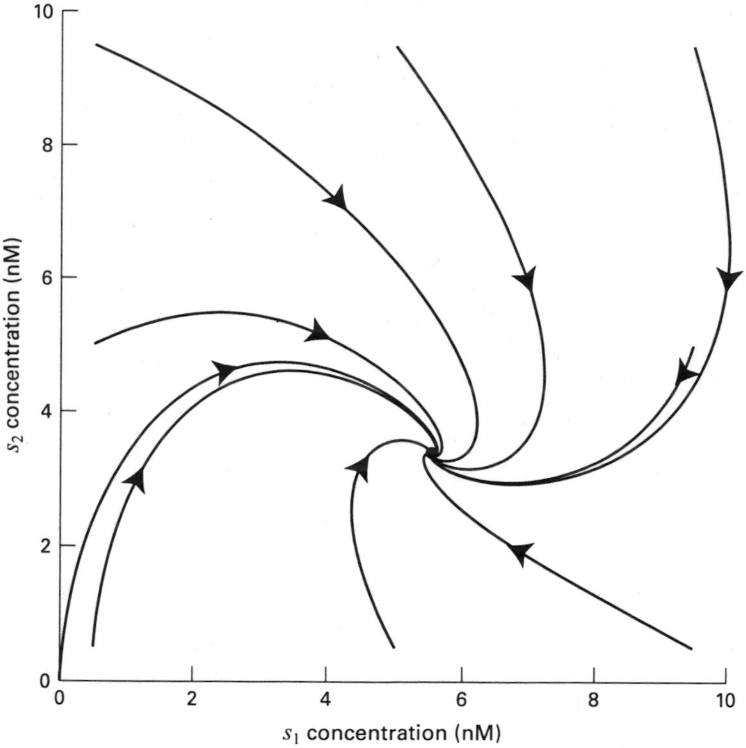

Figure 1.3
Phase-plane plot for the system of figure 1.1. Each trajectory begins at a different initial condition. Note that the trajectories cannot intersect; they can only come together at equilibria.

initial points in the s_1-s_2 plane and all end (in this case) at the steady state $(\bar{s}_1, \bar{s}_2)$. By capturing the behavior of multiple trajectories simultaneously, this single figure provides an overall impression of how the system behaves which would be difficult to achieve with a time-series plot.

An alternative to drawing many trajectories is to use short arrows to indicate the direction of motion at each point. The resulting plot, as shown in figure 1.4, is referred to as a *vector field* (or *direction field*). The trajectories lie parallel to (i.e., tangent to) the vector field at each point, and so can be constructed directly by "connecting the arrows." It is sometime useful to consider the analogy with particles suspended in a flowing fluid. The vector field describes the direction of motion of the fluid. The trajectories are the paths suspended particles would traverse as they are carried along with the flow.

The direction field can be derived directly from the differential equations—no explicit solution is needed. For a generic system involving two species s_1 and s_2,

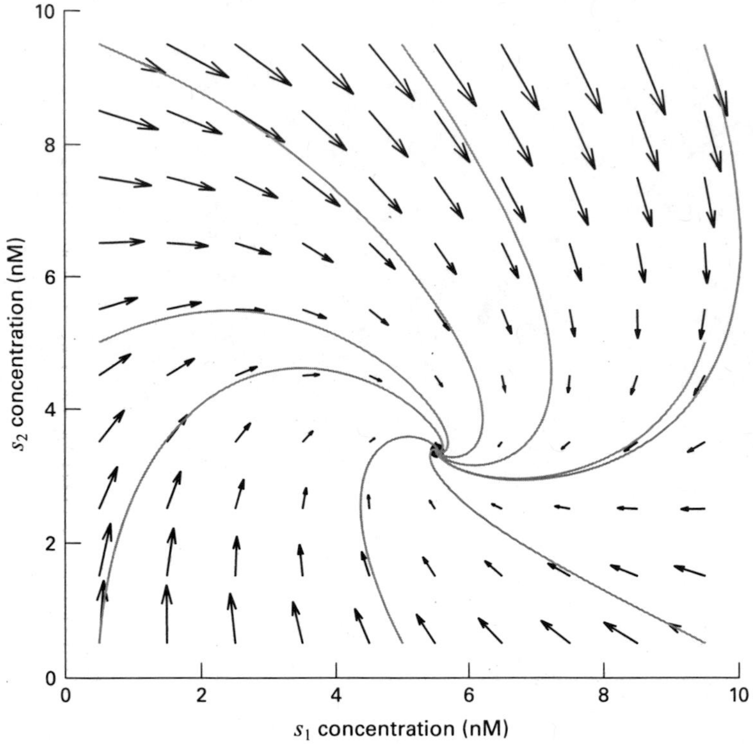

Figure 1.4
Vector field for the system defined by equation (1.2). The trajectories lie parallel to the arrows.

$$\frac{d}{dt}s_1(t) = f_1(s_1(t), s_2(t)),$$

$$\frac{d}{dt}s_2(t) = f_2(s_1(t), s_2(t)),$$

the slope of the (s_1, s_2) trajectory at any point is

$$\frac{ds_2}{ds_1} = \frac{ds_2/dt}{ds_1/dt} = \frac{f_2(s_1, s_2)}{f_1(s_1, s_2)}.$$

An arrow is drawn with this slope at each point. (If the denominator in this quotient is zero, then the arrow has an infinite slope, meaning that it points straight up or down.)

A key feature of the phase portrait is the set of points at which the trajectories "turn around," that is, change their direction with respect to one of the axes. These

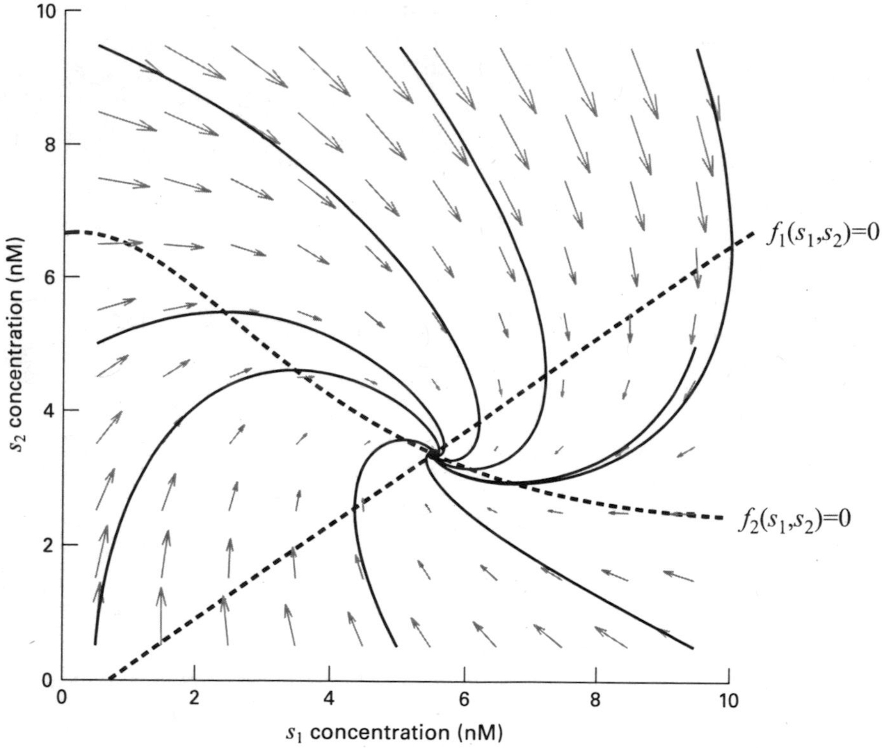

Figure 1.5
Trajectories for the system change direction as they cross the nullclines (*dotted lines*). Note also that the vector field points vertically on the s_1-nullcline and horizontally on the s_2-nullcline.

are the points at which one of the two variables $s_1(t)$ or $s_2(t)$ reaches a maximum or minimum, as shown in figure 1.2a. In this case, the maximum in s_2 occurs at time $t \approx 0.61$ h, at which point the concentrations are $(s_1, s_2) \approx (3.46, 4.75)$. Similarly, a maximum occurs in s_1 at time $t \approx 1.85$ h, at which point the concentrations are $(s_1, s_2) \approx (5.63, 3.42)$. These points can be identified on the same trajectory in the phase plane, as shown in figure 1.2b.

Turning points occur whenever the phase plane trajectory is directed either

1. vertically (vector points up or down): $\frac{d}{dt}s_1 = 0$; or
2. horizontally (vector points left or right): $\frac{d}{dt}s_2 = 0$.

The set of points (s_1, s_2) where the trajectory is vertical satisfies $f_1(s_1, s_2) = 0$, which defines the s_1-*nullcline* of the system. Likewise, the equation $f_2(s_1, s_2) = 0$ defines the s_2-*nullcline*. Figure 1.5 shows the nullclines for the system described by equation (1.2) along with a few trajectories and the vector field.

It is clear in figure 1.5 that the end point of the trajectories is marked by the intersection of the two nullclines. Note that the points (s_1, s_2) where the nullclines intersect satisfy $f_1(s_1, s_2) = f_2(s_1, s_2) = 0$ and thus are the equilibria of the system.

Because the nullclines can be generated without solving the differential equations, they allow direct insight into the dynamic behavior. This type of shortcut is a recurring theme in the analysis of dynamical systems.

1.3 Linear Systems and Linearization

If the function $\boldsymbol{f}$ in equation (1.1) is linear in $\boldsymbol{s}$, then the system can be written in the form

$$\frac{\mathrm{d}}{\mathrm{d}t}\boldsymbol{s}(t) = \boldsymbol{A}\boldsymbol{s}(t), \tag{1.3}$$

where $\boldsymbol{A}$ is an $n \times n$ matrix.

The assumption of linearity greatly simplifies the analysis of the system. In particular, analytic expressions for the solution of the differential equations are now available. In this case, the solution takes the form

$$\boldsymbol{s}(t) = e^{\boldsymbol{A}(t-t_0)}\boldsymbol{s}(t_0),$$

where $e^{\boldsymbol{A}}$ is the matrix exponential (Rugh, 1996).

Unfortunately, except for the simplest of systems, linear models are inappropriate for describing biochemical or genetic networks. Nevertheless, linear analysis can be used to understand the behavior of a nonlinear system in regions of the state space that are near a steady state $\bar{\boldsymbol{s}}$, as shown next. We approximate the function $\boldsymbol{f}$ in equation (1.1) by a linear function that coincides with $\boldsymbol{f}$ at the steady state:

$$\boldsymbol{f}(\boldsymbol{s}) \approx \boldsymbol{f}(\bar{\boldsymbol{s}}) + \frac{\partial \boldsymbol{f}}{\partial \boldsymbol{s}}(\boldsymbol{s} - \bar{\boldsymbol{s}}). \tag{1.4}$$

This approximation represents the first two terms in the Taylor series expansion of $\boldsymbol{f}$.

The matrix

$$\boldsymbol{A} = \frac{\partial \boldsymbol{f}}{\partial \boldsymbol{s}}$$

is known as the *Jacobian* of the system. If $\boldsymbol{s}$ (and hence $\boldsymbol{f}$) has n components, then the Jacobian is an $n \times n$ matrix where the i, jth element is the partial derivative of the ith component of $\boldsymbol{f}$ with respect to the jth element of $\boldsymbol{s}$, evaluated at the steady state $\bar{\boldsymbol{s}}$:

$$a_{i,j} = \left.\frac{\partial f_i(\boldsymbol{s})}{\partial s_j}\right|_{\boldsymbol{s}=\bar{\boldsymbol{s}}}.$$

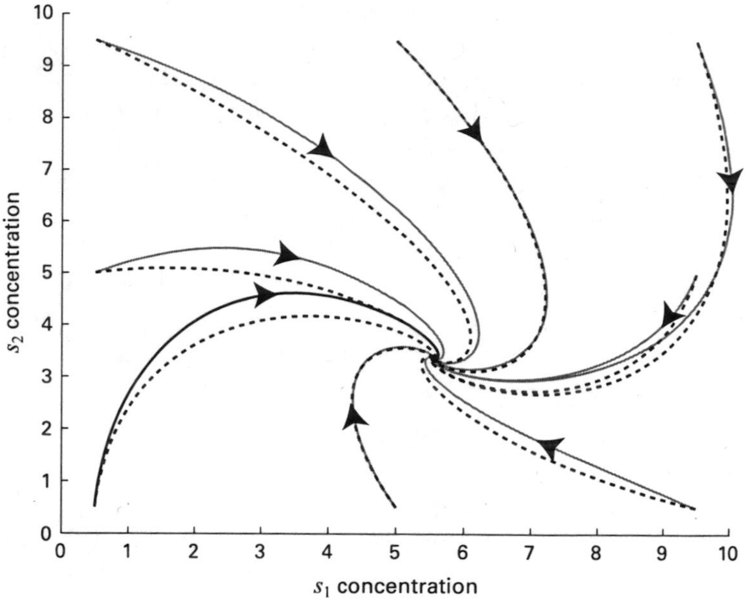

Figure 1.6
Phase portrait around a stable equilibrium for a nonlinear system (*solid lines*) and the linearization (*dotted lines*).

We introduce an auxiliary variable $\boldsymbol{x}(t)$, which represents the deviation of the species concentrations from their steady-state values:

$$\boldsymbol{x}(t) = \boldsymbol{s}(t) - \bar{\boldsymbol{s}}.$$

Then, because

$$\frac{\mathrm{d}}{\mathrm{d}t}\boldsymbol{x}(t) = \frac{\mathrm{d}}{\mathrm{d}t}\boldsymbol{s}(t) - \boldsymbol{0},$$

and $\boldsymbol{f}(\bar{\boldsymbol{s}}) = \boldsymbol{0}$, the system (1.1) can be rewritten as

$$\frac{\mathrm{d}}{\mathrm{d}t}\boldsymbol{x}(t) = \frac{\mathrm{d}}{\mathrm{d}t}\boldsymbol{s}(t) = \boldsymbol{f}(\boldsymbol{s}(t)) \approx \boldsymbol{f}(\bar{\boldsymbol{s}}) + \frac{\partial \boldsymbol{f}}{\partial \boldsymbol{s}}(\boldsymbol{s}(t) - \bar{\boldsymbol{s}}) = \boldsymbol{A}\boldsymbol{x}(t),$$

that is,

$$\frac{\mathrm{d}}{\mathrm{d}t}\boldsymbol{x}(t) = \boldsymbol{A}\boldsymbol{x}(t). \tag{1.5}$$

This is referred to as the *linearization* of equation (1.1).

The behavior of the linearized system (1.5) can provide considerable information about the nonlinear model (1.1). In particular, if the matrix A has no eigenvalues with zero real part and certain other technical conditions hold, then the behavior of the linear and nonlinear systems agrees whenever the trajectories remain near the equilibrium (Hartman, 1963; see figure 1.6 for an example).

Thus the linearization describes the qualitative behavior of the system when the state variables remain "close" to the equilibrium. Although there are results that give precise bounds on the error made in this approximation, they are generally intractable; we will not pursue them here. Because regulated systems often have the property that the state remains near the equilibrium, this approximation provides the central basis for many of the tools of control engineering.

1.4 Stability

The concept of stability is central to the analysis of dynamical systems. Depending on the particular type of system—for example, whether it is linear or nonlinear, whether it has inputs or not—there are numerous definitions of what it means for a system to be stable. In all cases, however, the general idea is that systems are stable if small perturbations, whether in the initial condition or due to external stimuli, do not give rise to large sustained changes in the behavior of the system.

Turning to stability in the context of the nonlinear system

$$\frac{\mathrm{d}}{\mathrm{d}t}\boldsymbol{x}(t) = \boldsymbol{f}(\boldsymbol{x}(t)),$$

let us suppose that the system has an equilibrium $\bar{s}$. We say that the equilibrium is *stable* if every initial condition s_0 that is near $\bar{s}$ gives rise to trajectories that stay close to the equilibrium. A precise definition of what it means to be close involves so-called δ-ϵ arguments (Khalil, 2002), which we will avoid here.

Note that stability does not specifically require that the trajectory tend to the equilibrium as $t \to \infty$, a separate property known as *attractivity*. When an equilibrium is both stable and attractive, we say that it is *asymptotically stable*. Because, in practice, it is rare for an equilibrium to be stable but not asymptotically stable (the case of *marginal* or *neutral* stability), in the computational biological community the two terms are often used interchangeably. Where an equilibrium is *unstable*, nearby trajectories will diverge from the steady state (figure 1.7).

It should be emphasized that stability is not a property of a system but rather of particular steady states. Nonlinear systems can have multiple equilibria, with varying stability properties (see chapter 7). In general, all linear systems have a single equilib-

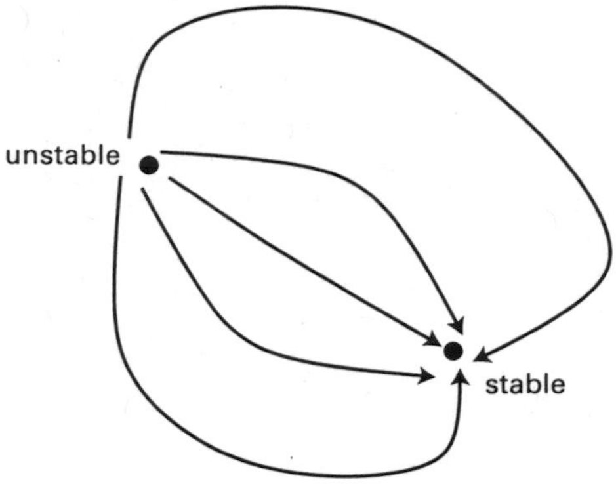

Figure 1.7
Two equilibria, one stable and the other unstable. Trajectories move away from the unstable and toward the stable equilibrium.

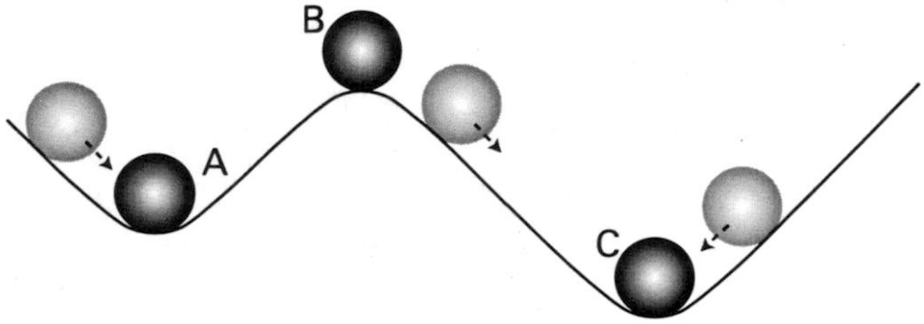

Figure 1.8
Stability and instability of equilibria likened to the behavior of balls on an undulating slope. For balls rolling on the slope depicted here, there are three possible equilibria, corresponding to the valley bottoms (stable) and the top of the hill (unstable). A ball balanced perfectly on the top of the hill (B) will remain there, but the slightest perturbation will cause it to roll away. Conversely, a ball in a valley bottom (A or C), if displaced by a small amount, will return to its resting place.

rium (at the origin) and so the term "stable system" is sometimes used to describe a linear system for which this equilibrium is stable.

In a standard analogy, stable and unstable states are likened to the valleys and hill tops of an undulating topography (figure 1.8). The valley bottoms represents stable equilibria: small perturbations will not let a ball escape and, in the presence of friction, the ball will return to a resting state at the bottom of the valley. Alternatively, the peak of a hill represents an unstable equilibrium: any disturbance will make a ball accelerate away from its rest state. A topography with two valleys would correspond to a bistable system. In this case, we can assign to each stable steady state a *basin of attraction*, which is the region in the phase space from which trajectories approach this point. The dividing line between the two is referred to as a *separatrix*.

1.4.1 Determining Stability for Linear Systems

How can we determine the stability of an equilibrium? For simple systems, we could draw the vector field and observe the direction of the arrows. For a more generally applicable method, however, let us first consider the linear case, where n linear differential equations are written in matrix form:

$$\frac{\mathrm{d}}{\mathrm{d}t}\boldsymbol{x}(t) = \boldsymbol{A}\boldsymbol{x}(t).$$

As mentioned earlier, the solution, given an initial state $\boldsymbol{x}(0) = \boldsymbol{x}_0$, is

$$\boldsymbol{x}(t) = e^{\boldsymbol{A}t}\boldsymbol{x}_0,$$

which can be expanded in the form

$$e^{\boldsymbol{A}t}\boldsymbol{x}_0 = \alpha_1 e^{\lambda_1 t}\boldsymbol{v}_1 + \cdots + \alpha_n e^{\lambda_n t}\boldsymbol{v}_n, \tag{1.6}$$

where the constants α_i depend on the initial condition $\boldsymbol{x}_0$, the λ_i are the *eigenvalues* of the matrix $\boldsymbol{A}$, and the $\boldsymbol{v}_i$ are the associated *eigenvectors*. We have assumed that the matrix $\boldsymbol{A}$ has n distinct eigenvalues, which is typically the case. The main stability results stated here hold in the general case (for more detailed discussions, see Rugh, 1996; Khalil, 2002).

Eigenvalues and Singular Values

The n eigenvalues of an $n \times n$ square matrix provide a valuable summary of the overall matrix structure and are used in a wide array of application areas (Horn and Johnson, 1985). The eigenvalues of a matrix $\boldsymbol{A}$ are the solutions λ of the *characteristic equation* $\det(\lambda\boldsymbol{I} - \boldsymbol{A}) = 0$. In the case of the 2×2 matrix

$$\boldsymbol{A} = \begin{bmatrix} a_{11} & a_{12} \\ a_{21} & a_{22} \end{bmatrix},$$

the characteristic equation is quadratic:

$$\lambda^2 - (a_{11} + a_{22})\lambda + (a_{11}a_{22} - a_{12}a_{21}).$$

Readers may recall that $a_{11} + a_{22}$ and $a_{11}a_{22} - a_{12}a_{21}$ are the *trace* and *determinant* of the matrix $\boldsymbol{A}$, respectively. Thus, in this case, the eigenvalues can be written explicitly in terms of these two values:

$$\lambda_{1,2} = \frac{1}{2}\left(\mathrm{trace}(\boldsymbol{A}) \pm \sqrt{\mathrm{trace}(\boldsymbol{A})^2 - 4\det(\boldsymbol{A})}\right).$$

For larger matrices, the characteristic equation is less tractable, and iterative numerical methods are typically employed to find eigenvalues. Closely related to the notion of eigenvalues are the *singular values*, which also provide a "summary" of a matrix. The singular values of an $n \times m$ matrix $\boldsymbol{A}$ are defined as the square roots of the eigenvalues of the square matrix $\boldsymbol{A}^T\boldsymbol{A}$. Although, in special cases, the singular values of a square matrix coincide with its eigenvalues, they generally provide a different measure of matrix properties. Singular values will be used to address system robustness in chapters 9, 10, and 11.

Stability Criteria

The stability of the linear system depends on the behavior of the time-varying terms $e^{\lambda_i t}$ that appear in the solution (1.6). In general, eigenvalues are complex numbers; thus we write

$$\lambda_k = \sigma_k + j\omega_k,$$

where $j = \sqrt{-1}$. (Note, outside of engineering, the more common notation is $i = \sqrt{-1}$.) Recalling Euler's formula for the exponential of a complex number:

$$e^{(\sigma+\omega j)t} = e^{\sigma t}(\cos(\omega t) + j\sin(\omega t)),$$

we see that the term $e^{\lambda_k t}\boldsymbol{v}_k$ will decay to zero asymptotically if and only if the real part of the eigenvalue (σ_k) is strictly less than zero. If this is true for all eigenvalues of $\boldsymbol{A}$, then all solutions tend to zero, and the system is asymptotically stable. In this case the matrix $\boldsymbol{A}$ is called *Hurwitz*. Similarly, we say that the polynomial

$$\lambda^n + a_{n-1}\lambda^{n-1} + \cdots + a_1\lambda + a_0 = 0$$

is Hurwitz if all the roots have negative real parts.

Alternatively, if *any* of the eigenvalues has real part σ_j greater than zero, then some trajectories will grow exponentially, thus the origin is unstable. Finally, in the special case that some of the eigenvalues have real part exactly equal to zero, the origin cannot be asymptotically stable. It may, however, be (neutrally) stable.

1.4.2 Determining Stability for Nonlinear Systems

We now consider the stability of an equilibrium $\bar{\boldsymbol{s}}$ of the nonlinear system

$$\frac{\mathrm{d}}{\mathrm{d}t}\boldsymbol{s}(t) = \boldsymbol{f}(\boldsymbol{s}(t)). \tag{1.7}$$

In this case, the stability properties of an equilibrium need to be characterized as *local* or *global*. An equilibrium is globally (asymptotically) stable if all trajectories converge to it, no matter what the initial condition (clearly this can only happen if there is a unique equilibrium), whereas it is locally stable if trajectories must start in some neighborhood of the equilibrium. In the special case of linear systems, global stability is implied by local stability.

To determine the local stability properties of an equilibrium, it is enough to consider the stability properties of the system's linearization. In particular, an equilibrium of equation (1.7) is asymptotically stable if all the eigenvalues of the Jacobian, evaluated at the equilibrium, have negative real part. Conversely, the equilibrium is unstable if the Jacobian has an eigenvalue with positive real part. If the linearization is stable but not asymptotically stable (implying that the Jacobian has eigenvalues on the imaginary axis), then we can draw no conclusion about the stability of the nonlinear system from the linearized analysis.

The characterization of local stability through linearization is known as Lyapunov's indirect method, named after the nineteenth-century Russian mathematician Aleksandr Mikhailovich Lyapunov, who developed many of the early concepts of stability. Lyapunov also provided tests of global stability that can be applied to nonlinear systems. In particular, suppose that a real-valued function $V(\mathbf{s})$ exists such that

1. $V(\mathbf{s}) \geq 0$ for all $\mathbf{s}$ and $V(\mathbf{s}) = 0$ if and only if $\mathbf{s} = \mathbf{0}$; and
2. $\dfrac{\mathrm{d}V(\mathbf{s}(t))}{\mathrm{d}t} < 0,$

that is, the value of V decreases along trajectories, then the equilibrium is locally asymptotically stable. If a third condition is also satisfied:

3. $V(\mathbf{s}) \to \infty$ as $\|\mathbf{s}\| \to \infty,$

then the equilibrium is globally asymptotically stable.

The function V is known as a *Lyapunov function*. Because energy dissipates through friction, it can be taken as a Lyapunov function when modeling mechanical systems. For biological systems, however, there is no obvious way of choosing a suitable Lyapunov function.

1.5 Oscillatory Behavior

Thus far, our analysis of long-term (asymptotic) behavior has been restricted to fixed-point steady states. We now extend our discussion to systems that give rise to persistent, oscillatory dynamics. Oscillatory systems are commonplace in biology, and there are numerous treatises dealing with them (see, for example, Goldbeter, 1996, and the references therein). As an example, we consider a simple metabolic model, illustrated in figure 1.9. This scheme is motivated by the "turbocharged" posi-

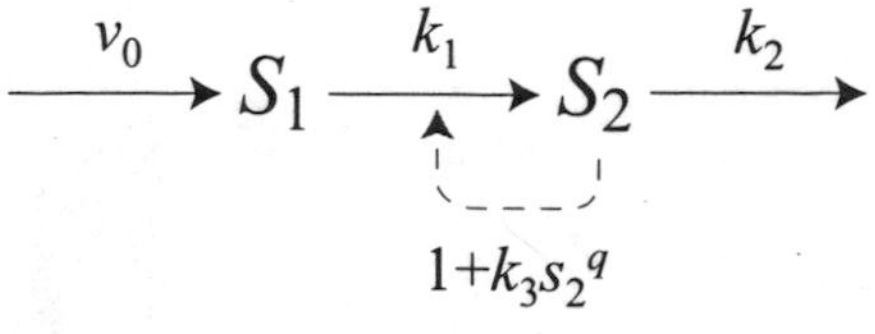

Figure 1.9
System involving positive feedback. Examples of such systems are found in catabolic pathways in which the first step involves coupling of ATP hydrolysis to activation of a substrate. Downstream, conversion of the substrate into product generates a surplus of ATP, which increases the available activated substrate (Teusink et al., 1998).

tive feedback aspect of the glycolytic chain in which the ATP output is used to produce more ATP (Teusink et al., 1998).

The potential for oscillations can be inferred from the model structure: the concentration of species S_2 builds up, causing further buildup until the pool of S_1 is depleted. The S_2 level then crashes until more S_1 is available, and so on. Although this intuitive argument indicates the potential for oscillatory behavior, it cannot predict the conditions under which oscillations will occur.

The equations describing the system are given by

$$\frac{d}{dt}s_1(t) = v_0 - k_1s_1(t)(1 + k_3s_2^q(t)), \tag{1.8a}$$

$$\frac{d}{dt}s_2(t) = k_1s_1(t)(1 + k_3s_2^q(t)) - k_2s_2(t). \tag{1.8b}$$

The system can exhibit stable behavior as shown in figure 1.10: both species concentrations converge to steady state. This same behavior can be observed over a wide range of initial conditions, as shown in the phase portrait, which indicates that there is a single steady state.

Note that the trajectories seem to be spiraling in as they approach the steady state. That behavior, apparent in the damped oscillations seen in the time-domain description, indicates that the system is somehow close to oscillatory behavior. Increasing the nonlinearity in the model leads to the sustained oscillatory behavior seen in figure 1.11a. Considering the accompanying phase portrait in figure 1.11b, we see that the trajectories are attracted to a cyclic track, called a *limit cycle*. When we compare this phase portrait with the one corresponding to the steady state (figure 1.10b), we see that the nullclines' structure has not changed significantly. What has changed is the stability of the single steady state, which is now unstable. For two-dimensional systems, there is a useful result known as the Poincaré-Bendixson theorem which guarantees the existence of a limit cycle in a case like this. With no stable equilibrium points and all trajectories bounded, the trajectories have to go *somewhere*, and a limit

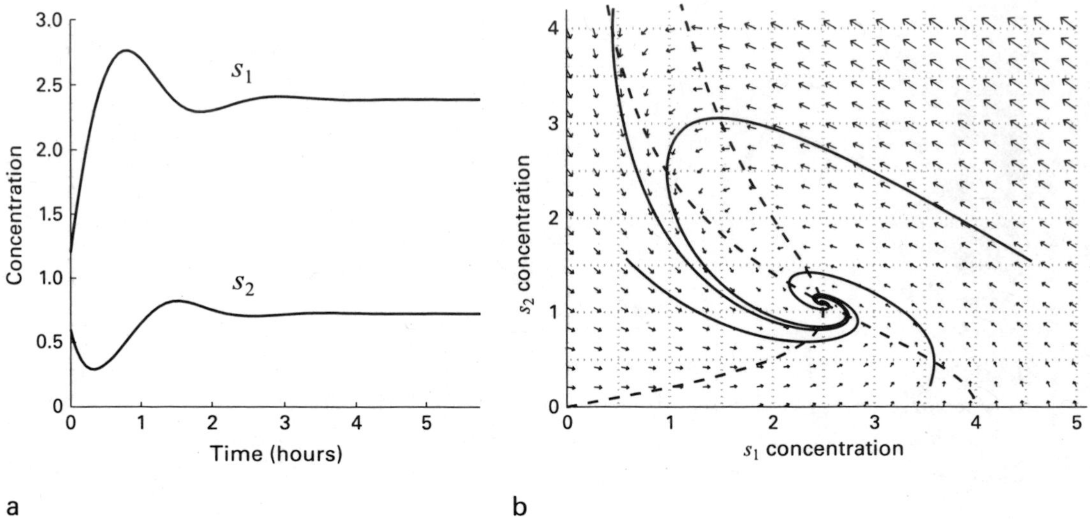

Figure 1.10
Time-domain (a) and phase-plane (b) plot of the system of figure 1.9, described by equation (1.8). Parameter values used are $v_0 = 8$, $k_1 = 1$, $k_2 = 5$, $k_3 = 1$ and $q = 2$.

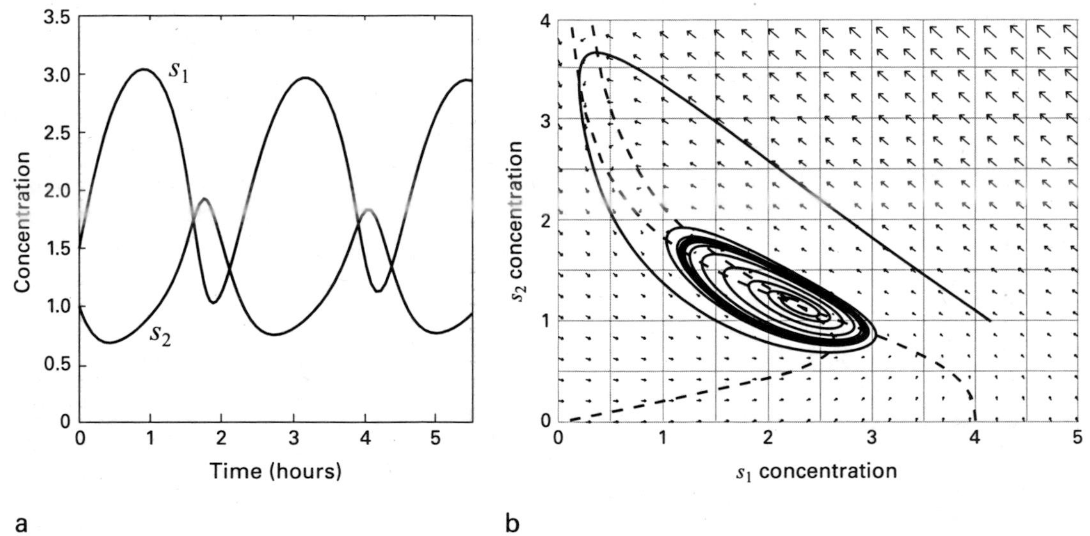

Figure 1.11
Time-domain (a) and phase-plane (b) plot of the system of figure 1.9, described by equation (1.8). The parameter values used are the same as in figure 1.10 except that the feedback strength has been increased to $q = 3$. This increase leads to sustained oscillatory behavior. In the phase plane, this can be seen as a closed curve.

cycle is the only remaining option (Khalil, 2002). Here we also see that the *qualitative* behavior of the system changes as the parameter values shift. The nature of those changes can be understood through bifurcation analysis.

1.6 Bifurcations

An important question when studying dynamical systems is whether the dynamic behavior of the system is retained as the system parameters change. This is of particular interest in biological systems, where the parameters may represent variables such as enzyme concentrations that are likely to vary significantly from one cell to another. When a system's properties do not undergo significant qualitative changes, we say that the system is *robust* or *structurally stable*. Robustness is the topic of chapters 9, 10, and 11. When qualitative changes do occur—for example, a stable equilibrium becomes unstable, or the system acquires a new equilibrium point—the system is said to undergo a *bifurcation*.

1.6.1 Bifurcation Diagrams

One way of studying bifurcations is by plotting the location and nature of the system's equilibria as a function of a parameter. These plots, which correspond to experimental dose-response curves, are known as *continuation diagrams*. Bifurcations occur at the points on a continuation diagram where a major change takes place. The specific value of the parameter where the change occurs is the *bifurcation point*. In these cases, the continuation diagram is known as a *bifurcation diagram*.

Figure 1.12 contains a bifurcation diagram showing the steady state of s_1 in system (1.8) as a function of the parameter q. In addition to the position of the equilibrium, the stability type is indicated by the line style: a stable (attracting) steady state is indicated by the solid line from $q = 1$ to $q \approx 2.86$; an unstable (repelling) fixed point, by the dashed line from that point up until $q = 5$. The value $q \approx 2.86$, at which the stability changes, is the bifurcation point. Recall that stability is dictated by the sign of the real part of the eigenvalues of the linearization of the system. These eigenvalues change as the parameter q changes; their real parts cross over from being negative to being positive at the bifurcation point.

In this particular system, the bifurcation point dictates not only the change in the stability of the fixed point, but also the appearance of a limit cycle. This particular type of bifurcation is known as a *Hopf bifurcation*. For such bifurcations, it is customary to denote the limit cycle by the maximum and minimum values reached in the oscillation (figure 1.13).

Knowledge of the bifurcation structure can be useful in assaying the robustness of system behavior. If the nominal conditions are near a bifurcation point, it is possible that environmental perturbations could push the system into a very different

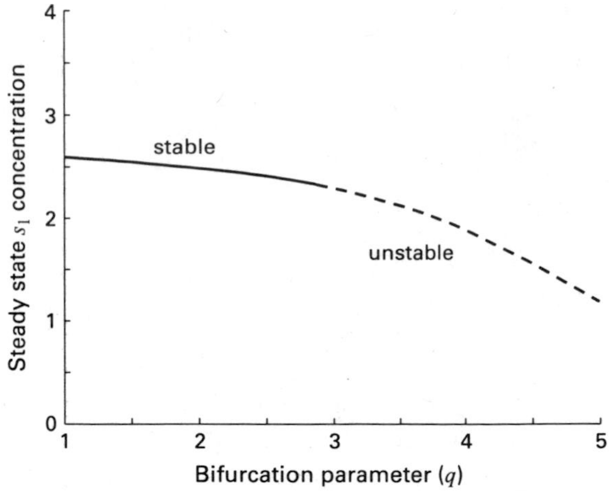

Figure 1.12
Bifurcation diagram for system (1.8), showing the steady state of s_1 as a function of the parameter q. Although one equilibrium exists for all these values of the parameter, the stability of the equilibrium changes at $q \approx 2.86$. For values smaller than this, the equilibrium is stable (*solid line*); for greater values, it is unstable (*dashed line*).

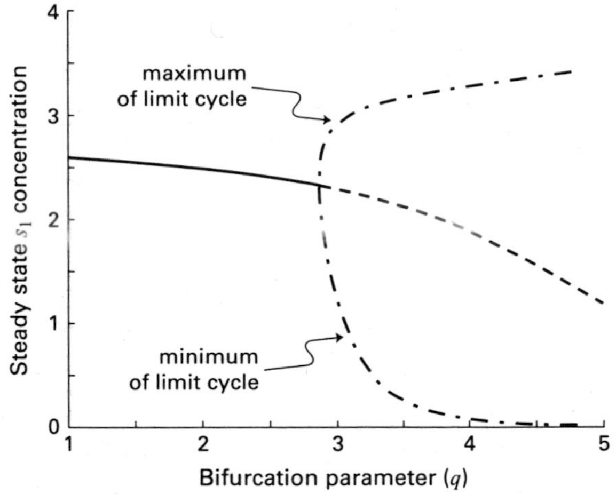

Figure 1.13
Bifurcation diagram for the system in figure 1.9, showing the steady state s_1 as a function of the parameter q and denoting the location and stability of the equilibrium as well as the maximum and minimum values achieved by s_1 during the steady-state oscillations (*dashed-dotted curve*).

behavior. On the other hand, if bifurcations are far away in parameter space, then the nominal behavior may be highly robust to changes in operating conditions.

1.7 Systems with Inputs and Outputs

Thus far, we have considered systems that evolve autonomously, and we have treated the whole state vector in our analysis. In control engineering, it is more common to consider systems that respond to external inputs and provide specific output signals to their environment.

We first consider systems of the form

$$\frac{\mathrm{d}}{\mathrm{d}t}\boldsymbol{s}(t) = \boldsymbol{f}(\boldsymbol{s}(t), \boldsymbol{u}(t)). \tag{1.9}$$

In this equation, the vector

$$\boldsymbol{u}(t) = \begin{bmatrix} u_1(t) \\ \vdots \\ u_m(t) \end{bmatrix}$$

represents species or other influences that act as external inputs to the system.

In a biological context, the inputs could represent any external parameters over which an experimenter has control or any regulatory signals coming from outside of the process of interest, for example, genetic control of enzyme abundance in a metabolic model. In these contexts, we might ask to what degree the input can influence the system dynamics (addressed in section 1.9.1). Alternatively, the input $\boldsymbol{u}(t)$ could be used to incorporate the effect of external disturbances on model behavior. In this case, the model allows us to analyze the robustness of system behavior in the face of these perturbations.

Additionally, control engineers consider a set of p observable outputs, denoted by

$$\boldsymbol{y}(t) = \begin{bmatrix} y_1(t) \\ \vdots \\ y_p(t) \end{bmatrix}.$$

The choice of output might be dictated by available experimental assays or by which components of the system interact with processes external to the network of interest. An alternative role for the output $\boldsymbol{y}(t)$ is to allow the analysis to focus on a particular aspect of system behavior. That might be the concentration of a particular component species or some function of overall behavior, for example, the flux through a metabolic pathway. For the simplest input-output systems, referred to as *memoryless* input-output maps, there is no system state and the output $\mathbf{y}$ at time t is given directly by a function of the input $\mathbf{u}$ at time t:

$$\mathbf{y}(t) = \mathbf{h}(\mathbf{u}(t)).$$

More generally, the effect of the input is not immediate, and so a dynamic description in which the output is a function of the system states and inputs is required:

$$\mathbf{y}(t) = \mathbf{h}(\mathbf{s}(t), \mathbf{u}(t)). \tag{1.10}$$

In this case, equations (1.9) and (1.10) represent a dynamic input-output system. Fixing a specific initial condition (typically at the origin), such systems can be thought of as maps from input signals $\mathbf{u}(t)$ to corresponding output signals $\mathbf{y}(t)$. In the specific case that the system has only one input ($m = 1$) and one output ($p = 1$) the system is said to be a *single-input, single-output* (SISO) system. Otherwise, the system is referred to as *multiple-input, multiple-output* (MIMO). For simplicity, we will assume that the systems are SISO.

1.7.1 Feedback

Central to control theory is the use of feedback control (figure 1.14). The idea is to arrange for the input variables to depend on the system response. When the input depends directly on the state variables:

$$u(t) = k_s(\mathbf{s}(t)),$$

for some suitable function k_s, known as a *control law*, the feedback is referred to as *static state feedback*. It is "state" feedback because the controller has access to the full state vector; and "static" because the feedback is implemented through a memoryless

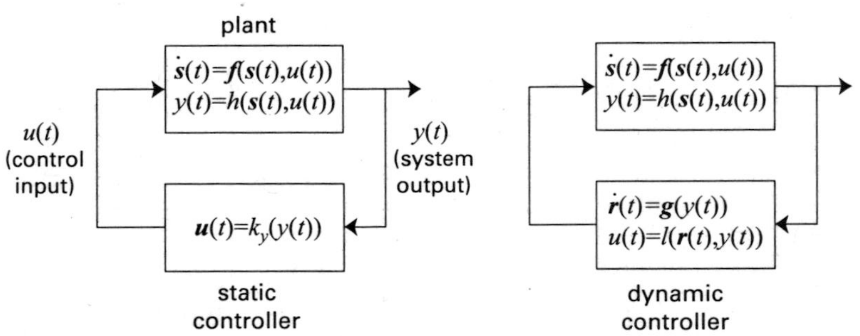

Figure 1.14
Typical feedback control system, consisting of two dynamical systems. One, referred to as the *plant*, is the system to be controlled. The other, known as the *controller*, acts in a feedback loop and is often designed to regulate the complete system. It can be a static (*left*) or a dynamic (*right*) function of the plant output. In biological implementations, the distinction between plant and controller is blurred.

map (i.e., a function). Alternatively, when the control depends only on the system output:

$$u(t) = k_y(y(t)),$$

the feedback is known as static *output feedback*.

In contrast to static feedback, the controller may itself involve an auxiliary dynamic system, which allows it to take into account past behavior of the system (as opposed to only acting on current system values). In this case, the feedback system is described by a vector of control states $\boldsymbol{r}(t)$. When such a feedback depends on output values, it takes the form

$$\frac{\mathrm{d}}{\mathrm{d}t}\boldsymbol{r}(t) = \boldsymbol{g}(y(t)), \tag{1.11a}$$

$$u(t) = l(\boldsymbol{r}(t), y(t)) \tag{1.11b}$$

and is referred to as *dynamic* output feedback. In this case, the feedback depends on a history of the output signals: the controller state $\boldsymbol{r}(t)$ retains a memory of the output $y(t)$. Incorporating this memory into the control input allows the system to cope with behavior that a direct (static) feedback cannot.

1.7.2 Linear Input-Output Systems

If the function $\boldsymbol{f}$ in equation (1.9) is linear in $\boldsymbol{s}$ and $\boldsymbol{u}$, then the system can be written in the form

$$\frac{\mathrm{d}}{\mathrm{d}t}\boldsymbol{s}(t) = \boldsymbol{A}\boldsymbol{s}(t) + \boldsymbol{B}\boldsymbol{u}(t). \tag{1.12}$$

Moreover, if the output depends linearly on the state and input, then we write

$$\boldsymbol{y}(t) = \boldsymbol{C}\boldsymbol{s}(t) + \boldsymbol{D}\boldsymbol{u}(t). \tag{1.13}$$

The system defined by equations (1.12) and (1.13) is referred to as a linear, time-invariant (LTI) system ("time-invariant" refers to the fact that the matrices do not change with time). If the system has m inputs, p outputs, and n states, then $\boldsymbol{A}$, $\boldsymbol{B}$, $\boldsymbol{C}$, and $\boldsymbol{D}$ are $n \times n$, $n \times m$, $p \times n$, and $p \times m$, matrices, respectively. The solution to this linear system can be written explicitly in the form

$$\boldsymbol{s}(t) = e^{\boldsymbol{A}(t-t_0)}\boldsymbol{s}(t_0) + \int_{t_0}^{t} e^{\boldsymbol{A}(t-\tau)}\boldsymbol{B}\boldsymbol{u}(\tau)\,\mathrm{d}\tau.$$

Linearization of a nonlinear input-output system follows in the manner described in section 1.3 and is centered at a steady state $\bar{\boldsymbol{s}}$ corresponding to a constant nominal input $\bar{\boldsymbol{u}}$. In addition to the state $\boldsymbol{x}$ and Jacobian $\boldsymbol{A}$, the linearization also includes the matrix $\boldsymbol{B}$ with i, jth element

$$b_{i,j} = \left. \frac{\partial f_i(\boldsymbol{s}, \boldsymbol{u})}{\partial u_j} \right|_{\boldsymbol{s}=\bar{\boldsymbol{s}},\, \boldsymbol{u}=\bar{\boldsymbol{u}}}.$$

In this case, the linearized equation is

$$\frac{\mathrm{d}}{\mathrm{d}t}\boldsymbol{x}(t) = \boldsymbol{A}\boldsymbol{x}(t) + \boldsymbol{B}\boldsymbol{v}(t), \tag{1.14}$$

where $\boldsymbol{v}(t) = \boldsymbol{u}(t) - \bar{\boldsymbol{u}}$.

Example Let us linearize the system described by equation (1.2). To define an input-output system, we consider the parameter k_2 as the input by setting $u(t) = k_2(t)$ and the concentration of S_2 as output: $y(t) = s_2(t) - \bar{s}_2$. We linearize about the equilibrium with input set to $\bar{u} = 20$ nMh^{-1}. The corresponding steady state is $(\bar{s}_1, \bar{s}_2) \approx (5.53, 3.37)$.

We now compute the Jacobian $\boldsymbol{A} = \begin{bmatrix} a_{11} & a_{12} \\ a_{21} & a_{22} \end{bmatrix}$:

$$a_{11} = \frac{\partial f_1}{\partial s_1} = -(k_{-1} + k_{-3}), \qquad a_{12} = \frac{\partial f_1}{\partial s_2} = k_3,$$

$$a_{21} = \frac{\partial f_2}{\partial s_1} = -\frac{k_4 \bar{u} q \bar{s}_1^{q-1}}{(1 + k_4 \bar{s}_1^q)^2} + k_{-3}, \qquad a_{22} = \frac{\partial f_2}{\partial s_2} = -(k_{-2} + k_3),$$

and the input matrix $\boldsymbol{B} = \begin{bmatrix} b_1 \\ b_2 \end{bmatrix}$:

$$b_1 = \frac{\partial f_1}{\partial u} = 0, \qquad b_2 = \frac{\partial f_2}{\partial u} = \frac{1}{(1 + k_4 \bar{s}_1^q)}.$$

The output equation has $\boldsymbol{C} = [0 \ \ 1]$ and $\boldsymbol{D} = 0$. ■

1.8 Frequency-Domain Analysis

The longtime (asymptotic) behavior of linear, time-invariant dynamical systems can be elegantly analyzed through an approach that makes use of a parallel, *frequency-domain* description of input and output signals. There are two features of LTI systems that are exploited in this analysis. The first is simply the linear nature of their input-output behavior that implies an *additive property*: provided the system starts with initial condition $\mathbf{x}(0) = \mathbf{0}$ (which corresponds to the nominal steady state of the biochemical network), the output produced by the sum of two inputs is the sum of the outputs produced independently by the two inputs. That is, if input $\mathbf{u}_1(t)$ elicits output $\mathbf{y}_1(t)$, and input $\mathbf{u}_2(t)$ yields output $\mathbf{y}_2(t)$, then input $\mathbf{u}_1(t) + \mathbf{u}_2(t)$ leads to output $\mathbf{y}_1(t) + \mathbf{y}_2(t)$.

The additive property allows a reductionist approach to the analysis of system response: if a complicated input can be written as a sum of simpler signals, the response to each of these simpler inputs can be addressed separately, and the original response can be found through a straightforward summation. This leads to a satisfactory procedure provided we are able to find a family of "simple" functions with the following two properties: (1) the family has to be "complete" in the sense that an arbitrary signal can be decomposed into a sum of functions chosen from this family; and (2) it must enjoy the property that the asymptotic response of a linear system to inputs chosen from the family is easily characterized. The family of sinusoids (sines and cosines) satisfies both of these conditions. The decomposition of a signal $y(t)$ into a combination of sinusoids is the foundation of *Fourier analysis* (Körner, 1988). This technique allows the description of a signal $y(t)$ in terms of its *Fourier transform* $Y(\omega)$, which provides a record of the frequency content of $y(t)$ and is an alternative characterization of the original function. In essence, the Fourier transform $Y(\omega)$ indicates the coefficients that appear in decomposing the original function $y(t)$ into a "sum" of sinusoids at different frequencies ω. (The sum is really an integral over a continuum of frequencies.)

In theory, the signal $y(t)$ can be recovered from $Y(\omega)$ by an inverse transform (which amounts to summing over the sinusoids). In practice, recovery of a signal from its transform is difficult to achieve. Nevertheless, important aspects of the signal can be gleaned directly from the graph of the transform. In particular, one can determine what sort of variations dominate the signal (for example, low-frequency or high-frequency) by comparing the content at various frequencies. Quickly varying signals have transforms with most of their content at high frequencies, whereas slowly varying functions show primarily low-frequency content. The second crucial property of linear, time-invariant systems is that, as mentioned above, their response to sinusoidal inputs can be easily described. For LTI systems, a sinusoidal input of frequency ω_0:

$$u(t) = \sin(\omega_0 t),$$

generates an output that is, after an initial transient, a sinusoid of the same frequency:

$$y(t) = A(\omega_0)\sin(\omega_0 t + \phi(\omega_0)).$$

This longtime response can be characterized by two frequency-dependent functions: $A(\omega_0)$, the amplitude of the oscillatory output, known as the *system gain*; and $\phi(\omega_0)$, the phase of the oscillatory output, referred to as the *phase shift*. As indicated, these depend on the particular frequency ω_0 of the input signal. The particular gain and phase shift that correspond to each frequency can be conveniently described by the assignment of a single complex number $A(\omega)e^{j\phi(\omega)}$ to each frequency ω. This complex-valued function is called the *frequency response* of the system.

1.8.1 The Laplace Transform

Although the Fourier transform provides a valuable description of a signal in terms of its frequency content, the definition is not a useful starting point for calculations. A more general tool, the *Laplace transform* (Körner, 1988), fills that role. The Laplace transform of a signal $y(t)$ is denoted $Y(s)$, where $s = \sigma + j\omega$ is a complex number with real part σ and imaginary part ω. The Fourier transform can be thought of as the special case of the Laplace transform in which $\sigma = 0$, although, in some cases, a function may have a Laplace transform but not have a well-defined Fourier transform (Körner, 1988).

The Transfer Function

For a linear, time-invariant system with zero initial conditions, input $\boldsymbol{u}(t)$, and output $\boldsymbol{y}(t)$, the transfer function is defined in terms of the Laplace transforms of the output $\boldsymbol{Y}(s)$ and the input $\boldsymbol{U}(s)$:

$$\boldsymbol{Y}(s) = \boldsymbol{G}(s)\boldsymbol{U}(s).$$

In the specific case that the system is defined by the linear differential equation (1.12) with output (1.13), the transfer function is given by

$$\boldsymbol{G}(s) = \boldsymbol{C}(s\boldsymbol{I} - \boldsymbol{A})^{-1}\boldsymbol{B} + \boldsymbol{D}. \tag{1.15}$$

This function will, in general, be matrix valued, but it is scalar valued when dealing with SISO systems. The frequency response of the system is the restriction of the transfer function to arguments of the form $s = j\omega$.

The complex-valued frequency response $G(j\omega) = A(\omega)e^{j\phi(\omega)}$ can be plotted in a number of ways. Perhaps the most useful of these visualizations is the *Bode plot*, in which the magnitude and argument of the frequency response are plotted separately (Bode, 1945). The magnitude $A(\omega)$, the system gain, is plotted on a log-log scale, where the gain is measured in decibels (dB) (defined by x dB $= 20 \log_{10} x$). The argument $\phi(\omega)$, the phase shift, appears on a semilog plot, with log frequency plotted against phase in degrees. An example is given in figure 1.15. The frequency-filtering properties of the system can be read directly from the magnitude plot. In this example, the higher-frequency content of an input signal will be attenuated (i.e., filtered)

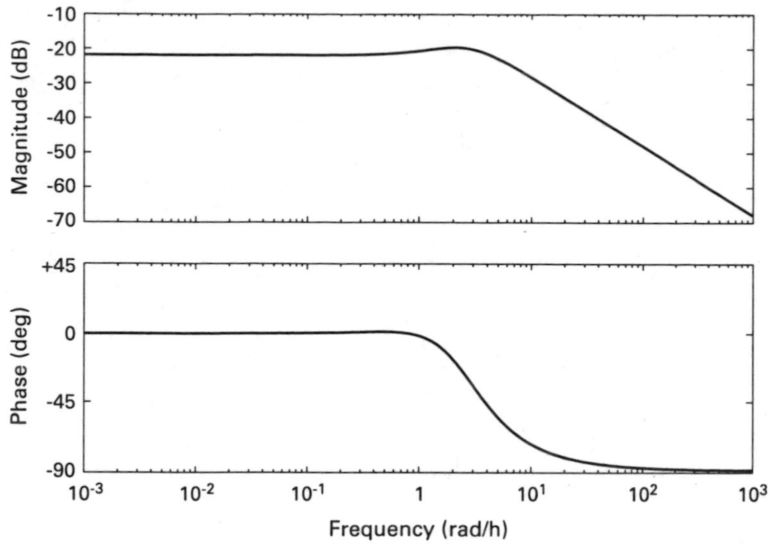

Figure 1.15
Frequency response of system described by equation (1.2), with $u = k_2$ and $y = s_2$. Using the linearization about the equilibrium, we computed the system's Bode magnitude (top) and phase (bottom) plots. From the magnitude plot, it can be seen that the system behaves as a low-pass filter with a cutoff frequency of approximately 1 rad/h.

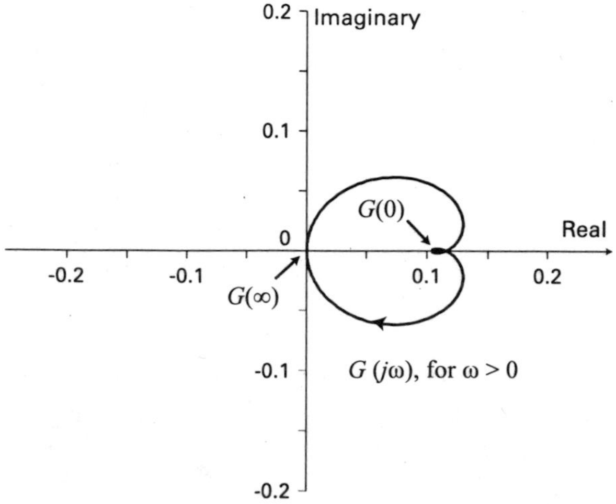

Figure 1.16
Nyquist plot. The frequency response of figure 1.15 can alternatively be represented as a Nyquist plot. The complex values of $G(j\omega)$ are plotted as the frequency ω ranges from $-\infty$ to ∞. The curve's arrowhead indicates the direction of increasing ω. The resulting curve is, in general, symmetric. Moreover, since the values of $G(j\omega)$ tend to zero for large frequencies ($\omega \to \pm\infty$), the plot generates a closed curve in the complex plane. The stability of a system under negative feedback can be determined by the number of times that the open-loop plot encircles the -1 point. In this case, there are no encirclements.

due to the low gain at frequencies above 1 rad/h. An input that consists solely of these high-frequency components (e.g., a highly variable noise signal) may be completely attenuated by the system (i.e., result in near-zero output), whereas other signals will have such highly variable noise filtered out by the system.

An alternative method for visualizing the frequency response is to plot it as a parametrized curve on the complex plane $\omega \mapsto A(\omega)e^{j\phi(\omega)}$. The resulting curve, known as a *Nyquist plot*, provides a valuable tool for addressing the stability of the system when the static output feedback with $k_y(y) = -y$ (called the *unity-feedback* closed loop) is implemented. An example is shown in figure 1.16. The Nyquist stability criterion will be taken up in chapters 9 and 10.

1.9 Controllability and Observability

There are two concepts that, though central to control engineering, are relatively unknown outside the field. Controllability and its counterpart observability deal, respectively, with the relationship between input and state and between state and output. To offer the simplest illustration, we will assume linear systems with one input or one output.

1.9.1 Controllability

Consider the system

$$\frac{\mathrm{d}}{\mathrm{d}t}\boldsymbol{x}(t) = \boldsymbol{A}\boldsymbol{x}(t) + \boldsymbol{B}u(t), \qquad \boldsymbol{x}(0) = \boldsymbol{x}_0, \tag{1.16}$$

where the state $\boldsymbol{x}(t)$ is an n-dimensional vector.

A system is *controllable* if the control input $u(t)$ is able to drive the state $\boldsymbol{x}(t)$ from any nonzero initial condition to the origin. Thus controllability is a measure of our ability to influence the system's state through $u(t)$. Specifically, we say that the system is "controllable over an interval $[0, t_f]$" if, for any initial condition $\boldsymbol{x}_0$, there exists an input $u(t)$, defined over the interval $[0, t_f]$ such that the solution $\boldsymbol{x}(t)$ of equation (1.16) satisfies $\boldsymbol{x}(t_f) = \boldsymbol{0}$.

Though the definition involves the dynamic behavior of the system, there is a simple algebraic test for controllability of linear, time-invariant systems. It relies only on the matrices $\boldsymbol{A}$ and $\boldsymbol{B}$. In particular, the system (1.16) is controllable over $[0, t_f]$ if and only if the matrix

$$\mathcal{C} = [\boldsymbol{B} \quad \boldsymbol{A}\boldsymbol{B} \quad \cdots \quad \boldsymbol{A}^{n-1}\boldsymbol{B}]$$

is invertible. Alert readers will note that the time interval $[0, t_f]$ does not play a role in the test. Consequently, if a system is controllable over one time interval, it is controllable over any time interval, and the time interval can be dropped. (The time interval plays a significant role if there is a bound on the input values.)

One of the more significant consequences of controllability arises when a linear state-feedback controller is implemented:

$$u(t) = -\boldsymbol{K}\boldsymbol{x}(t). \tag{1.17}$$

In this case, replacing $u(t)$ in equation (1.16) with the control law (1.17) gives rise to the following closed-loop dynamical system:

$$\frac{\mathrm{d}}{\mathrm{d}t}\boldsymbol{x}(t) = (\boldsymbol{A} - \boldsymbol{B}\boldsymbol{K})\boldsymbol{x}(t), \qquad \boldsymbol{x}(0) = \boldsymbol{x}_0. \tag{1.18}$$

The stability of this system is determined by the eigenvalues of the matrix $\boldsymbol{A} - \boldsymbol{B}\boldsymbol{K}$. It is a remarkable fact that, if the system is controllable, the location of these eigenvalues can be assigned arbitrarily by the appropriate choice of $\boldsymbol{K}$.

1.9.2 Observability

To address observability, we consider systems with a specified output signal, but no input. In particular:

$$\frac{\mathrm{d}}{\mathrm{d}t}\boldsymbol{x}(t) = \boldsymbol{A}\boldsymbol{x}(t), \qquad \boldsymbol{x}(0) = \boldsymbol{x}_0, \tag{1.19a}$$

$$y(t) = \boldsymbol{C}\boldsymbol{x}(t). \tag{1.19b}$$

The system is *observable* if, based on knowledge of the output $y(t)$ over some time interval, we can discern the state $\boldsymbol{x}$ of the system at the beginning of the interval. Recall that knowledge of the initial condition allows us to determine the state at subsequent times. Specifically, we say that the system (1.19) is observable over the time interval $[0, t_f]$ if any initial state $\boldsymbol{x}_0$ is uniquely determined by the ensuing output $y(t)$ for $t \in [0, t_f]$. As with controllability, an algebraic test can be used to determine observability. In particular, the system (1.19) is observable over $[0, t_f]$ if and only if the matrix

$$\mathcal{O} = \begin{bmatrix} \boldsymbol{C} \\ \boldsymbol{C}\boldsymbol{A} \\ \vdots \\ \boldsymbol{C}\boldsymbol{A}^{n-1} \end{bmatrix} \tag{1.20}$$

is invertible. As with controllability, the test is independent of the specific time interval, which implies that, if a system is observable over one time interval, it is observable over any time interval.

Observability guarantees that one can "look back" along the output signal and determine the state at a previous time. Typically, it is more useful to have real-time knowledge of the system's state. An *observer* is an auxiliary dynamical system that provides an asymptotically correct estimate for the current state. Define the system

$$\frac{\mathrm{d}}{\mathrm{d}t}\tilde{\boldsymbol{x}}(t) = \boldsymbol{A}\tilde{\boldsymbol{x}}(t) + \boldsymbol{L}(y(t) - \boldsymbol{C}\tilde{\boldsymbol{x}}(t)),$$

where the matrix $\boldsymbol{L}$ is yet to be chosen. The state $\tilde{\boldsymbol{x}}$ of the observer serves as an estimate of the state of the system (1.19). Defining the estimation error as

$$\boldsymbol{e}(t) = \boldsymbol{x}(t) - \tilde{\boldsymbol{x}}(t),$$

we see that

$$\frac{\mathrm{d}}{\mathrm{d}t}\boldsymbol{e}(t) = (\boldsymbol{A} - \boldsymbol{L}\boldsymbol{C})\boldsymbol{e}(t).$$

If this system is stable, then $\boldsymbol{e}(t)$ tends to zero or, equivalently, $\tilde{\boldsymbol{x}}$ tends to $\boldsymbol{x}$. This system will be stable provided that the eigenvalues of $\boldsymbol{A} - \boldsymbol{LC}$ have negative real parts. The matrix $\boldsymbol{L}$ can be chosen arbitrarily. This is analogous to the problem of choosing a control gain $\boldsymbol{K}$ so as to make $\boldsymbol{A} - \boldsymbol{BK}$ stable, considered above. As in that case, if the system is observable, then $\boldsymbol{L}$ can be chosen so that the system is stable. (In fact, there are no constraints on the placement of the eigenvalues of $\boldsymbol{A} - \boldsymbol{LC}$.) When the system equations are subject to stochastic disturbances, the corresponding observer is known as a *filter*. In the specific case that the matrix $\boldsymbol{L}$ minimizes the variance of the estimation error, the observer is known as the *Kalman filter* (Kalman, 1960). An application of the Kalman filter to the analysis of the signaling pathway regulating bacterial chemotaxis can be found in the work of Andrews et al. (2006).

2 Modeling and Analysis of Stochastic Biochemical Networks

Mustafa Khammash

The cellular environment is replete with noisy processes. A key source of this "intrinsic" noise is the randomness that characterizes the motion of cellular constituents at the molecular level. Cellular noise results not only in random fluctuations (over time) within individual cells, but also in phenotypic variability among clonal cellular populations. In some instances, fluctuations are suppressed downstream through intricate dynamical networks that act as noise filters. Yet, in other important instances, noise-induced fluctuations are exploited to the cell's advantage. Researchers are just now beginning to understand that the richness of stochastic phenomena in biology depends directly on the interactions of dynamics and noise and on the mechanisms through which these interactions occur. This chapter outlines some of the key approaches for the modeling and analysis of cellular noise and the resulting fluctuations in the copy numbers of cellular constituents.

2.1 Noise in Biological Networks: Origins and Implications

Events in biological networks follow from biochemical reactions at the molecular level. The random nature of such reactions can be traced back to the random collisions among reactant molecules whose trajectories are driven by thermal motion. Such randomness leads to fluctuations in the molecular copy numbers of reactants both among similar cells and within a single cell over time. These fluctuations (commonly referred to as noise) can propagate downstream and impact events and processes in accordance to the dynamics of the network interconnections. Cellular noise has been measured experimentally and classified according to its source (Elowitz et al., 2002; Swain et al., 2002): *intrinsic noise* refers to noise originating within the boundaries of the process under consideration and is due to the inherent stochastic nature of chemical reactions, whereas *extrinsic noise* has origins that are more global and affects all processes in the cell under consideration in a similar way (for example, regulatory protein copy numbers, RNAP numbers). Noise, both intrinsic and extrinsic, can play a critical role in biological processes.

a

b

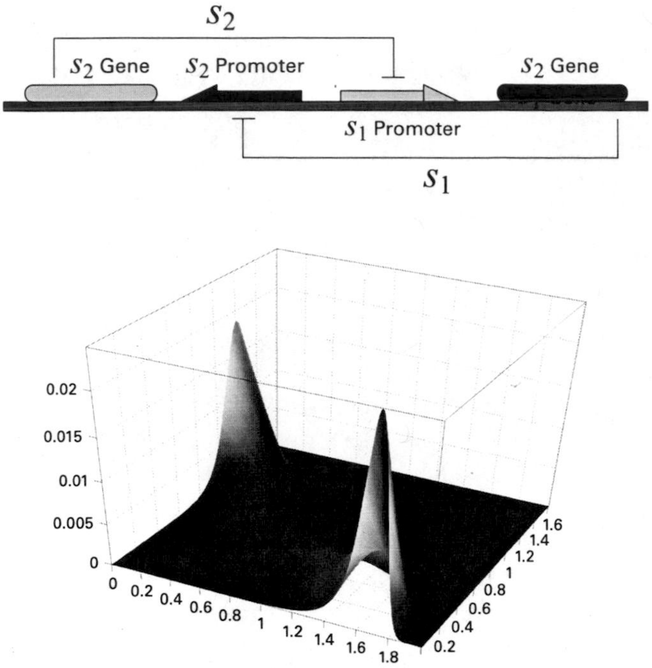

Figure 2.1
(a) Gardner-Cantor-Collins synthetic genetic toggle switch (Gardner et al., 2000). Protein s_1 suppresses the expression of the s_2 gene; protein s_2 suppresses the expression of the s_1 gene. (b) Bimodal nature of the distribution of proteins s_1 and s_2 that can arise for some parameter values of the toggle switch. The distribution shown was computed using the finite state projection method described in section 2.3.4.

McAdams and Arkin (1997, 1999) proposed that lysis-lysogeny fate decisions for phage λ are determined by a noise-driven stochastic switch, implying that the fate of a given cell can be determined only in a probabilistic sense. Another stochastic switch, which governs the piliation of *Escherichia coli*, has been modeled by Munsky et al. (2005). Aside from endogenous switches, bistable genetic switches have been constructed and tested (Gardner et al., 2000; Hasty et al., 2002). Depending on their parameter values, such switches are driven by stochastic noise to exhibit states with a bimodal probability distribution (figure 2.1). Elowitz and Leibler (2000) reported on the first synthetic oscillator, called the "repressilator," a novel circuit of three genes, each expressing a product that represses the next gene, thereby creating a feedback loop of three genes. A deterministic model and a discrete stochastic model that captures the effect of noise guided their design. The role of noise in the operation of the repressilator was recently studied by Yoda et al. (2007). Noise also appears to play an important role in the noise-enhanced robustness of oscillations in relaxation oscillators, for example, in the circadian rhythm. This effect, which is sometimes referred to as coherence resonance, has been studied by Vilar et al. (2002) and El-Samad and Khammash (2006a). Yet another curious effect of noise can be seen in the fluctuation-enhanced sensitivity of intracellular regulation referred to as "stochastic focusing" (Paulsson et al., 2000). In gene expression, noise-induced fluctuations in gene products have been the subject of considerable interest (El-Samad and Khammash, 2006b; Isaacs et al., 2003; Paulsson, 2004; Rosenfeld et al., 2002; Swain, 2004; Thattai and van Oudenaarden, 2001). Many of these studies look at the propagation of noise in gene networks, as well as the impact and limitations of various types of feedback in suppressing such fluctuations.

2.1.1 Deterministic versus Stochastic Modeling

One approach to modeling reactions in biochemical networks, discussed in chapter 1, uses the law of mass action, which results in a set of differential equations that describe the evolution of concentrations of species adopted by the network over time. As an example, consider the reaction

$$A + B \xrightarrow{k} C.$$

A deterministic formulation of chemical kinetics yields the following description:

$$\frac{\mathrm{d}[C]}{\mathrm{d}t} = k[A] \times [B],$$

where $[\cdot]$ denotes the concentration, which is considered to be a continuous variable. In contrast, a discrete stochastic formulation of the same reaction describes the *probability* that the numbers of molecules of species A and B take certain integer values at a given time t. In this way, populations of the species within the network of interest are treated as random variables. In this description, reactions take place randomly according to certain probabilities determined by several factors including reaction rates and species populations. For example, given certain integer populations of A and B, say N_A and N_B, at time t, the probability that the above reaction takes place within the interval $[t, t + \mathrm{d}t)$ is proportional to $(N_A \times N_B/V)\,\mathrm{d}t$, where V is the volume of the space containing the molecules of A and B and $\mathrm{d}t$ is a small time increment.

Thus, in this mesoscopic stochastic formulation of chemical kinetics, molecular species are characterized by their probability density function, which quantifies the amount of fluctuations around a certain mean value. When molecule numbers get large (while maintaining the same initial concentration), fluctuations become negligible, and the mesoscopic description converges to the macroscopic description. In typical cellular environments where small volumes and molecule copy numbers are the norm, mesoscopic stochastic descriptions offer a more accurate representation of chemical reactions and their accompanying fluctuations. Because they can generate distinct phenomena that simply cannot be captured by deterministic descriptions, such fluctuations need to be accounted for. The next section gives a more detailed description of the stochastic framework for modeling chemical reactions and provides a connection between the stochastic and deterministic descriptions.

2.2 Stochastic Chemical Kinetics

Consider a chemically reacting system of volume Ω containing N molecular species $S_1, \dots, S_N$ that react through M allowable reactions $R_1, \dots, R_M$. We assume that the system is well stirred and is in thermal equilibrium, thus the reaction volume is at a constant temperature T and the molecules move due to the thermal energy. Let $\boldsymbol{X}(t) = [X_1(t) \dots X_N(t)]^T$ be the state vector, where $X_i(t)$ is a random variable that describes the number of molecules of species S_i in the system at time t. Elementary reactions may be either monomolecular: $S_i \to$ Products, or bimolecular: $S_i + S_j \to$ Products. Each reaction channel R_k defines a transition from some state $\boldsymbol{X} = \boldsymbol{x}_i$ to some other state $\boldsymbol{X} = \boldsymbol{x}_i + \boldsymbol{s}_k$, which reflects the change in the state after the reaction has taken place. The variable $\boldsymbol{s}_k$ is termed the *stoichiometric vector*, and the set of all M reactions define the *stoichiometry matrix*:

$$\boldsymbol{S} = [\boldsymbol{s}_1 \dots \boldsymbol{s}_M].$$

Associated with each reaction R_k is a *propensity function*, $w_k(\boldsymbol{x})$, which captures the rate of the reaction k. Specifically, $w_k(\boldsymbol{x})\,\mathrm{d}t$ is the probability that, given the system is in state $\boldsymbol{x}$ at time t, the kth reaction will take place exactly once in the time interval $[t, t + \mathrm{d}t)$. The propensity function for various reaction types is given in table 2.1.

2.2.1 The Chemical Master Equation

Also known as the forward Kolmogorov equation, the *chemical master equation* (CME) describes the time evolution of the probability that the system is in a given

Table 2.1
Propensity functions for the various elementary reaction types

Reaction type		Propensity function
$S_i \to$ Products		$c\boldsymbol{x}_i$
$S_i + S_j \to$ Products	$(i \neq j)$	$c'\boldsymbol{x}_i\boldsymbol{x}_j$
$S_i + S_i \to$ Products		$c''\boldsymbol{x}_i(\boldsymbol{x}_i - 1)/2$

Note: If we denote by k, k', and k'' the reaction rate constants from deterministic mass action kinetics for the first, second, and third reaction types shown in the table, it can be shown that $c = k$, $c' = k'/\Omega$, and $c'' = 2k''/\Omega$.

state $\boldsymbol{x}$. The CME can be derived based on the Markov property of chemical reactions. Suppose the system is in state $\boldsymbol{x}$ at time t. Within an error of order $\mathcal{O}(\mathrm{d}t^2)$, the following statements apply:

- The probability that an R_k reaction fires exactly once in the time interval $[t, t+\mathrm{d}t]$ is given by $w_k(\boldsymbol{x})\,\mathrm{d}t$.
- The probability that no reactions fire in the time interval $[t, t+\mathrm{d}t]$ is given by $1 - \sum_k w_k(\boldsymbol{x})\,\mathrm{d}t$.
- The probability that more than one reaction fires in the time interval $[t, t+\mathrm{d}t]$ is zero.

Let $P(\boldsymbol{x}, t)$ denote the probability that the system is in state $\boldsymbol{x}$ at time t. We can express $P(\boldsymbol{x}, t+\mathrm{d}t)$ as follows:

$$P(\boldsymbol{x}, t+\mathrm{d}t) = P(\boldsymbol{x}, t)\left(1 - \sum_{k=1}^{M} w_k(\boldsymbol{x})\,\mathrm{d}t\right) + \sum_{k=1}^{M} P(\boldsymbol{x} - s_k, t) w_k(\boldsymbol{x} - s_k)\,\mathrm{d}t + \mathcal{O}(\mathrm{d}t^2).$$

On the right-hand side, the first term is the probability that the system is already in state $\boldsymbol{x}$ at time t and no reactions occur in the next $\mathrm{d}t$. In the second term, the kth term in the summation is the probability that the system at time t is an R_k reaction away from being at state $\boldsymbol{x}$, and that an R_k reaction takes place in the next $\mathrm{d}t$.

Moving $P(\boldsymbol{x}, t)$ to the left-hand side, dividing by $\mathrm{d}t$, and taking the limit as $\mathrm{d}t$ goes to zero yields the chemical master equation:

$$\frac{\mathrm{d}P(\boldsymbol{x}, t)}{\mathrm{d}t} = \sum_{k=1}^{M} (w_k(\boldsymbol{x} - \boldsymbol{s}_k) P(\boldsymbol{x} - \boldsymbol{s}_k, t) - w_k(\boldsymbol{x}) P(\boldsymbol{x}, t)). \tag{2.1}$$

2.2.2 Deterministic and Stochastic Models: A Connection

We now establish a connection between the stochastic process $\boldsymbol{X}(t)$ and the solution of the deterministic reaction rate equations arising from conventional mass-action kinetics. The latter corresponds to the trajectories of the concentrations of species $S_1, \ldots, S_N$. Let these concentrations be denoted by $\Phi(t) = [\Phi_1(t) \ldots \Phi_N(t)]^T$. Accordingly, $\Phi(\cdot)$ satisfies the mass-action ordinary differential equation:

$$\frac{\mathrm{d}\Phi}{\mathrm{d}t} = \boldsymbol{S} f(\Phi(t)), \qquad \Phi(0) = \Phi_0.$$

For a meaningful comparison with the stochastic solution, we shall compare the function $\Phi(t)$ with the volume-normalized stochastic process $\boldsymbol{X}^{\Omega}(t) = \boldsymbol{X}(t)/\Omega$. We now ask, how does $\boldsymbol{X}^{\Omega}(t)$ relate to $\Phi(t)$? The answer is given by the following fact, which is a consequence of the law of large numbers (Ethier and Kurtz, 1986):

Fact 2.1 Let $\Phi(t)$ be the deterministic solution to the reaction rate equations

$$\frac{d\Phi}{dt} = \boldsymbol{S}f(\Phi(t)), \qquad \Phi(0) = \Phi_0.$$

Let $\boldsymbol{X}^{\Omega}(t)$ be the *stochastic* representation of the same chemical system with $\boldsymbol{X}^{\Omega}(0) = \Phi_0$. Then, for every $t \geq 0$,

$$\lim_{\Omega \to \infty} \sup_{s \leq t} |\boldsymbol{X}^{\Omega}(s) - \Phi(s)| = 0, \quad \text{almost surely.} \qquad \blacksquare$$

Put into words, this fact states that over any finite time interval, the stochastic description *converges* to the deterministic one *in the thermodynamic limit.* Although this result is reassuring, in practice, the large volume assumption cannot be justified: the cell volume is fixed. Indeed, a stochastic description could differ appreciably from its large volume limit—a fact that makes stochastic models necessary.

2.3 Stochastic Analysis Tools

Stochastic analysis tools may be broadly divided into four categories: (1) Monte Carlo methods, which compute sample paths whose statistics are used to extract information about the system; (2) methods that approximate the stochastic process $\boldsymbol{X}(t)$ by solutions of certain stochastic differential equations; (3) methods that compute the trajectories of various moments of $\boldsymbol{X}(t)$; and (4) methods that compute the evolution of probability densities of the stochastic process $\boldsymbol{X}(t)$.

2.3.1 Monte Carlo Simulations

Because the chemical master equation is typically infinite dimensional with no obvious analytical solution, most analyses at the mesoscopic scale have been conducted using Monte Carlo algorithms. The most widely used of these algorithms is Gillespie's stochastic simulation algorithm (SSA; Gillespie, 1976) and its variants. These are described next.

The Gillespie Algorithm

Each step of the stochastic simulation algorithm begins at a time t and at a state $\boldsymbol{X}(t) = \boldsymbol{x}$ and comprises three substeps:

1. Generate the time until the next reaction;
2. Determine which reaction occurs at that time; and
3. Update the time and state to reflect the previous two choices.

The SSA approach is exact in the sense that it results in a random variable with a probability distribution exactly equal to the solution of the corresponding chemical master equation. However, each run of the SSA provides only a single trajectory. Thus numerous trajectories need to be generated that are then used to compute statistics of interest.

We now describe these substeps in more detail.

To each of the reactions $\{R_1, \ldots, R_M\}$, we associate a random variable $\mathcal{T}_i$ that describes the time for the next firing of reaction R_i. A key fact is that $\mathcal{T}_i$ is exponentially distributed with parameter w_i. From these, we can define two additional random variables, one continuous and the other discrete:

$$\mathcal{T} = \min_i\{\mathcal{T}_i\} \quad \text{(time to the next reaction)},$$

$$\mathcal{R} = \arg\min_i\{\mathcal{T}_i\} \quad \text{(index of the next reaction)}.$$

It can be shown that

1. $\mathcal{T}$ is exponentially distributed with parameter $\sum_i w_i$.

2. $\mathcal{R}$ has the discrete distribution $P(\mathcal{R} = k) = \dfrac{w_k}{\sum_i w_i}$.

Gillespie's simulation algorithm relies on samples of the random variables $\mathcal{T}$ and $\mathcal{R}$ to advance through each sample path.

Gillespie's Stochastic Simulation Algorithm:

- *Step 0* Initialize time t and state population $\boldsymbol{x}$.
- *Step 1* Draw a sample τ from the distribution of $\mathcal{T}$ (figure 2.2).
- *Step 2* Draw a sample μ from the distribution of $\mathcal{R}$ (figure 2.2).
- *Step 3* Update time $t \leftarrow t + \tau$. Update the state $\boldsymbol{x} \leftarrow \boldsymbol{x} + \boldsymbol{s}_\mu$.

Leaping Methods

The biggest drawback of the stochastic simulation algorithm is that it must step through one reaction at a time and thus is often prohibitively slow. One approximate accelerated simulation strategy is known as *tau leaping* (Gillespie, 2001), which advances the system by a *preselected* time τ that encompasses more than one reaction event. Roughly speaking, tau leaping requires that τ be chosen small enough so that the following *leap condition* is satisfied: the expected state change induced by the leap must be sufficiently small that no propensity function changes "appreciably" during that time. Tau leaping has been shown to speed up the simulation of certain systems significantly (Gillespie, 2001; Gillespie and Petzold, 2003; Rathinam et al., 2003, 2005), although it is not as foolproof as the SSA. If one takes leaps that are too large, the tau leap assumptions may be violated and the results may be inaccurate

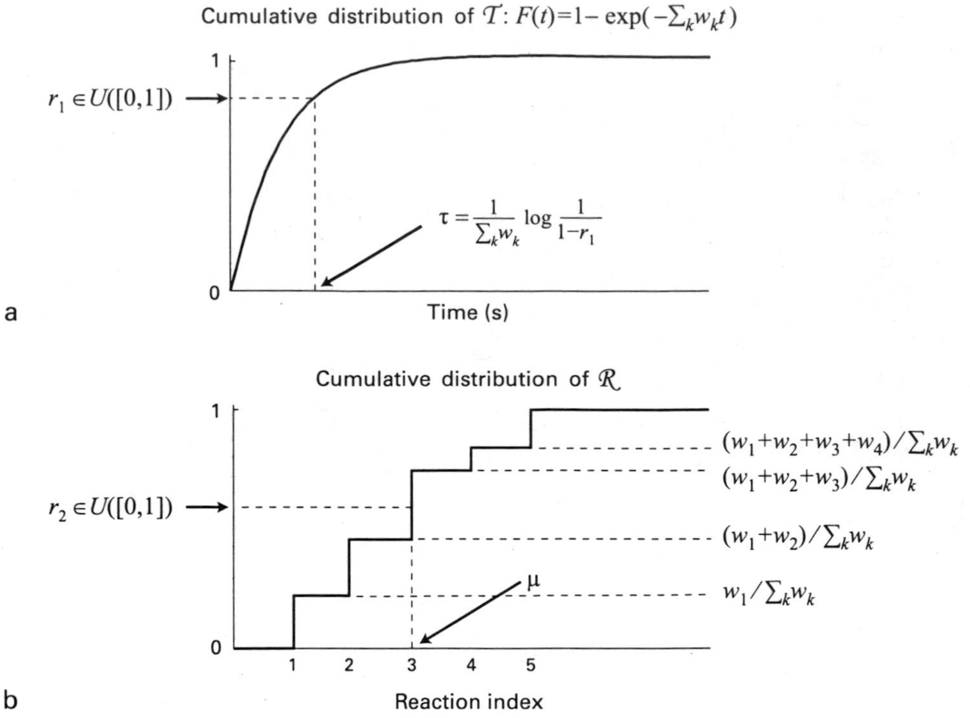

Figure 2.2
Cumulative distribution of the two random variables $\mathcal{T}$ and $\mathcal{R}$. (a) A sample of $\mathcal{T}$ is drawn by first drawing a uniformly distributed random number r_1 and then finding its inverse image under F, the cumulative distribution of $\mathcal{T}$. (b) A sample from the distribution of $\mathcal{R}$ is drawn using a similar procedure.

or even nonsensical. For example, some species populations might be driven negative. Moreover, if the system is stiff, meaning that it has widely varying time scales with the fastest mode being stable, the leap condition will generally limit the size of τ to the time scale of the fastest mode, with the result that large leaps cannot be taken.

2.3.2 Stochastic Differential Equation Approximations

The Chemical Langevin Equation: Diffusion Approximation

One approximation to the chemical master equation, the *chemical Langevin equation*, can be obtained through the tau-leaping multireaction update formula: with $\boldsymbol{X}(t) = \boldsymbol{x}$, suppose that the leap time τ can be taken small enough to satisfy the leap condition, but large enough that $w_k(\boldsymbol{x})\tau \gg 1$ for every reaction. If the leap condition holds, the number of reactions in the interval τ is a Poisson random variable with mean and variance equal to $w_k(\boldsymbol{x})\tau$. When this quantity is much larger than one, this random variable is well approximated by a normal random variable with the same mean and variance. This leads to the *Langevin leaping formula*:

$$X(t+\tau) \approx x + \sum_{k=1}^{M} s_k w_k(x)\tau + \sum_{k=1}^{M} s_k \sqrt{w_k(x)}\mathcal{N}_k(0,1)\sqrt{\tau}, \tag{2.2}$$

which expresses the state increment $X(t+\tau) - x$ as the sum of two terms: a deterministic "drift" term proportional to τ and a fluctuating "diffusion" term proportional to $\sqrt{\tau}$, where the $\mathcal{N}_k(0,1)$ denote independent standard normal random variables (Gillespie, 2000, 2002). From equation (2.2), one can approximate the stochastic process X by another stochastic process V, which is described by the following nonlinear stochastic differential equation:

$$dV(t) = \sum_{k=1}^{M} s_k w_k(V(t))\,\mathrm{d}t + \sum_{k=1}^{M} s_k \sqrt{w_k(V(t))}\,\mathrm{d}B_k(t),$$

where the $B_k(t)$ are independent standard Brownian motion processes. This approximation is sometimes called the *chemical Langevin equation approximation* or the *diffusion approximation* (Gillespie, 2000, 2002; Kurtz, 1978).

van Kampen's Linear Noise Approximation

Another approximation that leads to a stochastic differential equation is van Kampen's linear noise approximation (LNA; Elf and Ehrenberg, 2003; Khammash and El-Samad, 2005; Tomioka et al., 2004; van Kampen, 1981). It is essentially an approximation to the process $X(t)$ that takes advantage of the fact that in the large volume limit ($\Omega \to \infty$), the process $X^{\Omega}(t) = X(t)/\Omega$ converges to the solution $\Phi(t)$ of the deterministic reaction rate equation:

$$\frac{\mathrm{d}\Phi}{\mathrm{d}t} = Sf(\Phi(t)), \qquad \Phi(t_0) = \Phi_0.$$

Defining a scaled "error" process,

$$V^{\Omega}(t) = \sqrt{\Omega}(X^{\Omega}(t) - \Phi(t)),$$

and using the central limit theorem, it can be shown (Ethier and Kurtz, 1986) that $V^{\Omega}(t)$ converges in distribution to the solution $V(t)$ of the linear stochastic differential equation:

$$\mathrm{d}V(t) = J_F(\Phi(t))V(t)\,\mathrm{d}t + \sum_{k=1}^{M} s_k \sqrt{w_k(\Phi(t))}\,\mathrm{d}B_k(t),$$

where J_F denotes the Jacobian of $F(\cdot) = Sf(\cdot)$. Hence the linear noise approximation results in a process $X(t) \approx \Omega\Phi(t) + \sqrt{\Omega}V(t)$, which can be viewed as the sum of a deterministic term given by the solution to the deterministic reaction rate equation, and a zero mean stochastic term given by the solution to a linear stochastic differential equation. Though it is reasonable for systems with sufficiently large numbers of molecules (and volume), examples show that the LNA can yield poor results when this assumption is violated. This includes scenarios where the mean of a stochastic representation does not coincide with the solution of the corresponding reaction rate equation (see Paulsson et al., 2000 on *stochastic focusing*).

2.3.3 Moment Computations

When studying stochastic fluctuations that arise in biochemical networks, one is often interested in computing moments and variances of biochemical species. The moment dynamics can be described using the chemical master equation. To compute the first moment $E[X_i]$, we multiply the CME by x_i and then sum over all $(x_1, \ldots, x_N) \in \mathbf{N}^N$ to get

$$\frac{\mathrm{d}E[X_i]}{\mathrm{d}t} = \sum_{k=1}^{M} s_{ik} E[w_k(\boldsymbol{X})].$$

Similarly, to get the second moments $E[X_i X_j]$, we multiply the CME by $x_i x_j$ and sum over all $(x_1, \ldots, x_N) \in \mathbf{N}^N$, which gives

$$\frac{\mathrm{d}E[X_i X_j]}{\mathrm{d}t} = \sum_{k=1}^{M} s_{ik} E[X_j w_k(\boldsymbol{X})] + E[X_i w_k(\boldsymbol{X})] s_{jk} + s_{ik} E[w_k(\boldsymbol{X})] s_{jk}.$$

These last two equations can be expressed more compactly in matrix form. Defining $w(x) = [w_1(x) \ldots w_M(x)]^T$, the moment dynamics become

$$\frac{\mathrm{d}E[\boldsymbol{X}]}{\mathrm{d}t} = \boldsymbol{S}E[\boldsymbol{w}(\boldsymbol{X})],$$

$$\frac{\mathrm{d}E[\boldsymbol{X}\boldsymbol{X}^T]}{\mathrm{d}t} = \boldsymbol{S}E[\boldsymbol{w}(\boldsymbol{X})\boldsymbol{X}^T] + E[\boldsymbol{w}(\boldsymbol{X})\boldsymbol{X}^T]^T \boldsymbol{S}^T + \boldsymbol{S}(\text{diag } E[\boldsymbol{w}(\boldsymbol{X})])\boldsymbol{S}^T.$$

In general, this set of moment equations cannot be solved explicitly because they will not always be closed: depending on the form of the propensity vector $\boldsymbol{w}(\cdot)$, the dynamics of the first moment $E(\boldsymbol{X})$ may depend on the second moments $E(\boldsymbol{X}\boldsymbol{X}^T)$, the second-moment dynamics may in turn depend on the third moments, and so on, resulting in an infinite system of ordinary differential equations. The following subsections will elaborate on the important special case when moment equations *are* closed, and then discuss approaches to deal with scenarios when they are not.

Special Case: Affine Propensities

If the propensity function is affine:

$$\boldsymbol{w}(\boldsymbol{x}) = \boldsymbol{W}\boldsymbol{x} + \boldsymbol{w}_0 \quad (\boldsymbol{W} \text{ is an } N \times N \text{ matrix; } \boldsymbol{w}_0 \text{ is } N \times 1),$$

then

$$E[\boldsymbol{w}(\boldsymbol{X})] = \boldsymbol{W}E[\boldsymbol{X}] + \boldsymbol{w}_0,$$

and

$$E[\boldsymbol{w}(\boldsymbol{X})\boldsymbol{X}^T] = \boldsymbol{W}E[\boldsymbol{X}\boldsymbol{X}^T] + \boldsymbol{w}_0 E[\boldsymbol{X}^T].$$

This gives us the following moment equations:

$$\frac{\mathrm{d}}{\mathrm{d}t}E[\boldsymbol{X}] = \boldsymbol{S}\boldsymbol{W}E[\boldsymbol{X}] + \boldsymbol{S}\boldsymbol{w}_0,$$

$$\frac{\mathrm{d}}{\mathrm{d}t}E[\boldsymbol{X}\boldsymbol{X}^T] = \boldsymbol{S}\boldsymbol{W}E[\boldsymbol{X}\boldsymbol{X}^T] + E[\boldsymbol{X}\boldsymbol{X}^T]\boldsymbol{W}^T\boldsymbol{S}^T + \boldsymbol{S}\,\mathrm{diag}(\boldsymbol{W}E[\boldsymbol{X}] + \boldsymbol{w}_0)\boldsymbol{S}^T$$
$$+ \boldsymbol{S}\boldsymbol{w}_0 E[\boldsymbol{X}^T] + E[\boldsymbol{X}]\boldsymbol{w}_0^T\boldsymbol{S}^T.$$

Clearly, this is a closed system of linear ODEs that can be solved directly for the first and second moments.

Defining the covariance matrix

$$\boldsymbol{\Sigma} = E[(\boldsymbol{X} - E[\boldsymbol{X}])(\boldsymbol{X} - E[\boldsymbol{X}])^T],$$

we can also compute covariance equations:

$$\frac{\mathrm{d}}{\mathrm{d}t}\boldsymbol{\Sigma} = \boldsymbol{S}\boldsymbol{W}\boldsymbol{\Sigma} + \boldsymbol{\Sigma}\boldsymbol{W}^T\boldsymbol{S}^T + \boldsymbol{S}\,\mathrm{diag}(\boldsymbol{W}E[\boldsymbol{X}] + \boldsymbol{w}_0)\boldsymbol{S}^T.$$

The steady-state moments and covariances can be obtained by solving linear algebraic equations. Let

$$\bar{\boldsymbol{X}} = \lim_{t\to\infty} E[\boldsymbol{X}(t)]$$

and

$$\bar{\boldsymbol{\Sigma}} = \lim_{t\to\infty} \boldsymbol{\Sigma}(t).$$

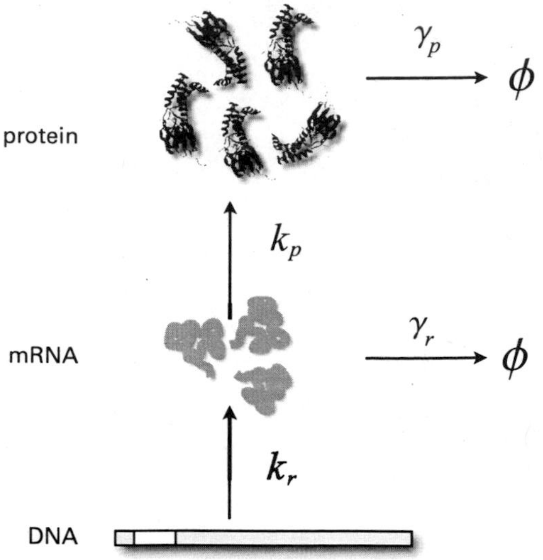

Figure 2.3
Simple model for gene expression. mRNA is transcribed at a rate k_r and is degraded at a rate γ_r. Protein is translated from mRNA at a rate k_p and is degraded at a rate γ_p.

Then

$$\boldsymbol{S}\boldsymbol{W}\overline{\boldsymbol{X}} = -\boldsymbol{S}\boldsymbol{w}_0,$$

and

$$\boldsymbol{S}\boldsymbol{W}\overline{\boldsymbol{\Sigma}} + \overline{\boldsymbol{\Sigma}}\boldsymbol{W}^T\boldsymbol{S}^T + \boldsymbol{S}\,\mathrm{diag}(\boldsymbol{W}\overline{\boldsymbol{X}} + \boldsymbol{w}_0)\boldsymbol{S}^T = \boldsymbol{0}.$$

The latter is an algebraic Lyapunov equation. Such equations arise in many applications in control theory, and efficient numerical techniques exist for their solution.

Example: Application to Gene Expression The above stationary covariance equations can be applied to compute exact expressions for the mean and coefficient of variation of a simple gene expression circuit. Consider the gene expression model shown in figure 2.3.

Let $X_1(t)$ and $X_2(t)$ be random variables describing the number of mRNA and protein molecules, respectively. The stoichiometry matrix for the gene expression reactions is

$$\boldsymbol{S} = \begin{bmatrix} 1 & -1 & 0 & 0 \\ 0 & 0 & 1 & -1 \end{bmatrix},$$

while the propensity vector is given by

$$w(X) = \begin{bmatrix} k_r \\ \gamma_r X_1 \\ k_p X_1 \\ \gamma_p X_2 \end{bmatrix} = \begin{bmatrix} 0 & 0 \\ \gamma_r & 0 \\ k_p & 0 \\ 0 & \gamma_p \end{bmatrix} \begin{bmatrix} X_1 \\ X_2 \end{bmatrix} + \begin{bmatrix} k_r \\ 0 \\ 0 \\ 0 \end{bmatrix}$$

$$= \boldsymbol{W}\boldsymbol{X} + \boldsymbol{w}_0.$$

Defining $\boldsymbol{A} = \boldsymbol{S}\boldsymbol{W}$, we can compute the vector of stationary means:

$$\overline{\boldsymbol{X}} = -\boldsymbol{A}^{-1}\boldsymbol{S}\boldsymbol{w}_0 = \begin{bmatrix} \dfrac{k_r}{\gamma_r} \\ \dfrac{k_p k_r}{\gamma_p \gamma_r} \end{bmatrix}.$$

Defining

$$\boldsymbol{B}\boldsymbol{B}^T = \boldsymbol{S}\,\mathrm{diag}(\boldsymbol{W}\overline{\boldsymbol{X}} + \boldsymbol{w}_0)\boldsymbol{S}^T,$$

we see that the stationary covariance matrix is the solution to the Lyapunov equation:

$$\boldsymbol{A}\overline{\boldsymbol{\Sigma}} + \overline{\boldsymbol{\Sigma}}\boldsymbol{A}^T + \boldsymbol{B}\boldsymbol{B}^T = \boldsymbol{0},$$

which is given by

$$\overline{\boldsymbol{\Sigma}} = \begin{bmatrix} \dfrac{k_r}{\gamma_r} & \dfrac{k_p k_r}{\gamma_r(\gamma_r + \gamma_p)} \\ \dfrac{k_p k_r}{\gamma_r(\gamma_r + \gamma_p)} & \dfrac{k_p k_r}{\gamma_p \gamma_r}\left(1 + \dfrac{k_p}{\gamma_r + \gamma_p}\right) \end{bmatrix}.$$

A common measure of the degree of variability is the *coefficient of variation*, C_v, defined as the standard deviation divided by the mean. The coefficient of variation for the mRNA and protein can now be computed easily:

$$C_{vr} = \left(\frac{1}{k_r/\gamma_r}\right)^{1/2}, \quad \text{and} \quad C_{vp} = \left(\frac{\gamma_r \gamma_p}{k_r k_p}\right)^{1/2}\left(1 + \frac{k_p}{\gamma_r + \gamma_p}\right)^{1/2}.$$

In the case of mRNA, we have just shown the well-known result that the coefficient of variation equals $1/\sqrt{\text{mean}}$. ■

Moment Closure

An important property of the Markov processes that describe chemical reactions is that when one constructs a vector μ with all the first- and second-order statistical uncentered moments of the process state X, this vector evolves according to a *linear* equation of the form

$$\frac{d\mu}{dt} = \boldsymbol{A}\mu + \boldsymbol{B}\bar{\mu}. \tag{2.3}$$

Unfortunately, as pointed out earlier, equation (2.3) is not in general a closed system because the vector $\bar{\mu}$ may contain moments of order larger than two, whose evolution is not provided by equation (2.3). In fact, this will always be the case when bimolecular reactions are involved. To overcome this difficulty, one can approximate the *open linear system* (2.3) by the following *closed nonlinear system*:

$$\frac{d\nu}{dt} = \boldsymbol{A}\nu + \boldsymbol{B}\varphi(\nu), \tag{2.4}$$

where ν is an approximation to the solution μ to (2.3) and $\varphi(\cdot)$ is a *moment closure function* that attempts to approximate the moments in $\bar{\mu}$ based on the values of the moments in μ. The construction of $\varphi(\cdot)$ often relies on postulating a given type for the distribution of X and then expressing the higher-order moments in $\bar{\mu}$ by a nonlinear function $\varphi(\mu)$ of the first- and second-order moments in μ. Postulating a normal distribution is quite popular (Gomez-Uribe and Verghese, 2007; Nasell, 2003b; Whittle, 1957), but when the population standard deviations are not much smaller than the means, choosing $\varphi(\cdot)$ based on a normal distribution assumption often leads to bad approximations. Other authors construct moment closure functions $\varphi(\cdot)$ based on different assumed distributions for X, which include lognormal (Keeling, 2000), Poisson, and binomial (Nasell, 2003a) distributions.

Hespanha (2005) proposed a new technique for moment closure that does not require a priori assumptions on the shape of the distribution for X. Instead, the moment closure $\varphi(\cdot)$ is computed by trying to match all (or a large number of) the time derivatives of the exact solution to equation (2.3) with the corresponding time derivatives of the approximate solution to equation (2.4), for a given set of initial conditions. With this approach, one can indeed match derivatives between equation (2.3) and equation (2.4) with small error (Singh and Hespanha, 2006, 2007a,b). Moreover, this can be done with moment closure functions $\varphi(\cdot)$ that do not depend on the (most often poorly known) parameters of the chemical reactions. This leads to an automated methodology to construct the approximate closed systems (2.4). A set of Matlab scripts that constructs truncated moment dynamics given a set of chemical reactions may be found in Hespanha (2006).

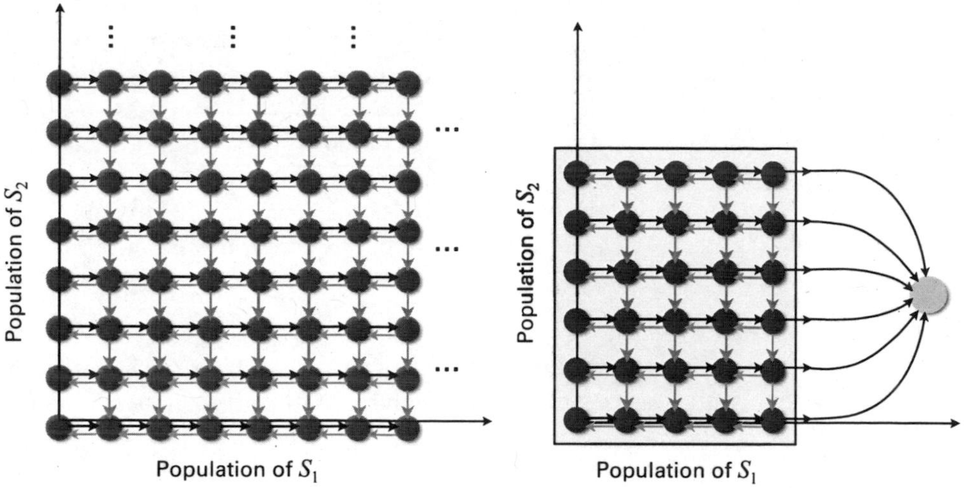

Figure 2.4
Finite-state projection. (Left) State space for a system with two species. The corresponding process is a continuous-time, discrete-state Markov process whose state space is typically quite large or infinite. Arrows indicate possible transitions within states. (Right) Projected system for a specific projection region (*gray box*). The projected system is obtained as follows. Transitions within the projection region are kept unchanged. Transitions that emanate from states within the region and end at states outside (in the original system) are routed to a single absorbing state in the projected system. Transitions into the projection region are deleted. As a result, the projected system is a finite-state Markov process, and the probability of each state can be computed exactly.

2.3.4 Density Computations

Another approach used to analyze models described by the chemical master equation aims to compute the probability density functions for the random variable X. This is achieved by approximate solutions of the CME, using a new analytical approach called the finite-state projection (FSP) (Munsky and Khammash, 2006; Peles et al., 2006). The FSP approach relies on a projection that preserves an important subset of the state space (for example, that supporting the bulk of the probability distribution), while projecting the remaining large or infinite states onto a single "absorbing" state (figure 2.4).

Probabilities for the resulting finite-state Markov chain can be computed exactly, and can be shown to give a lower bound for the corresponding probability for the original full system. The finite-state projection algorithm provides a means of systematically choosing a projection of the chemical master equation that satisfies any prespecified accuracy requirement. The basic idea of the FSP is as follows. In matrix form, the CME may be written as

$$\frac{\mathrm{d}\boldsymbol{P}(t)}{\mathrm{d}t} = \boldsymbol{A}\boldsymbol{P}(t), \tag{2.5}$$

where $\boldsymbol{P}(t)$ is the (infinite) vector of probabilities corresponding to each possible state in the configuration space. The generator matrix $\boldsymbol{A}$ embodies the propensity functions for transitions from one configuration to another and is defined by the reactions and the enumeration of the configuration space. A projection can now be made to achieve an arbitrarily accurate approximation as outlined next. Given an index set of the form $J = \{j_1, j_2, j_3, \ldots\}$ and a vector $\boldsymbol{v}$, let $\boldsymbol{v}_J$ denote the subvector of $\boldsymbol{v}$ chosen according to J, and for any matrix $\boldsymbol{A}$, let $\boldsymbol{A}_J$ denote the submatrix of $\boldsymbol{A}$ whose rows and columns have been chosen according to J. With this notation, we can restate the result from Munsky and Khammash (2006):

Fact 2.2: Finite-State Projection Consider any distribution that evolves according to equation (2.5). Let $\boldsymbol{A}_J$ be a principal submatrix of $\boldsymbol{A}$ and $\boldsymbol{P}_J$ be a subvector of $\boldsymbol{P}$, both corresponding to the indexes in J. If, for a given $\varepsilon > 0$ and $t_f \geq 0$, we have

$$\mathbf{1}^T \exp(\boldsymbol{A}_J t_f)\boldsymbol{P}_J(0) \geq 1 - \varepsilon,$$

then

$$\|\exp(\boldsymbol{A}_J t_f)\boldsymbol{P}_J(0) - \boldsymbol{P}_J(t_f)\|_1 \leq \varepsilon. \tag{2.6}$$

■

Inequality (2.6) provides a bound on the error between the exact solution $\boldsymbol{P}_J$ to the (infinite) chemical master equation and the matrix exponential of the (finite) reduced system with generator $\boldsymbol{A}_J$. This result is the basis for an algorithm to compute the probability density function with guaranteed accuracy. The FSP approach and various improvements on the main algorithm are described by Munsky and Khammash (2008).

2.4 Conclusions and Future Directions

This chapter has presented several approaches for the stochastic analysis of chemical reactions arising in gene networks. Although, in isolation, none of these techniques provides a solution that is both computationally effective and accurate, they are clearly complementary in scope. The density computation methods (for example, finite-state projection) are especially suitable for very low molecule counts, moment closure methods for medium to large counts, and the linear noise approximation for very large counts. Combining these different types of approximations in a unified framework to devise models that are both accurate and computationally efficient remains an important research goal. Essentially, one would like to follow the limiting volume argument for chemical species that have a large number of molecules (resulting in linear models for the covariances), but use Monte Carlo or finite-projection techniques for chemical species that do not.

3 Spatial Modeling

Pablo A. Iglesias

Many, if not most, of the mathematical models in cell biology assume a homogeneous spatial distribution of the biochemical entities involved. This assumption facilitates the use of ordinary differential equations to describe the reactions governing the species concentrations. In practice, however, cells are not well-stirred biochemical reactors where proteins and other species are uniformly distributed. Instead, they consist of highly complex environments in which species are segregated to different spatial domains. In many cases, the spatial arrangement and regulation are as important to cellular function as the temporal behavior. For example, spatial segregation is crucial during development, giving rise to polarity in cells and, subsequently, in organisms. This chapter shows how spatial models in biology arise; it also presents some important properties regarding the control of spatial heterogeneities in cell biology.

3.1 Spatial Heterogeneities in Biological Models

Implicit in any ordinary-differential-equation (ODE) model of biological systems is the assumption that chemical concentrations are spatially homogeneous. This assumption is valid for systems in which the reactions are confined to small volumes or when the mobility of the interacting chemical species is restricted. In many systems, however, spatial heterogeneities are present and are important for proper cell function. Specifically, diffusion and transport of biochemical molecules can play a significant role in biological regulation.

This spatial segregation can be treated with compartment models, in which different parts of the cells are modeled separately using ODEs and transport between the compartments is incorporated. A good example of this type of model is found in Görlich et al. (2003), in which the cytoplasm and nucleus are modeled as well-stirred compartments and transport through the nuclear pore complexes is modeled explicitly.

Alternatively, partial-differential-equation (PDE) models can be used. In these models, systems of differential equations of the form

$$\frac{\mathrm{d}c_i(t)}{\mathrm{d}t} = f_i(c_1, \ldots, c_n),$$

where the f_i refer to the reaction terms, are replaced by equations that depend on both time, t, and spatial location, $\boldsymbol{x}$. This can be inside a domain, Ω, or on the domain's boundary, $\partial\Omega$. The former can represent the cytoplasm, nucleus, or some other intracellular compartment; the latter can be, for example, the plasma membrane or nuclear envelope.

Taking transport into account, the concentration of species i obeys

$$\frac{\partial c_i(\boldsymbol{x}, t)}{\partial t} = -\nabla J_i(\boldsymbol{x}, t) + f_i(c_1(\boldsymbol{x}, t), \ldots, c_n(\boldsymbol{x}, t)). \tag{3.1}$$

In this equation, the function $J_i(\boldsymbol{x}, t)$ specifies the flux of molecules of species c_i into an infinitesimal volume at spatial location $\boldsymbol{x}$. It has units of molecules per unit area per unit time. The differential operator ∇ depends on the spatial dimension and the representation (Cartesian, polar, etc.) used to describe it. For simplicity, we mostly assume one-dimensional systems in this chapter, in which case $\nabla = \partial/\partial x$. The general principles presented here apply to more realistic spatial domains.

3.1.1 Diffusion

There are several ways in which molecular flux can arise. We first consider diffusion, the random motion of a molecules from areas of high concentration to areas of low concentration. Arising from the Brownian motion of individual molecules, diffusion can be described statistically on a microscale (Berg, 1993). Instead, we will employ a macroscale description based on Fick's law of diffusion, which states that the flux of a species into a region is proportional to the concentration gradient:

$$J_i(\boldsymbol{x}, t) = -D_i(\boldsymbol{x})\nabla c_i(\boldsymbol{x}, t). \tag{3.2}$$

The proportionality constant $D_i(\boldsymbol{x})$ is known as the *diffusion coefficient*. In general, it can depend on the location in the domain. The "$-$" sign is used since diffusion takes molecules from regions of higher concentration into regions of lower concentration. If we substitute equation (3.2) into equation (3.1), we get

$$\frac{\partial c_i(\boldsymbol{x}, t)}{\partial t} = \nabla(D_i(\boldsymbol{x})\nabla c_i(\boldsymbol{x}, t)) + f_i(c_1(\boldsymbol{x}, t), \ldots, c_n(\boldsymbol{x}, t)). \tag{3.3}$$

It is common to assume that diffusion is homogeneous—that is, independent of the spatial variable—in which case equation (3.3) simplifies to

$$\frac{\partial c_i(\boldsymbol{x},t)}{\partial t} = D_i \nabla^2 c_i(\boldsymbol{x},t) + f_i(c_1(\boldsymbol{x},t),\ldots,c_n(\boldsymbol{x},t)). \tag{3.4}$$

3.1.2 Transport

Diffusion represents passive transport. Biochemical species inside cells also move because of active transport processes. For example, actin or microtubule motors carry cargo along these polarized filaments in different directions (Bray, 2000). To model this transport, we will use the advection equation

$$J_i(\boldsymbol{x},t) = v_i(\boldsymbol{x}) c_i(\boldsymbol{x},t), \tag{3.5}$$

where $v_i(\boldsymbol{x})$ is the velocity of the particles of species i. This velocity may depend on the spatial location $\boldsymbol{x}$.

3.1.3 Boundary Conditions

When specifying an ordinary differential equation, one must define the initial condition. For partial differential equations, one must also specify *boundary conditions.* Three types of boundary conditions are commonly used in models of biochemical reactions. We will assume that the partial differential equation is defined in the domain Ω with boundary $\partial\Omega$. For Dirichlet boundary conditions, the first type, the values of the concentration are fixed:

$$c_i(\boldsymbol{x},t) = c_{0,i}(\boldsymbol{x}), \qquad \boldsymbol{x} \in \partial\Omega.$$

In practice, a more common situation in biological systems is to specify the flux of the species at the boundaries. For Neumann boundary conditions, the second type, the equation

$$\nabla c_i(\boldsymbol{x}) \cdot \vec{\mathbf{n}} = g_i(c_1(\boldsymbol{x},t),\ldots,c_n(\boldsymbol{x},t)), \qquad \boldsymbol{x} \in \partial\Omega,$$

specifies the boundary conditions, where $\vec{\mathbf{n}}$ is the outward normal direction to $\partial\Omega$.

The third type, Robin boundary conditions, arises as a linear combination of the first and second types:

$$k_1 c_i(\boldsymbol{x}) + k_2 \nabla c_i(\boldsymbol{x}) \cdot \vec{\mathbf{n}} = g_i(c_1(\boldsymbol{x},t),\ldots,c_n(\boldsymbol{x},t)), \qquad \boldsymbol{x} \in \partial\Omega.$$

The functions g_i that arise in Neumann and Robin boundary conditions can include reactions that take place on the boundary, for example, reactions of membrane-bound particles that drive intracellular dynamics.

3.1.4 The Diffusion Equation

When there are no reaction terms, the partial differential equation reduces to the diffusion equation, which is also known as the heat equation. In one dimension:

$$\frac{\partial c(x,t)}{\partial t} = D\frac{\partial^2 c(x,t)}{\partial x^2}. \tag{3.6}$$

In a specific case that proves to be useful, the species diffuses in an infinite medium: $x \in (-\infty, \infty)$, and all the molecules are initially at the origin:

$$c(x,0) = c_0\delta(x),$$

where $\delta(x)$ refers to the Dirac delta function and the boundary conditions are such that $\lim_{x\to\infty} c(\pm x, t) < \infty$. Although the solution can be obtained in a straightforward manner using the method of *separation of variables* (Haberman, 1983), it is easier just to verify that

$$c(x,t) = \frac{c_0}{\sqrt{4\pi Dt}} e^{-x^2/(4Dt)} \tag{3.7}$$

satisfies equation (3.6). The profile of $c(x,t)$ in equation (3.7) is clearly that of a Gaussian curve with mean zero and variance $2Dt$. We can view this as the probability that a particle, initially at the origin, will have moved a distance x in time t seconds. In particular, we can get the mean square displacement:

$$\langle x^2 \rangle = \int_{-\infty}^{\infty} x^2 c(x,t)\,\mathrm{d}x = 2Dt, \tag{3.8}$$

directly from the fact that equation (3.7) specifies a Gaussian distribution.

3.2 Morphogen Gradients

All biological systems exhibit spatial polarity at some point or another. In fact, even spherical bacteria (known as *cocci*) develop spatial asymmetry at the time of division (Cabeen and Jacobs-Wagner, 2007). How biological species polarize is one of the great questions in cellular and developmental biology (Arkowitz and Iglesias, 2008).

In developmental biology, morphogen gradients are believed to effect differentiation. A morphogen is a diffusible molecule that is produced or secreted at one end of an organism. Diffusion away from the localized source forms a gradient of concentration. This spatial pattern then controls the activity of downstream effectors, which may be other signaling molecules or other cells. Because the response of these effectors depends on the local concentration of the morphogen, which itself depends on the position, different cells along the length of the gradient experiencing different concentrations yield different responses (figure 3.1). The idea that gradients of diffusible particles could effect spatial patterns is about 100 years old (Ephrussi and St Johnston, 2004). Nobelist Thomas Hunt Morgan (1901) proposed that gradients

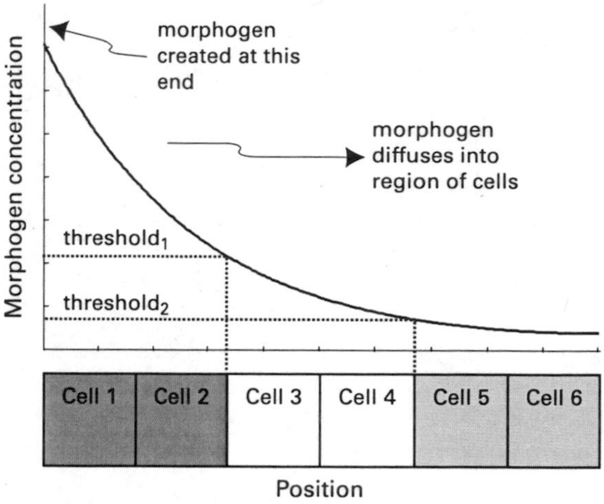

Figure 3.1
Morphogen gradients and the French flag model. A morphogen is secreted at the left end of a domain and diffuses into the area, where it is degraded or deactivated, creating a spatial gradient. The morphogen triggers downstream effectors in a concentration-dependent manner. In this example, the six cells exhibit three distinct phenotypes, depending on whether the morphogen concentration is above one or both of the thresholds. The resemblance of the resultant pattern to the French tricolor gives this model its name (Wolpert, 1969).

could be responsible for regeneration. That same year, Theodor Boveri (1901) suggested that the patterns in sea urchin larvae could be achieved by gradients.

The concept of morphogen gradients as a means of controlling spatial patterning and positional information during development was particularly championed by Lewis Wolpert (1969), who developed the so-called French flag model (figure 3.1). The existence of morphogens and their role in development have now been established experimentally; the first morphogen discovered was the transcription factor bicoid, in the fruit fly *Drosophila melagonaster* (Driever and Nüsslein-Volhard, 1988).

3.2.1 A Simple One-Dimensional Model

To see how a morphogen gradient can arise, we consider the case of a system consisting of a single species, whose concentration is denoted by $c(x, t)$, diffusing in a one-dimensional finite environment: $x \in \Omega = [0, L]$.

We assume that $c(x, t)$ is being produced or activated at one boundary ($x = 0$). From there, it diffuses into the environment, where it is inactivated at a rate proportional to its concentration. The concentration of $c(x, t)$ is described by

$$\frac{\partial c(x,t)}{\partial t} = D\frac{\partial^2 c(x,t)}{\partial x^2} - k_- c(x,t). \tag{3.9}$$

We assume that, initially, the concentration is zero everywhere: $c(x,0) = 0$. The production appears as a boundary condition in terms of flux

$$D\frac{\partial c}{\partial x}\bigg|_{x=0} = -k_+.$$

The minus sign corresponds to the fact that the flux is into the domain. The other condition we impose is that there is no flux at the other end of the domain:

$$\frac{\partial c}{\partial x}\bigg|_{x=L} = 0.$$

The solution of this differential equation, written in terms of the steady state ($c_\infty(x)$) and transient ($\tilde{c}(x,t)$), is given by

$$c(x,t) = \underbrace{\frac{k_+\lambda}{D}\frac{\cosh([x-L]/\lambda)}{\sinh(L/\lambda)}}_{c_\infty(x)} + \underbrace{\frac{k_+}{D}\sum_{n=-\infty}^{\infty}\tau_n \cos\left(\frac{n\pi x}{L}\right)e^{-t/\tau_n}}_{\tilde{c}(x,t)}, \tag{3.10}$$

where $\lambda = \sqrt{D/k_-}$ and $\tau_n = 1/[k_- + n^2\pi^2 D/L^2]$. The functions $\cosh(z) = (e^z + e^{-z})/2$ and $\sinh(z) = (e^z - e^{-z})/2$ are the hyperbolic cosine and sine, respectively. The time-dependent spatial profile of this solution is illustrated in figure 3.2.

Although this chapter focuses on the steady-state solution, it is important to keep in mind that, in biology, the transient term may be crucially important (Bergmann et al., 2007).

3.2.2 Controlling the Shape of the Gradient

The parameter λ is known as the *dispersion*. The steady-state spatial gradient of the diffusing species is controlled by the nondimensional parameter

$$\Phi = \frac{L}{\lambda} = \sqrt{\frac{L^2 k_-}{D}},$$

known as the Thiele modulus. We will focus on the concentration at the two ends. First, we look at the end from which the species is being produced or activated ($x = 0$):

$$c_\infty(0) = \frac{k_+\lambda}{D}\coth(\Phi).$$

Then we look at the other end ($x = L$):

$$c_\infty(L) = \frac{k_+\lambda}{D\sinh(\Phi)}.$$

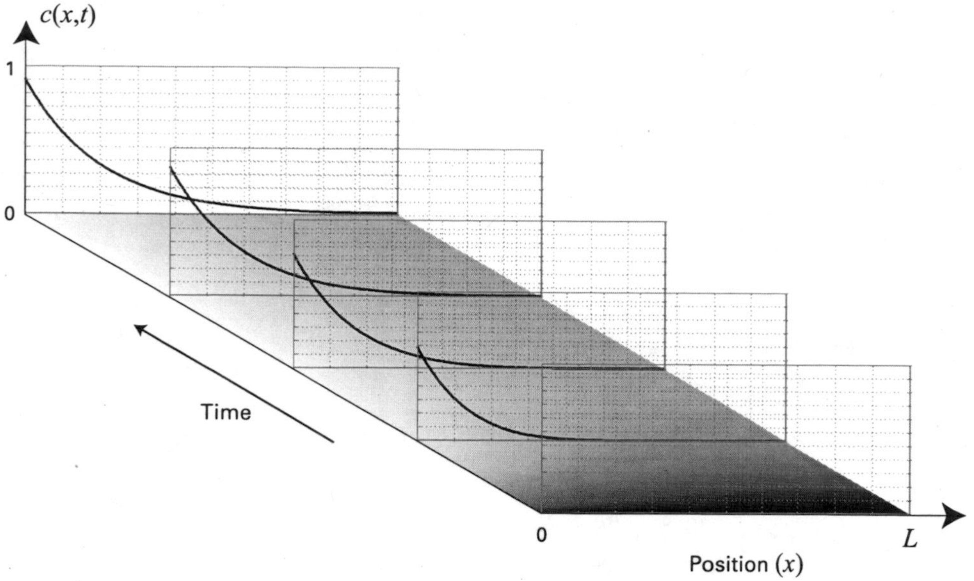

Figure 3.2
Time-dependent solution of equation (3.9), with zero initial condition for a region of length $L = 1$ μm. Coefficients used are $k_- = 1\ \mathrm{s}^{-1}$, $D = 10\ \mu\mathrm{m}^2/\mathrm{s}$, and $k_+ = 1\ \mu\mathrm{M}(\mu\mathrm{m}\ s)^{-1}$. Shown along the plane is the time-dependent concentration, with darker shades representing lower concentrations. Also shown is the concentration at three intermediate points and at steady state as a function of the position (x).

Finally, as a measure of the spatial distribution, we look at the gain, defined as the ratio between the two:

$$\frac{c_\infty(0)}{c_\infty(L)} = \cosh(\Phi).$$

These three curves are plotted in figure 3.3 as a function of the Thiele modulus.

Clearly, there are three ways that the Thiele modulus can be regulated: L, k_- and D. We now turn to how each of these can be used in biology to control cell function.

Size and Shape of the Cell

Cells can use the dimension and shape of the environment to regulate the concentration of $c(x,t)$ (Meyers et al., 2006). As the dimension L is increased, two things are apparent. First, the concentration $c_\infty(0)$ is a monotonically decreasing function of L, reaching the asymptotic value of $k_+\lambda/D$ as $L \uparrow \infty$ (figure 3.3a). Hence the larger the environment, the smaller the peak concentration of $c_\infty(x)$, and this is relatively constant once $\Phi \simeq 1$. Consequently, the concentration throughout the region is also lowered. Second, at the other end ($x = L$), $c_\infty(L)$ decreases exponentially for $\Phi \gg 1$. Thus, in addition to the overall level decreasing, the gradient gain increases substantially (figure 3.3b).

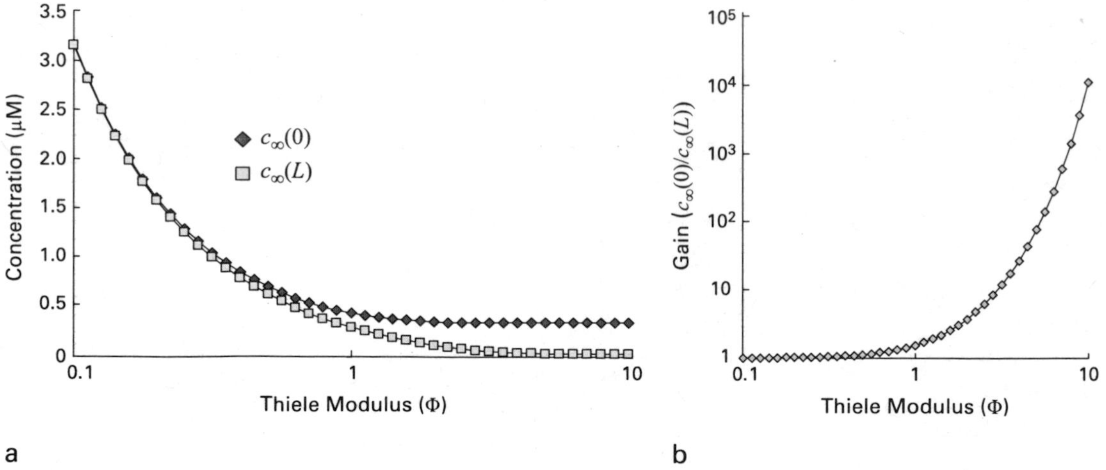

Figure 3.3
Concentration gradient (a) and gain (b) as a function of the Thiele modulus. The coefficients used are $D = 10\ \mu m^2/s$, $k_+ = 1\ \mu M(\mu m\ s)^{-1}$ and $k_- = 1\ s^{-1}$.

This form of control can be used in two ways. First, assume that the environment is growing (for example, a rod-shaped bacterium that is lengthening, or a limb that is growing during development). How does the organism know that it has reached the correct size? One way that has been proposed is through the use of morphogen gradients (Day and Lawrence, 2000). In this case, the source of the morphogen is away from the region of growth. As shown in the model above, as the size of the organism increases, the concentration $c_\infty(x)$ decreases uniformly. Once this reaches a certain threshold, cues stopping growth can be triggered. There is considerable evidence for such a mechanism in wing size determination in *Drosophila* (Hufnagel et al., 2007).

Size can also be used to control the concentration of the diffusing species and related downstream signaling. Consider a situation in which a cell transduces a signal sensed at its membrane into a nuclear signal. This transfer of information takes place by producing an intracellular molecule at the membrane, and allowing it to diffuse toward the nucleus. Figure 3.4 contrasts the signal sensed at the nucleus as a function of cell shape. In the ellipsoidal cell, because the distance between the membrane and the nucleus is smaller, the overall concentration is higher than in the spherical cell. It is clear that the shape of the cell can greatly affect the signal received at the nucleus.

3.2.3 Controlling the Degradation Rate

The degradation rate k_- appears in the numerator of the dispersion λ. Hence, k_-'s effect on determining the shape of the gradient is similar to that of L. The parameter k_- specifies the average lifetime of a molecule. In particular, if we assume that the

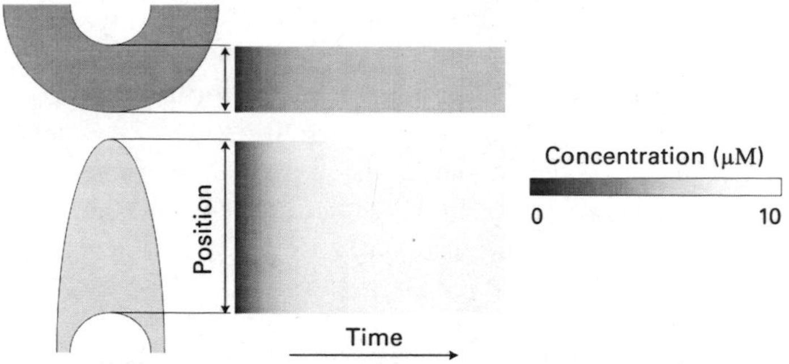

Figure 3.4
Effect of domain shape on concentration of the diffusing species. In this simple model showing only half the domain and the steady-state concentration, a species is released at the cell membrane (*outer circle* or *ellipse*) and diffuses in the domain, having zero flux at the nucleus (*inner circle*). Also shown is the concentration as a function of time along the radial axis. It is clear that the steady-state concentration in the ellipsoidal cell is higher than that of the spherical cell, even though the cell area is the same.

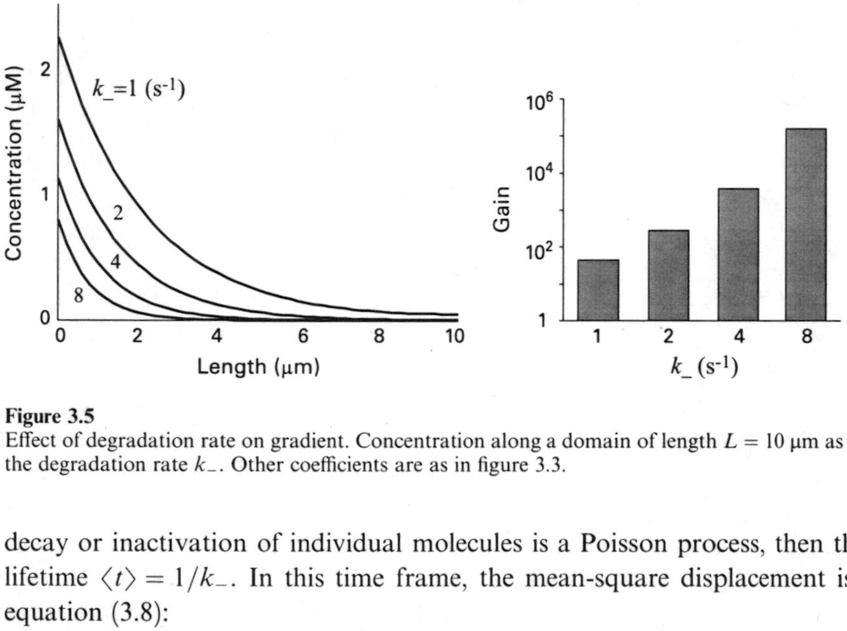

Figure 3.5
Effect of degradation rate on gradient. Concentration along a domain of length $L = 10$ μm as a function of the degradation rate k_{-}. Other coefficients are as in figure 3.3.

decay or inactivation of individual molecules is a Poisson process, then the average lifetime $\langle t \rangle = 1/k_{-}$. In this time frame, the mean-square displacement is given by equation (3.8):

$$\langle x^2 \rangle = 2D\langle t \rangle = \frac{2D}{k_{-}}.$$

Thus larger values of k_{-} signify shorter lifetimes over which the molecule diffuses. Thus, fewer activated molecules reach the other end: $x = L$. This translates to fewer molecules throughout the environment, as well as steeper gradients (figure 3.5).

The Ran (Ras-related nuclear protein) system is an example of how k_- is regulated to control the gradient of a diffusible protein (Caudron et al., 2005). Ran is a small GTPase, found in both inactive (RanGDP) and active (RanGTP) forms inside cells, with its activity being dictated by whether it is bound to GDP or GTP. Its spatial regulation follows closely the simple model described above. RanGDP is converted to RanGTP through the action of an enzyme, RCC1 (regulator of chromosome condensation 1), that is found on the surface of the chromatin. Thereafter, RanGTP diffuses into the cytoplasm, where it is hydrolyzed (converted to RanGDP) through the action of a second enzyme, RanGAP (Ran GTPase-activating protein).

Inside the cytoplasm, RanGTP binds to a third molecule, known as importin-β. Through in vitro studies, the compound of RanGTP:Importin-β has been found to be extremely stable: its dissociation rate (k_-) is on the order of 4.5×10^{-4} s^{-1}, meaning that, on average, it stays around for approximately 2,000 seconds. With this lifetime, the dispersion of the RanGTP:Importin-β complex is approximately 165 μm, which is considerably larger than the typical cell dimension.

In live cells, another protein, RanBP1 (Ran Binding Protein 1) helps to break down the RanGTP:Importin-β compound, allowing RanGTP to be hydrolyzed. In the presence of RanBP1, the dissociation rate of RanGTP and importin-β increases about 1,000-fold, making the dispersion length about the radius of the cell. This simple calculation suggests that, in a cell with reduced levels of RanBP1, the abundance of RanGTP:Importin-β would be considerably reduced; this, indeed, has been observed experimentally (Li et al., 2007).

3.2.4 Controlling the Diffusion Coefficient

The third way in which the gradient can be controlled is through the diffusion coefficient. In practice, this can be achieved by changing the local environment in which the molecules are expected to diffuse. Cellular particles have to diffuse in a crowded environment that may have considerable number of physical barriers preventing easy passage. Moreover, the diffusing molecules may interact with other species, hindering their movement. These effects can lead to a greatly reduced effective diffusion coefficient. As an example, the effective diffusion coefficient of a fluorescently tagged morphogen known as Decapentaplegic (Dpp) in the developing fruit fly wing has been measured to be 0.1 $\mu m^2/s$, which is three orders of magnitude less than would be expected according to its size when diffusing freely in water (section 3.6.3).

An interesting and somewhat paradoxical property of regulation through the diffusion coefficient is that, because the concentration of the species near the source increases as the diffusion coefficient is decreased (figure 3.6), it may lead to activation farther away from the source (Lander, 2007). Depending on where the threshold for activation is set, decreased diffusion may lead to a greater portion of the environment being above the threshold. This behavior was seen in a recent model of the Sonic

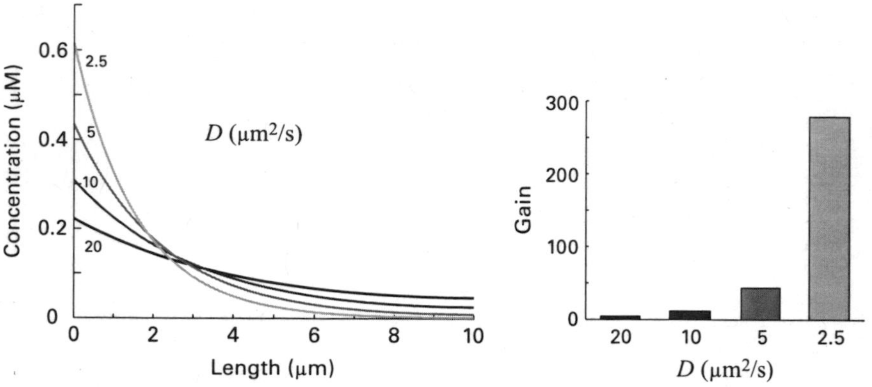

Figure 3.6
Effect of diffusion coefficient. Reducing the diffusion coefficient increases the gradient (gain). Although this reduces the concentration far away from the source, it also increases the concentration near the source.

hedgehog (Shh) gradient that patterns the ventral neural tube of the chick embryo (Saha and Schaffer, 2006).

3.3 Turing Patterns

The concept of the morphogen gradient, though now a well-accepted principle in developmental biology, is not the only model that has been used to account for pattern formation during development. In his influential 1952 paper, Alan Turing proposed that systems could provide the basis for morphogenesis—the formation of shapes in biology. Turing reasoned that, however devoid of shape biological systems might be in their early stages, during development, patterns appeared and that these patterns might be due to an instability of the spatially homogeneous solution. These diffusion-driven bifurcations are now named Turing instabilities. To someone not familiar with this class of systems, the fact that diffusion can destabilize a stable system may be somewhat surprising. After all, alone, diffusion acts to minimize spatial heterogeneities.

The Turing mechanism provides an elegant model of how pattern formation may arise and has been used to account for a large number of patterns seen in biology, from pigmentation patterns in large cats (Liu et al., 2006) to vertebrate limb development (Newman et al., 2008). Because of the lack of experimental data supporting it, and because data from some of the better-studied developmental systems have called its validity into question (Maini et al., 2006), it has not been widely accepted in the biological community. Recently, however, experimental evidence supporting the Turing mechanism has emerged. Thomas Schlake and coworkers identified WNT and its inhibitor Dickkopf (DKK) as primary determinants of murine hair follicle spacing and confirmed predictions of a WNT/DKK-specific mathematical model based on the Turing mechanism (Sick et al., 2006).

3.3.1 Turing Instabilities in Two-Component Systems

To see how these instabilities arise, we consider a system with two interacting species (say U and V) that are capable of diffusing. The system is described by

$$\frac{\partial}{\partial t}\begin{bmatrix} U(x,t) \\ V(x,t) \end{bmatrix} = \begin{bmatrix} f(U,V) \\ g(U,V) \end{bmatrix} + \begin{bmatrix} D_u \nabla^2 U(x,t) \\ D_v \nabla^2 V(x,t) \end{bmatrix}.$$

We assume that the system has a spatially homogeneous solution with $\bar{U}$ and $\bar{V}$:

$$f(\bar{U},\bar{V}) = 0, \quad \text{and} \quad g(\bar{U},\bar{V}) = 0,$$

with $\nabla^2 \bar{U} = 0$ and $\nabla^2 \bar{V} = 0$. We linearize the system around this equilibrium. Defining

$$u(x,t) = U(x,t) - \bar{U} \quad \text{and} \quad v(x,t) = V(x,t) - \bar{V},$$

we obtain

$$\frac{\partial}{\partial t}\begin{bmatrix} u(x,t) \\ v(x,t) \end{bmatrix} = \underbrace{\begin{bmatrix} a_{11} & a_{12} \\ a_{21} & a_{22} \end{bmatrix}}_{A} \begin{bmatrix} u(x,t) \\ v(x,t) \end{bmatrix} + \underbrace{\begin{bmatrix} D_u & 0 \\ 0 & D_v \end{bmatrix}}_{D} \begin{bmatrix} \nabla^2 u(x,t) \\ \nabla^2 v(x,t) \end{bmatrix}, \tag{3.11}$$

where $\boldsymbol{A}$ is the Jacobian and $\boldsymbol{D}$ is a diagonal matrix describing the species diffusion rates.

We seek conditions under which

1. without diffusion, the spatially homogeneous equilibrium is stable; and
2. with diffusion, it becomes unstable.

The first condition requires that the matrix $\boldsymbol{A}$ be Hurwitz. Because this is a system of two interacting species, we can turn to section 1.4.1 to note that stability is equivalent to the following two inequalities:

$$\text{trace}(\boldsymbol{A}) = a_{11} + a_{22} < 0, \tag{3.12}$$

and

$$\det(\boldsymbol{A}) = a_{11}a_{22} - a_{12}a_{21} > 0. \tag{3.13}$$

We now consider the second condition. We assume that the perturbations about the spatially homogeneous solutions are of the form

$$\begin{bmatrix} u(x,t) \\ v(x,t) \end{bmatrix} = \begin{bmatrix} u_0 \\ v_0 \end{bmatrix} e^{\lambda t} e^{jqx}.$$

(Recall that we use the notation $j = \sqrt{-1}$.) These solutions oscillate in space (with wavelength $q/2\pi$), and their stability is determined by the sign of λ. As we will see later, the size of the spatial domain determines which values of q are admissible. For now, we do not restrict q.

Substituting these solutions into the linear equation (3.11) leads to

$$(\lambda \boldsymbol{I} - \boldsymbol{A} + \boldsymbol{D}q^2) \begin{bmatrix} u_0 \\ v_0 \end{bmatrix} = \boldsymbol{0}.$$

These solutions are unstable if the real part of λ is positive. Thus the system is unstable because of diffusion if the matrix $\boldsymbol{A} - \boldsymbol{D}q^2$ has eigenvalues with positive real part; that is, if

$$\begin{aligned} \operatorname{trace}(\boldsymbol{A} - \boldsymbol{D}q^2) &= a_{11} + a_{22} - q^2(D_u + D_v) \\ &= \operatorname{trace}(\boldsymbol{A}) - q^2(D_u + D_v) > 0, \end{aligned} \tag{3.14}$$

or

$$\begin{aligned} \det(\boldsymbol{A} - \boldsymbol{D}q^2) &= (a_{11} - q^2 D_u)(a_{22} - q^2 D_v) - a_{12}a_{21} \\ &= a_{11}a_{22} - a_{12}a_{21} - q^2(a_{11}D_v + a_{22}D_u) + q^4 D_u D_v \\ &= \det(\boldsymbol{A}) - q^2(a_{11}D_v + a_{22}D_u) + q^4 D_u D_v < 0. \end{aligned} \tag{3.15}$$

We can immediately make several observations.

1. *If the diffusion coefficients are equal ($D_u = D_v = d$), then the homogeneous equilibrium is stable.* To see this, note that

$$\lambda \boldsymbol{I} - \boldsymbol{A} + \boldsymbol{D}q^2 = (\lambda + dq^2)\boldsymbol{I} - \boldsymbol{A}.$$

Thus the eigenvalues of $\boldsymbol{A} - \boldsymbol{D}q^2$ are those of $\boldsymbol{A}$, shifted to the left by $dq^2 > 0$.

2. *Diffusive instabilities can arise only if inequality (3.15) holds.* Because $\operatorname{trace}(\boldsymbol{A}) < 0$, and $q^2(D_u + D_v) > 0$, it is clear that

$$\operatorname{trace}(\boldsymbol{A} - \boldsymbol{D}q^2) = \operatorname{trace}(\boldsymbol{A}) - q^2(D_u + D_v) < 0.$$

This inequality implies that inequality (3.14) cannot hold. Thus, if a diffusive instability exists, then it must be because inequality (3.15) holds.

3. *If the system has a diffusive instability, then the diagonal coefficients of* $\boldsymbol{A}$ *must have opposite signs, as must the off-diagonal coefficients.* To see why this assertion holds, note that inequality (3.12) requires that at least one of the two diagonal elements be negative. If, however, both $a_{11} < 0$ and $a_{22} < 0$, then

$$-q^2(a_{11}D_v + a_{22}D_u) > 0,$$

and each of the three terms in inequality (3.15) is positive, implying that the inequality cannot hold. It follows that $a_{11}a_{22} < 0$. However, this inequality, along with $\det(\boldsymbol{A}) > 0$, means that

$$a_{12}a_{21} < a_{11}a_{22} < 0.$$

Thus, without loss of generality, we may assume that diffusive instabilities arise only in systems where the Jacobian has one of two sign patterns:

$$\boldsymbol{A} = \begin{bmatrix} + & - \\ + & - \end{bmatrix}, \quad \text{or} \quad \boldsymbol{A} = \begin{bmatrix} + & + \\ - & - \end{bmatrix}. \tag{3.16}$$

4. *If the system has a diffusion-driven instability, then the dispersion of v must be greater than that of u.* We have already established that diffusion-driven instability requires that inequality (3.15) hold. Because both the constant and q^4 terms are positive, however, this means that the q^2 term must be negative:

$$-q^2(a_{11}D_v + a_{22}D_u) < 0.$$

Dividing by $-q^2 D_u D_v$, we obtain

$$\underbrace{\frac{a_{11}}{D_u}}_{1/\lambda_u} > \underbrace{\frac{(-a_{22})}{D_v}}_{1/\lambda_v} > 0.$$

Hence $\lambda_v > \lambda_u$, as claimed. Note that the dispersion of species v requires a minus sign because the coefficient $a_{22} < 0$. Normally, this would have been written as $a_{22} = -k_{v,-}$ to match the format of section 3.2.

Condition (3.16) implies that two interacting patterns are possible for a system with diffusion-driven instability. In the first case, species u contributes both to its own production ($a_{11} > 0$) and to that of v ($a_{21} > 0$). Similarly, v negatively regulates both u and v ($a_{12} < 0$ and $a_{22} < 0$, respectively). For this reason, u and v are referred to as the activator and inhibitor, respectively (figure 3.7). Moreover, because the dispersion of v is greater than that of u, activator-inhibitor systems are said to require local enhancement and long-range inhibition.

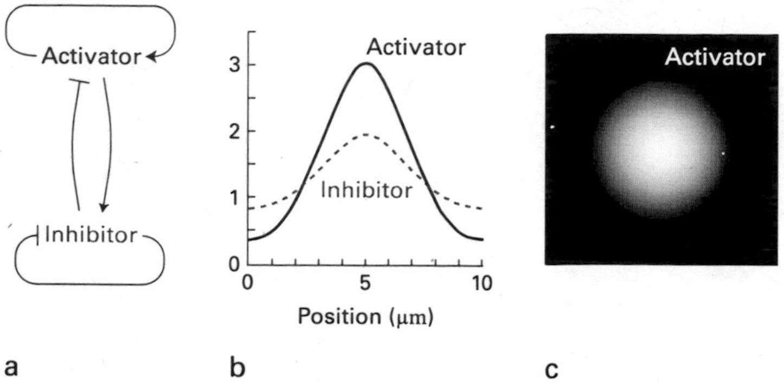

Figure 3.7
Activator-inhibitor system. (a) Activator acts to increase the concentrations of both itself and the inhibitor, whereas the inhibitor acts to decrease the concentration of both. (b) Typical profile for the activator and inhibitor in a one-dimensional problem. (c) Typical profile for the activator in a two-dimensional system. The model used for the latter is that of section 3.4.

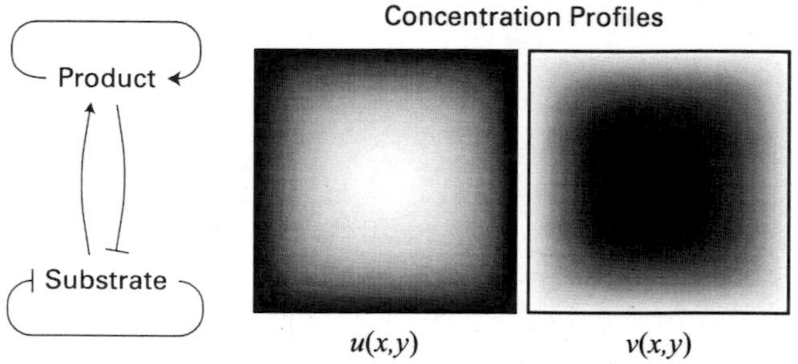

Figure 3.8
Substrate depletion. In these systems, the substrate is needed for the production of the product. As product is formed, substrate is depleted. Shown is the spatially dependent concentration of both the product ($u(x, y)$) and substrate ($v(x, y)$) for the system described in section 3.5.

In the second case, production of u is activated both by its own presence ($a_{11} > 0$) and by the presence of v ($a_{12} > 0$). However, an increase in u reduces the concentration of v ($a_{21} < 0$). In this mechanism, species v can represent a substrate required for the formation of u. As u is formed, the amount of v is depleted; as more substrate is produced, more u can also be produced. For this reason, systems with this type of sign pattern are referred to as substrate-depletion systems (figure 3.8).

We now consider inequality (3.15), a quadratic function of q^2, in greater detail. As stated above, because the coefficients in front of the q^4 and q^0 terms are both positive, the quadratic function can only take negative values if the discriminant is positive:

$$(a_{11}D_v + a_{22}D_u)^2 > 4D_uD_v(a_{11}a_{22} - a_{12}a_{21}).$$

Dividing by $(D_uD_v)^2$ yields

$$\left(\frac{a_{22}}{D_v} + \frac{a_{11}}{D_u}\right)^2 > 4\left(\frac{a_{11}}{D_u}\frac{a_{22}}{D_v} - \frac{a_{12}}{D_u}\frac{a_{21}}{D_v}\right),$$

which is equivalent to

$$\left(\frac{a_{22}}{D_v} - \frac{a_{11}}{D_u}\right)^2 > 4\left|\frac{a_{12}}{D_u}\frac{a_{21}}{D_v}\right| > 0. \tag{3.17}$$

Because $a_{11} > 0$ and $a_{22} < 0$,

$$\frac{a_{22}}{D_v} - \frac{a_{11}}{D_u} = -\left(\left|\frac{a_{22}}{D_v}\right| + \left|\frac{a_{11}}{D_u}\right|\right) < 0.$$

Thus, by taking square roots of equation (3.17), we obtain the inequality

$$\frac{a_{11}}{D_u} - \frac{a_{22}}{D_v} > 2\sqrt{\left|\frac{a_{12}}{D_u}\right|\left|\frac{a_{21}}{D_v}\right|} > 0. \tag{3.18}$$

3.3.2 Effects of Spatial Dimension on the Appearance of a Pattern

If there is no restriction on the spatial parameter q, then inequality (3.18) provides a necessary and sufficient condition for the existence of a Turing instability. In general, however, q cannot be chosen arbitrarily, but is instead constrained by the dimensions of the environment over which the solution evolves.

As an example, consider the case of a finite, one-dimensional domain $\Omega = [0, L]$ with Dirichlet or Neumann boundary conditions. The spatial dimension restricts q to one of the discrete values, known as *wave numbers*:

$$q \in \{q_n\}, \qquad \text{where} \qquad q_n = \frac{2\pi n}{L}.$$

Moreover, the general solution of the linearized equation (3.11) is

$$\begin{bmatrix} u(x,t) \\ v(x,t) \end{bmatrix} = \sum_n \begin{bmatrix} u_{0,n} \\ v_{0,n} \end{bmatrix} e^{\lambda_n t + jq_n x}.$$

The individual modes in this sum are unstable if and only if

$$\det(A) - q_n^2(a_{11}D_v + a_{22}D_u) + q_n^4 D_u D_v < 0$$

for the discrete values of q_n.

As shown in figure 3.9, it may happen that inequality (3.18) holds for some q, but not for the admissible values of q_n. We also see that, as the size of the environment grows, an ever-increasing number of modes can become unstable. This observation fits the general intuition of how spatial patterns may arise during development. It is possible that in a small organism, no discrete mode may satisfy inequality (3.15). As the organism grows, however, a greater number of unstable modes may arise, leading to different patterns. As an example, models have been used to explain the coloration patterns of the growing angelfish *Pomacanthus semicirculatus* (Painter et al., 1999). When young, the fish displays three white stripes on a dark background. As it grows, new stripes develop and insert themselves between the preexisting stripes. The appearance of these extra stripes can be attributed to the extra modes that become unstable.

3.4 Activator-Inhibitor Systems

In 1972, Alfred Gierer and Hans Meinhardt rediscovered that differences in the diffusion properties of interacting species could lead to unstable homogeneous solutions. More important, by developing nonlinear models incorporating local enhancement and long-range inhibition, they moved away from the linear determination of instability that Turing carried out to the development of models that displayed the patterns. Over the years, they have developed a large number of models describing patterns arising in biology (Meinhardt, 1982). Let us consider their original model of an activator-inhibitor system, which was postulated as a means of explaining tentacle formation in hydra (Gierer and Meinhardt, 1972).

The nonlinear system of equations is given by

$$\frac{\partial u}{\partial t} = \frac{\alpha u^2}{v} - \beta u + \epsilon D \frac{\partial^2 u}{\partial x^2},$$

$$\frac{\partial v}{\partial t} = \gamma u^2 - \delta v + D \frac{\partial^2 v}{\partial x^2},$$

where all of the parameters α, β, γ, δ, ϵ, and d are positive constants. The homogeneous solution is given by the pair

$$(\bar{u}, \bar{v}) = \frac{\beta\delta}{\alpha\gamma}\left(1, \frac{\gamma}{\delta}\right)$$

and the associated Jacobian

$$A = \begin{bmatrix} 2\alpha\bar{u}/\bar{v} - \beta & -\alpha\bar{u}^2/\bar{v}^2 \\ 2\gamma\bar{u} & -\delta \end{bmatrix} = \begin{bmatrix} \beta & -\beta^2/\alpha \\ 2\alpha\delta/\beta & -\delta \end{bmatrix}.$$

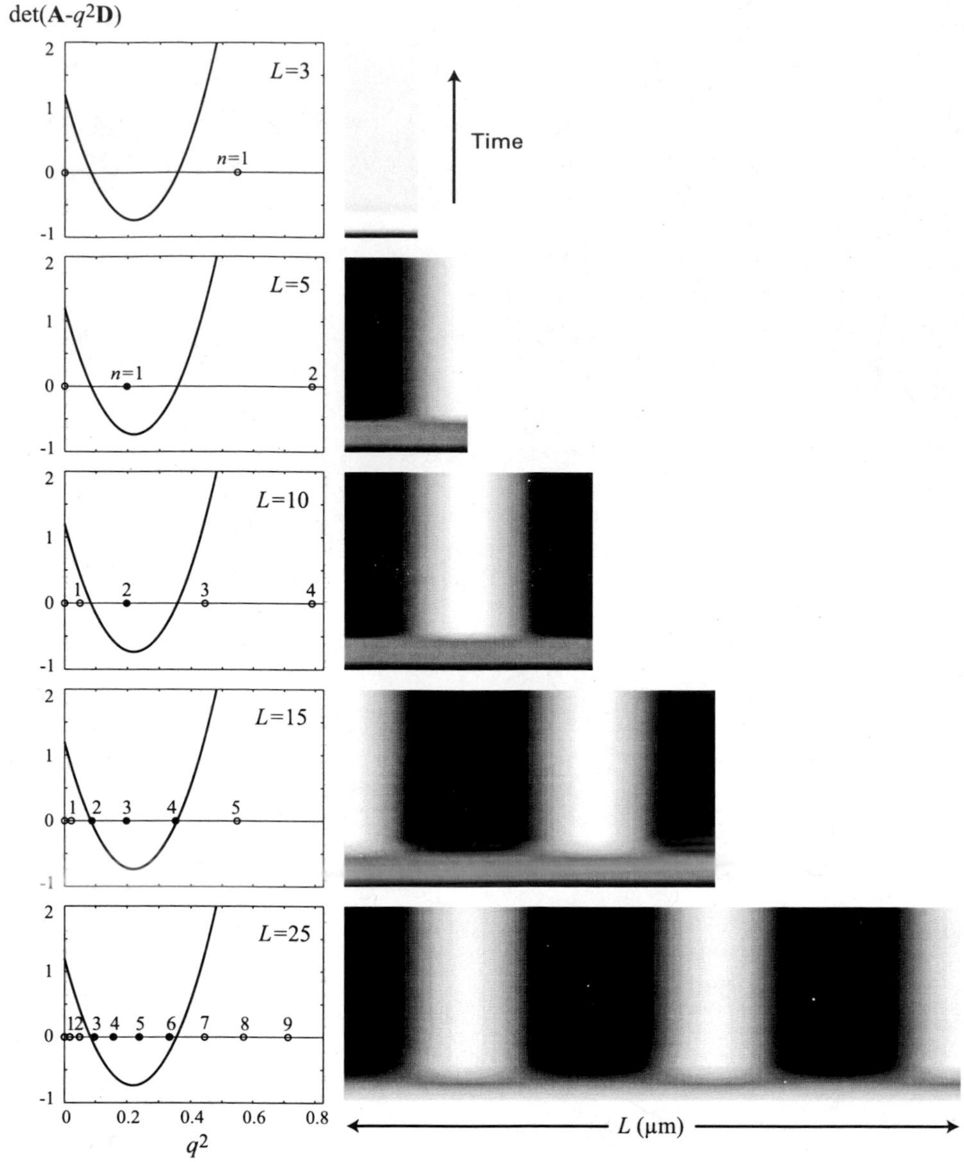

Figure 3.9
Effect of spatial dimension on the existence of instabilities. The spatial dimension specifies the allowable wave numbers that become unstable. The system of figure 3.7 is simulated on a series of one-dimensional domains of different size. In all these cases, the parameters are such that instabilities may arise. Shown is the quadratic term from inequality (3.15). If the length is 3 μm, however, no wave number is unstable; for increasing lengths, different wave numbers become unstable. Which pattern dominates will be determined by the initial conditions and the real value of the wave numbers' eigenvalues.

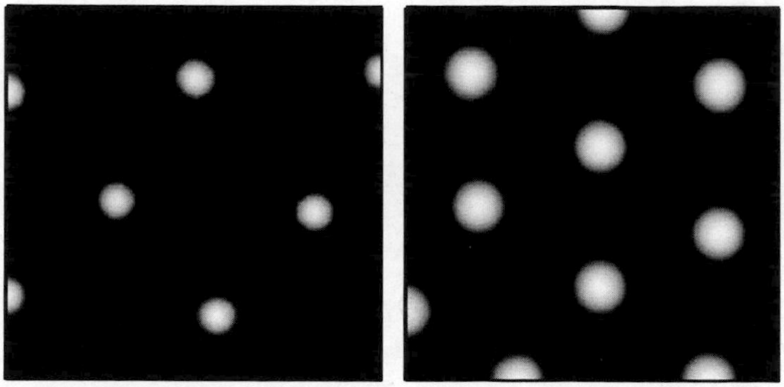

Figure 3.10
Turing instabilities. Shown are different patterns obtained by the activator-inhibitor system described in section 3.4. Parameters used are $\alpha = 1$, $\beta = 1$, $\gamma = 1$, $\delta = 1.1$, $D = 10$, and $\epsilon = 0.1$.

The requirement that $\det(A) = 2\beta\delta > 0$ is satisfied automatically. The second stability requirement: $\mathrm{trace}(A) = \beta - \delta < 0$, is satisfied provided that $\delta > \beta$. From inequality (3.18), for diffusion-driven instabilities to arise, we require that

$$(\beta - \delta\epsilon)^2 > 4\beta\delta\epsilon,$$

which is satisfied provided that

$$\epsilon < (3 - \sqrt{8})\frac{\beta}{\delta} \approx 0.17\frac{\beta}{\delta}.$$

Some of the patterns that arise from these equations are shown in figure 3.10.

3.5 Substrate Depletion

In equation (3.16), instabilities can also arise from the sign pattern corresponding to a substrate-depletion system. An example of such a system is given by the following set of equations (Barrio et al., 1999):

$$\frac{\partial u}{\partial t} = \alpha u + v + r_1 uv + r_2 uv^2 + \epsilon D \frac{\partial^2 u}{\partial x^2},$$

$$\frac{\partial v}{\partial t} = \gamma u - \beta v - r_1 uv - r_2 uv^2 + D \frac{\partial^2 v}{\partial x^2}.$$

In addition to the linear component, there are two terms that denote exchange from v to u. The two parameters r_1 and r_2 dictate whether the exchange is linear or quadratic in v. Thus the corresponding nonlinearities are either quadratic or cubic.

The homogeneous solutions for this systems are given by

$$v = -\frac{\alpha+\gamma}{1-\beta}u, \qquad \beta \neq 1.$$

By setting $\gamma = -\alpha$, we restrict the steady-state value to the origin: $(u, v) = (0, 0)$. In this case, the Jacobian is

$$A = \begin{bmatrix} \alpha & 1 \\ -\alpha & -\beta \end{bmatrix},$$

which fits the sign pattern. The equilibrium is stable if $\beta > \alpha > 0$ and $\alpha\beta + \gamma < 0$. Inequality (3.18) predicts that diffusion will destabilize the system if

$$\alpha + \epsilon\beta > 2\sqrt{\alpha\epsilon}.$$

Following Liu et al. (2006), we select

$$\alpha = 0.899, \quad \beta = 0.91, \quad D = 6, \quad \epsilon = 0.075, \quad r_1 = 2, \quad \text{and} \quad r_2 = 3.5.$$

A spatially heterogeneous pattern obtained using this system is shown in figure 3.8.

3.6 Determining Diffusion Coefficients

The reaction diffusion equations, together with the initial and boundary conditions specify the system describing the concentration of molecules in the system. To determine the system's evolution requires that we know the correct value of the diffusion coefficient inside living cells. This section describes how diffusion coefficients can be measured.

3.6.1 Fluorescence Recovery after Photobleaching

We first consider the diffusion of molecules that move along the surface of a membrane, which can be that of the cell or an intracellular object, such as the nucleus. Because of this restriction, the molecules diffuse in two dimensions. We are also going to assume that the diffusing species is not interacting or, more likely, that its concentration is at steady state, so the number of molecules is not changing.

The idea behind *fluorescence recovery after photobleaching* (FRAP) is to create in the living cell a discontinuous initial condition in the spatial distribution of the fluorescence of the molecule of interest. This is done by taking advantage of a property of biological fluorophores known as photobleaching. When the fluorescent molecules are exposed to light, they fluoresce, and this response can be detected in the microscope. When the light power from the laser is sufficiently strong, however, irrevers-

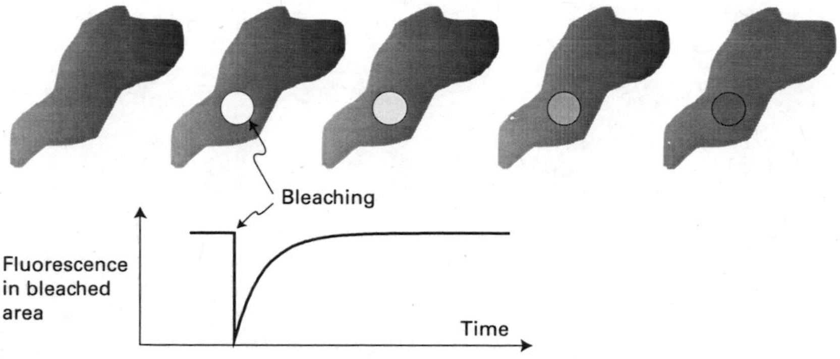

Figure 3.11
Fluorescence recovery after photobleaching. Using a high-powered laser, the fluorescence in a small region (*white circle*) can be eliminated. Diffusion replenishes the area of fluorescent fluorophores; the rate of recovery can be used to measure the diffusion coefficient.

ible photochemical bleaching of the fluorophore in that region ensues. After this photobleaching, the light and dark molecules diffuse into and out of the initially dark area leading to a blurring of the dark area's boundary (figure 3.11).

The recovery rates of this brightness can be used to determine the diffusion coefficients (Axelrod et al., 1976). If $c(r,t)$ denotes the concentration of the unbleached fluorophores, then

$$\frac{dc(r,t)}{dt} = -aI(r)c(r,t),$$

where $I(r)$ denotes the intensity of the laser light that induces the photobleaching. Solving this equation leads to the function

$$c(r,t) = c(r,0)\exp[-aI(r)t].$$

Typically, it is assumed that the laser power has a Gaussian profile. For simplicity, however, we assume a perfectly circular profile of radius r_0:

$$I(r) = \begin{cases} P_0/\pi r_0^2, & r < r_0, \\ 0, & r \geq r_0, \end{cases}$$

where P_0 is the total laser power.

In the absence of interactions, equation (3.4) reduces to the simpler diffusion equation

$$\frac{\partial c(\boldsymbol{x},t)}{\partial t} = D\nabla^2 c(\boldsymbol{x},t).$$

Under the assumption that the bleached area is small, the boundary condition is $c(r = \infty) = c_0$, and the initial condition is

$$c(r,0) = \begin{cases} c_0 \exp[-aI(r)T], & r < r_0, \\ c_0, & r \geq r_0, \end{cases}$$

where T is the time over which the laser is used. Integrating over the bleached spot, we define

$$\bar{c}(t) = \int c(r,t)\,\mathrm{d}^2r.$$

The solution to this diffusion equation leads to an expression describing the fractional recovery:

$$f(t) = \frac{\bar{c}(t) - \bar{c}(0)}{\bar{c}(\infty) - \bar{c}(0)},$$

given by

$$f(t) = 1 - \frac{\tau_D}{t} e^{-2\tau_D/t}[I_0(2\tau_D/t) + I_1(2\tau_D/t)],$$

where I_0 and I_1 are modified Bessel functions and $\tau_D = r^2/4D$ (Soumpasis, 1983). Based on this equation, one way of determining τ_D, and hence D, is to measure $f(t_{1/2}) = 1/2$, which leads (Axelrod et al., 1976) to

$$D \approx 0.224\frac{r_0^2}{t_{1/2}}.$$

3.6.2 Fluorescence Correlation Spectroscopy

For molecules that are free to diffuse in three dimensions, the preferred technique is *fluorescence correlation spectroscopy*, in which the laser light is focused on an extremely small sample (approximately one femtoliter). Molecules in this volume fluoresce, and the intensity of the emitted light (denoted $I(t)$) is measured and tracked over time. As the molecules diffuse, the measured fluorescence intensity fluctuates. One way of determining the amount of diffusion is to evaluate the normalized autocorrelation function:

$$G(\tau) = \frac{\langle I(t)I(t+\tau)\rangle}{\langle I^2(t)\rangle} - 1.$$

For particles diffusing in three dimensions, the autocorrelation is given (Haustein and Schwille 2007) by

$$G(\tau) = G(0)\frac{1}{1+\tau/\tau_D}\frac{1}{\sqrt{1+(r_0^2/z_0^2)\cdot(\tau/\tau_D)}},$$

where r_0 and z_0 describe the radius and height of the volume onto which the laser beam is directed. After fitting experimental data to this equation, the diffusion coefficient can be obtained from

$$D = \frac{r_0^2}{4\tau_D}.$$

3.6.3 The Einstein-Stokes Relationship

Particles subject to Brownian motion in a liquid experience force. In 1905, Einstein used statistical mechanics to demonstrate a relationship between this force and the diffusion coefficient. In particular,

$$D = \text{mob}\ k_B T,$$

where k_B is Boltzmann's constant, T is the absolute temperature, and "mob" refers to the particle's mobility, the ratio between the force applied to the particle and its drift velocity:

$$\text{mob} = \frac{v}{F}.$$

Particles diffusing in a liquid at low Reynolds number, which is the case in biology at the cellular level, are said to be undergoing viscous flow. In this case, the relationship between velocity and the resultant drag force is given by

$$F = \gamma v,$$

where γ is the drag coefficient. For a sphere of radius r, the drag coefficient can be computed as

$$\gamma = 6\pi r\eta,$$

where η is the viscosity of the fluid. For water at room temperature, this equals 8.94×10^{-4} Pa·s. Inside a cell, however, this can be 5–10 times larger (Caudron et al., 2005). Together, this means that

$$D = \frac{k_B T}{6\pi r\eta}, \qquad (3.19)$$

which is known as the Einstein-Stokes relationship.

Note that equation (3.19) requires that one know the effective radius of the particle. Although in practice unknown, it can be estimated from the protein's mass by assuming that the molecule is a sphere. From the gene's sequence, the chemical composition is known, and hence the mass of the resultant gene product can be easily calculated. The typical unit of mass used in biochemistry is the dalton, which is defined as one-twelfth the mass of a carbon atom, and equals approximately 1.66×10^{-24} g. The conversion factor used from mass to volume is 0.74 ml/gram (Caudron et al., 2005). For example, the green fluorescent protein (GFP) commonly used to tag proteins in living cells is a 26.9 kDa protein. Thus we estimate that it occupies a volume equal to

$$
\begin{aligned}
V &\approx (26.9 \times 10^3 \text{ kDa})(1.66 \times 10^{-24} \text{ g})(0.74 \text{ ml/gram}) \\
&\approx 3.3 \times 10^{-23} \text{ ml} \\
&= 3.3 \times 10^{-8} \ (\mu\text{m})^3.
\end{aligned}
$$

Under the assumption that the protein is a sphere, the radius is

$$r = \left(\frac{3V}{4\pi}\right)^{1/3} \approx 1.99 \text{ nm}.$$

The Einstein-Stokes relationship would predict a diffusion coefficient inside the cell equal to

$$D = \frac{k_B T}{6\pi r \eta} \approx 24.4 \ \mu\text{m}^2/\text{s},$$

at 298 K and assuming a viscosity five times higher than water. By way of comparison, the diffusion coefficient of fluorescent proteins inside living cells has been measured to be 22–25 μm^2/s (Maertens et al., 2005; Potma et al., 2001).

3.7 Conclusions

This chapter has highlighted the need for including spatial dynamics in models of biochemical reactions. As we saw in sections 3.2 and 3.3, diffusion and transport can make significant differences to the response and even stability of systems. An area where more work is needed is in the use of stochastic methods, such as those presented in chapter 2, for explaining spatial phenomena. Andrews and Bray (2004), Elf and Ehrenberg (2004) and Altschuler et al. (2008) provide some of the first results in this important area.

4 Quantifying Properties of Cell Signaling Cascades

Simone Frey, Olaf Wolkenhauer, and Thomas Millat

Cells use transduction pathways in order to respond to signals from their environment. These pathways are composed of specific biochemical networks, which have a characteristic structural design. In recent years, several mathematical models have been proposed to describe signaling processes. When considering the dynamics of a signal transduction pathway, questions about the magnitude of average activation time or duration of a signal arise. Quantitative measures provide one approach for answering such questions. In this chapter, we analyze features such as the signaling time and signal duration of linear kinase-phosphatase cascades. We apply three commonly used mathematical models to show that one main purpose for the specific design of the MAPK (mitogen-activated protein kinase) cascade structure could be to trigger slow signals of short duration. Thus our results may help to explain the design of a specific pathway structure. Furthermore, we examine the deviations that occur when applying different model approximations to describe the same signaling pathway.

4.1 Why We Need Quantitative Measures of Dynamic Properties

The functioning of cells requires the transfer of information between and within cells. The transmission of signals is realized through interacting proteins, groups of which are organized into signaling pathways. Because signal processing and cellular responses are highly dynamic biological processes, the investigation of steady-state behavior is not sufficient to characterize them. As reported in recent publications, changes in the temporal profile of a stimulus can result in very different cellular processes. In PC12 cells, for example, the duration of signaling through extracellular signal-regulated kinases (ERKs) may result in proliferation or in differentiation. The two hormones EGF (epidermal growth factor) and NGF (nerve growth factor) use the same pathway to trigger either process. Stimulation with EGF results in transient ERK phosphorylation, which causes proliferation, whereas stimulation with NGF results in sustained ERK activation, which triggers differentiation (Marshall, 1995; Sabbagh et al., 2001; Vaudry et al., 2002). It can be seen here that for the cellular response and therefore for the specific function of the pathway, the properties of a signal (e.g., duration) are crucial. The meaningful determination of quantitative properties of dynamic signals is not simple. Over recent years, mathematical theories have been developed to describe the regulation of signaling pathways as a function of key parameters. A measure frequently used to determine signal properties is *characteristic time*. A well-known example is the *relaxation time* of a system (Atkins and de Paula, 2002; Schwarz, 1968), which is related to the eigenvalues of the system Jacobian at steady state (defined in section 1.3). An advantage of this approach is that such a characteristic time is independent of the initial or final state of the system and depends only on intrinsic systems properties (e.g., biochemical interaction and kinetic parameters). This measure is well defined from a theoretical point of view but difficult to determine from data and for complex systems. For this reason, Llorens et al. (1999) introduced a so-called geometrical approach to provide an experimentally measurable analog; it considers the average time taken by any observable of a metabolic pathway to transition from one given state to another, when either a perturbation or a persistent variation is applied.

A similar approach was developed in control systems theory to characterize the signal transition to a new steady state, where the rise time t_r is defined as the time required for a response to rise from 10% to 90% of its final value. Analogously, the delay time t_d is the time to reach 50% of the final value. Additionally, characteristic quantities for oscillatory signals have been introduced, for example, the settling time t_s, which determines how long it takes to reach and stay within a specific tolerance band (see Franklin et al., 2006; Nagrath and Gopal, 1986).

In this chapter, we demonstrate how quantitative measures can be used to investigate features of signaling cascades. We apply an approach in which integrals over dynamic signals determine the features of the signaling pathway. In contrast to the characteristic times used above, here signals do not tend to a new steady state but instead return to their initial value. We make use of quantitative signaling measures first introduced by Heinrich et al. (2002) to analyze linear kinase-phosphatase cascades (figure 4.1). Their study focused on a number of questions including: How do the magnitudes of signal output and signal duration depend on the kinetic properties of pathway components, such as kinases or phosphatases (Hornberg et al., 2005)? Can high signal amplification be coupled with fast signaling? How are signaling pathways designed to ensure that they are safely off in the absence of stimulation, yet display high signal amplification following receptor activation? How can different agonists stimulate the same pathway in distinct ways to elicit either a sustained or a transient response? These examples show that the quantitative measures introduced above can be used to investigate various intrinsic properties of signaling cascades.

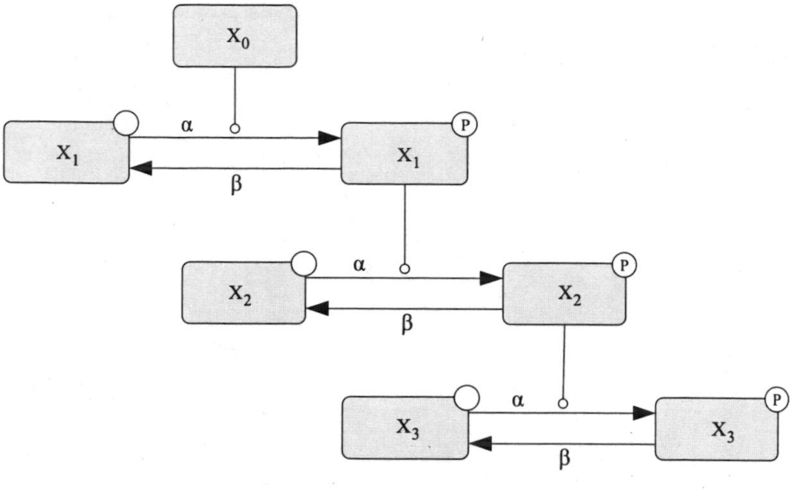

Figure 4.1
Sequential kinase-phosphatase cascade model (Heinrich et al., 2002). The dephosphorylated form of a protein is denoted as X_i and the phosphorylated as X_i^P. X_0, α, and β denote the receptor concentration, and the kinase and phosphatase rate constants, respectively.

The motivation for characterizing a signal can also arise from a more structural point of view: we may want to unravel features that make one pathway structure more favorable than others. Or, assuming the same network structure but different model approximations, we may want to describe quantitative and qualitative differences in model behavior.

Section 4.2 introduces and explains quantitative signaling measures. The analysis based on quantitative measures is performed with respect to a multivariate objective function. This function is not trivial but is necessary since the design of a complex signal transduction pathway most likely depends on numerous factors. Section 4.3 briefly introduces MAPK cascades and provides references to the literature. Section 4.4 presents three different model approximations, two of them proposed for modeling the MAPK cascade. Section 4.5 demonstrates the application of the quantitative measures of section 4.2 to the models of section 4.4. Finally, section 4.6 provides a discussion of and the outlook for our work.

4.2 Quantitative Signaling Measures

We follow the approach of Heinrich et al. (2002), which uses integrals of different order to characterize signaling cascades, and which describes the signal in question in terms of signaling time, duration, strength, and amplification. Consider the linear kinase-phosphatase cascade shown in figure 4.1, where *linear* means that there is no

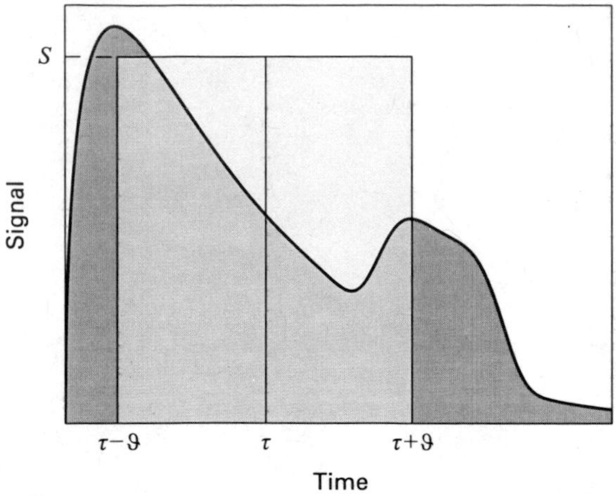

Figure 4.2
Graph of hypothetical time-dependent signal, showing signaling time τ, signal duration ϑ, and signal strength S (Heinrich et al., 2002). Note that the signaling time τ does not necessarily coincide with the time where the maximum level of the signal is reached. This would only be the case for a symmetric profile. By definition of S, the area under the curve is equal to the area of the rectangle spanned by S and 2ϑ.

cross-talk and no feedback or feedforward loop. Each protein X_i exists in either a phosphorylated (X_i^{P}) or a nonphosphorylated (X_i) state. Here the phosphorylated state is presumed to be the active state; the nonphosphorylated state is presumed to be inactive. Assuming, for example, that at time $t_0 = 0$, every protein is found only in its inactive state, the occurrence of a stimulus triggers the phosphorylation of protein X_1, which then stimulates the phosphorylation-dephosphorylation cycle of its successor, with the corresponding kinase and phosphatase rate constants.

We now introduce the quantitative measures of Heinrich and coworkers that we used to characterize the temporal profile of a protein X_i^{P}, schematically shown in figure 4.2. As can be seen in this figure, the approach can also be used for more complex biochemical signaling pathways.

The *signaling time* is a quantitative measure defined as the average time to activate protein X_i. It is defined as an average, analogous to the mean value of a statistical distribution:

$$\tau_i = \frac{\int_0^\infty t X_i^{\mathrm{P}}\,\mathrm{d}t}{\int_0^\infty X_i^{\mathrm{P}}\,\mathrm{d}t}. \tag{4.1}$$

The denominator is the integral over the entire phosphorylated protein signal and corresponds to the total amount of all X_i^{P} molecules generated during the signaling period. More generally, the area under the curve can be interpreted as the total amount of activated protein X_i^P. Note that, although the protein concentration $X_i^P(t)$ is time dependent, we suppress an explicit indication in this chapter to make the formulas more clearly arranged.

The *signal duration* is defined as the average time during which protein X_i^P remains activated:

$$\vartheta_i = \sqrt{\frac{\int_0^\infty t^2 X_i^P \, dt}{\int_0^\infty X_i^P \, dt} - \tau_i^2}, \tag{4.2}$$

and is similar to the standard deviation of a statistical distribution. By definition, the signal duration depends on the signaling time defined in equation (4.1).

To develop a measure of the average signal concentration S_i, Heinrich and co-workers considered the area under the curve $X_i^P(t)$ (see again figure 4.2), that is, $\int_0^\infty X_i^P \, dt$. They defined the average concentration S_i so that $2\vartheta_i \times S_i$ spans a rectangle whose area is the same as the area under the curve. Thus combining the quantitative measures as defined in equation (4.1) and equation (4.2), we arrive at the *signal amplitude* or signal strength, given by

$$S_i = \frac{\int_0^\infty X_i^P \, dt}{2\vartheta_i}, \tag{4.3}$$

where S_i defines the average concentration of protein X_i^P.

The *amplification factor*, defined as

$$A_{ij} = \frac{S_j}{S_i}, \quad \text{for } i < j; \; i, j = 0, \ldots, n, \text{ where } n = \text{cascade length}, \tag{4.4}$$

compares the signal amplitude of the jth component with one of its precursors i. For $A_{ij} > 1$, the signal is amplified; for $A_{ij} < 1$, it is damped; for $A_{ij} = 1$, it remains constant. Additional amplification factors can be defined in the same way to compare other elements of the cascade (e.g., double phosphorylation of proteins, as shown in figure 4.3). The overall amplification of a signaling pathway can thus be measured as

$$A_n = \frac{S_n}{S_0}, \tag{4.5}$$

where S_0 denotes the signal amplitude of the stimulus, and S_n the signal amplitude of the most downstream and phosphorylated protein. The signal amplitude or the amplification factor can be used as a consistent baseline for comparing different pathway structures. Thus one can compare signaling time and signal duration of

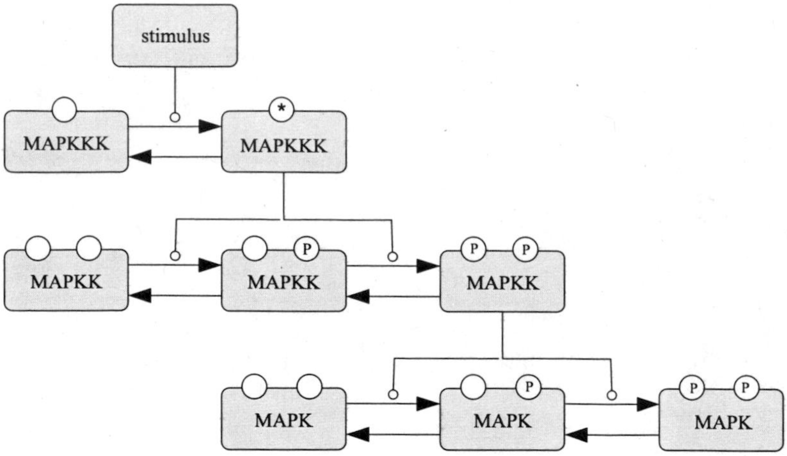

Figure 4.3
Schematic representation of a three-level MAPK cascade using CellDesigner (Funahashi et al., 2003). Circles with a "P" inside denote phosphate groups involved in phosphorylation. Here the second and third levels involve double phosphorylation. The circle with a "*" inside denotes activation.

pathways differing in length or number of phosphorylations, for example, while keeping these measures constant (Frey et al., 2008). For a model with numerous parameters, however, determining a parameter set that results in the desired amplitude is more complicated, as shown in section 4.5. On the other hand, for a systematic comparison, one does not need to use the same signal amplitude, but can relate the rate constants, as shown at the end of section 4.4.

Application of the quantitative signaling measures of equations (4.1)–(4.3) is restricted to a special class of signal shapes. The signal has to be finite at initial time t_0 (in our investigations, we assume a value of zero). In addition, the term $\int_0^\infty t^2 X_i^{\mathrm{P}}\,\mathrm{d}t$ in equation (4.2) requires that, in the limit as $t \to \infty$, the signal disappears faster than t^{-2}. According to these limitations, the formalism introduced can be applied to a cascade where the initial concentrations of the phosphorylated proteins, the nonphosphorylated proteins, intermediates and the external stimulus are finite. The MAPK cascades satisfy these constraints, allowing the formalism to be applied to this family of pathways, as the following section demonstrates.

4.3 Cellular Signaling Cascades: The MAPK Cascade

The mitogen-activated protein kinase (MAPK) cascades are evolutionarily well conserved from yeast to mammals (Widmann et al., 1999). They relay signals from diverse stimuli and receptors to the nucleus and are involved in many crucial developmental cellular processes (Avruch, 2007; Pearson et al., 2001), including proliferation, differentiation (Raman et al., 2007) and apoptosis (Lamkanfi et al., 2007). Many oncogenes have been shown to encode proteins that transmit mitogenic signals upstream of this cascade, so that the MAPK pathway provides a simple unifying explanation for the mechanism of action of many nonnuclear oncogenes (Seger and Krebs, 1995). Consequently, these signaling pathways play an important role in many diseases (Kyriakis and Avruch, 2001; Lawrence et al., 2008); their role in cancer is of special interest.

The ubiquitous MAPK pathway consists of a cascade of at least three activation steps (Bardwell, 2005), as shown schematically in figure 4.3. It is assumed that the corresponding kinase and phosphatase activities do not depend on the degree of phosphorylation. The first protein kinase (MAPKKK) activates the second kinase (MAPKK). The second kinase then activates the third kinase (MAPK) through double phosphorylation (Alessi et al., 1994; Anderson et al., 1990). Active MAPK then enters the nucleus and regulates transcription factors that control gene expression. For about a decade, the MAPK cascade has been the subject of mathematical modeling. Two questions arise:

1. What are the consequences of different model assumptions assessed by quantitative signaling measures?
Recent studies show that approximations can strongly influence the response of the system (Blüthgen et al., 2006; Flach and Schnell, 2006; Millat et al., 2007).

2. Do certain structures realize a design principle with respect to the quantitative measures (see Frey et al., 2008)?

To investigate these questions, we chose three commonly used models of MAPK cascades, representing different levels of approximation. The mathematical analysis of MAPK cascades reveals many interesting features of this fundamental structure in cellular signaling. The sequential alignment of (de)phosphorylation cycles amplifies their inherent ultrasensitivity (Goldbeter and Koshland, 1981) in such a way that the cascades are presumed to act as very sensitive, highly nonlinear biochemical switchs (Ferrell, 1996; Ferrell and Machleder, 1998). Additional combination with positive or negative feedback, or both, brings about other interesting nonlinear behavior, such as bistability and oscillations (Kholodenko, 2000; Wang et al., 2006; Xiong and Ferrell, 2003).

4.4 Models of the MAPK Cascade

During the last decade, various models of the MAPK cascade have been developed. Some of them were generated to focus on the investigation of the core module and its principal features (e.g., Bhalla et al., 2002; Blüthgen and Herzel, 2003; Huang and Ferrell, 1996). Others incorporate growth-factor receptors and adaptor proteins (e.g., Bhalla and Iyengar, 1999; Pircher et al., 1999; Sasagawa et al., 2005; Schoeberl et al., 2002) or describe the effect of feedback or feedforward loops and cross-talk with consequences on dynamic and steady-state properties (e.g., Asthagiri and Lauffenburger, 2001; Kholodenko et al., 1999; Markevich et al., 2004; McClean et al., 2007; for reviews, see Orton et al., 2005; Vayttaden et al., 2004).

Here we focus on the core structure of the MAPK cascade as shown in figure 4.3. For our analysis, we chose three well-established models with different levels of approximation. The model by Huang and Ferrell (1996; hereafter denoted HF) is the most detailed one with respect to the kinetic representation, whereas the models by Heinrich et al. (2002; denoted Hg) and Kholodenko (2000; denoted K) apply some approximations to simplify the mathematical representation.

The mechanistic HF model assumes that each activation or deactivation of a protein follows an enzyme kinetic reaction (Cornish-Bowden, 2004; Segel, 1993). For a protein X_i and its upstream active kinase X_{i-1}^{P} that catalyzes the phosphorylation of X_i to X_i^{P}, we have

$$X_i + X_{i-1}^{\mathrm{P}} \underset{d_i}{\overset{a_i}{\rightleftharpoons}} (X_i X_{i-1}^{\mathrm{P}}) \overset{k_i}{\rightarrow} X_i^{\mathrm{P}} + X_{i-1}^{\mathrm{P}}. \tag{4.6}$$

The bimolecular reaction of X_i and X_{i-1}^{P} results in an intermediary complex $(X_i X_{i-1}^{\mathrm{P}})$, which can dissociate into the reactants or the activated form X_i^{P} and the unchanged kinase X_{i-1}^{P}. The three reaction rates involved are determined by rate coefficients: association constant a_i, dissociation constant d_i, and catalytic constant k_i. X_{i-1}^{P} is assumed to be an ideal catalyst. Furthermore, it is assumed that all other participants (e.g., H_2O, ATP) are constant over the whole observation period, which leads to a reaction scheme that is formally equivalent to a standard enzymatic reaction (Millat et al., 2007). Additionally, there are some conservation laws for proteins and enzymes (Huang and Ferrell, 1996). The resulting system of coupled nonlinear ordinary differential equations describes the time course of the proteins and their different activation states as well as the time course of the intermediary complexes as in equation (4.6):

$$\frac{\mathrm{d}}{\mathrm{d}t}(X_i X_{i-1}^{\mathrm{P}}) = a_i X_i \times X_{i-1}^{\mathrm{P}} - (d_i + k_i)(X_i X_{i-1}^{\mathrm{P}}).$$

The term $(X_i X_{i-1}^{\mathrm{P}})$ denotes a complex. Note that, for our investigation, we introduce an additional equation for the time-dependent stimulus:

$$X_0(t) = \mathcal{X}_0 \mathrm{e}^{-\lambda t},$$

in comparison to the original system of Huang and Ferrell (1996), where the stimulus is a constant external parameter. This equation describes receptor deactivation as a consequence of a pulselike stimulus. For the sake of simplicity, we choose a simple monomolecular deactivation mechanism and assume a receptor concentration of $\mathcal{X}_0$ at the initial time $t_0 = 0$. This stimulus profile guarantees that the final signal of the phosphorylated protein approaches zero for $t \to \infty$, as required for calculating the quantitative signaling measures.

The mechanistic model can be simplified if one assumes that the intermediate complexes operate at a quasi–steady state. This assumption converts the differential equation for the complex into an algebraic equation, for example:

$$0 = a_i X_i \times X_{i-1}^{\mathrm{P}} - (d_i + k_i)(X_i X_{i-1}^{\mathrm{P}}), \tag{4.7}$$

which can be solved for the complex concentration $(X_i X_{i-1}^{\mathrm{P}})$. Inserting this into the enzyme kinetic reaction reduces it formally to a bimolecular reaction (Millat et al., 2007). If one further assumes a constant phosphatase concentration, the dephosphorylation reaction reduces to a pseudo-monomolecular reaction (Millat et al., 2007). The change of concentration of a protein can then be described by the differential equation

$$\frac{\mathrm{d}}{\mathrm{d}t} X_i^{\mathrm{P}} = \tilde{\alpha}_i X_i \times X_{i-1}^{\mathrm{P}} - \beta_i X_i^{\mathrm{P}}, \tag{4.8}$$

where $\tilde{\alpha}_i = \alpha_i / X_i^{\mathrm{tot}}$ and X_i^{tot} denotes the total concentration of protein X_i (Heinrich et al., 2002). The rate equation (4.8) can be further simplified if we apply the conservation laws for the proteins and additionally assume that the concentration of intermediate complexes is negligible with respect to the concentration of proteins that are not bound in a complex. Under that assumption, the conservation law reduces to a simple relation between the protein concentrations:

$$X_i^{\mathrm{tot}} = X_i + X_i^{\mathrm{P}}. \tag{4.9}$$

Inserting this into equation (4.8) yields the Hg model:

$$\frac{\mathrm{d}}{\mathrm{d}t} X_i^{\mathrm{P}} = \alpha_i X_{i-1}^{\mathrm{P}} \left(1 - \frac{X_i^{\mathrm{P}}}{X_i^{\mathrm{tot}}}\right) - \beta_i X_i^{\mathrm{P}}. \tag{4.10}$$

Finally, we introduce a third MAPK model by Kholodenko (2000), which we use for our investigation. As in the previous models, all of the biochemical reactions involved follow the enzyme kinetic reaction of equation (4.6). Again, we use the quasi-steady-state approximation to simplify the system of coupled differential equations. In contrast to the HF and Hg models discussed above, the K model uses the

Table 4.1
Overview of the three models used for analysis in chapter 4

Model (Reference)	Example equation	Parameters
HF (Huang and Ferrell, 1996)	$\frac{d}{dt}X_i^P = (d_{i+1}+k_{i+1})(X_{i+1}X_i^P) + k_i(X_iX_{i-1}^P) - a_{i+1}X_{i+1}X_i^P$	a_i, d_i, k_i
K (Kholodenko, 2000)	$\frac{d}{dt}X_i^P = \frac{k_iX_{i-1}^PX_i}{K_M+X_i} - \frac{V_iX_i^P}{K_M+X_i^P}$	$K_M^i = \frac{d_i+k_i}{a_i}$ $V_i = k_iP$
Hg (Heinrich et al., 2002)	$\frac{d}{dt}X_i^P = \alpha_iX_{i-1}^P(1-X_i^P/X_i^{tot}) - \beta_iX_i^P$	$\alpha_i = \frac{k_i}{K_M^i}$ $\beta_i = \frac{k_i}{K_M^i}P$

Note: The third column shows the specific parameters and their derivation from the mechanistic HF model.

conservation law for the proteins to express the level of upstream kinase. Owing to this explicit transformation, the K model is a simplification of the Hg model (see equation (4.8); Millat et al., 2007). The resulting rate equation for activation and deactivation of a protein,

$$\frac{d}{dt}X_i^P = \frac{k_iX_{i-1}^P \times X_i}{K_M+X_i} - \frac{V_iX_i^P}{K_M+X_i^P}, \tag{4.11}$$

contains two Michaelis-Menten-like expressions and is similar to the model of Goldbeter and Koshland (1981). The limiting rate V_i results again from the assumption of a constant phosphatase concentration. Note that neglecting the complex concentration within the model is crucial (Blüthgen et al., 2006; Millat et al., 2007). The structure of the differential equations, the required kinetic parameters, and their relation to the mechanistic constants are summarized in table 4.1. Section 4.5 applies the quantitative signaling measures to compare the HF, Hg, and K models and discusses the consequences of the approximations used in the latter two models.

We now derive the *kinetic parameters* uniquely related within the models. For simplicity, the number of parameters is reduced by examining only cascades with identical cycles. The starting point is again the mechanistic HF model, where the activation and deactivation of a protein X_i is determined by the association constant a_i, the dissociation constant d_i, and the catalytic constant k_i. Owing to the quasi-steady-state assumption used in the Hg model, we obtain new kinetic parameters that depend on the mechanistic parameters of the HF model. The activation rate constant α_i follows from the balance equation (4.7) as

$$\alpha_i = \frac{k_i}{K_M^i}. \tag{4.12}$$

Since the concentration of phosphatase is assumed to be constant, β_i is an effective deactivation rate constant (Atkins and de Paula, 2002; Millat et al., 2007):

$$\beta_i = \frac{k_i}{K_M^i} P, \tag{4.13}$$

which is the product of the phosphatase concentration P and the ratio of the catalytic rate constant k_i and the Michaelis constant:

$$K_M^i = \frac{d_i + k_i}{a_i}. \tag{4.14}$$

Additionally, the Kholodenko model contains the limiting rate V_i, which is defined as usual as

$$V_i = k_i P, \tag{4.15}$$

where P again is the phosphatase concentration (for an overview, see table 4.1).

As the sequential derivation shows, the parameters of the simplified models are determined by the mechanistic parameters a_i, d_i, and k_i. Note that this derivation is not reversible (Millat et al., 2007). Thus, if the limiting rate V_i and the Michaelis-Menten constant for an enzyme kinetic reaction are known, it is not possible to evaluate uniquely all three mechanistic kinetic parameters. Since all model parameters can be determined from the mechanistic parameters, all models are well defined if the mechanistic parameters are known. To simplify the analysis, we use identical activation and deactivation cycles to compose the MAPK cascades. Furthermore, we assume that the coefficients of deactivation scale with the kinetic coefficients of activation. In this way, we reduce the number of system parameters to three coefficients for activation and a factor that determines the coefficients for deactivation. We assume that, at the initial time t_0, all proteins are inactive and thus in their nonphosphorylated form.

4.5 Comparison of Models Using Quantitative Measures

If the mechanistic interactions, approximations, or intermediate steps of a signaling pathway are unknown or uncertain, the modeling of the pathway becomes precarious. Quantitative measures can be used to characterize the effect of the different approaches as well as to compare different network structures within a single model approach. In comparing models of different mechanistic interactions, approximations, and intermediate steps, we use the model approximations introduced before. The different intermediate steps include single, double, and triple phosphorylations of a three-level cascade. We also consider steps that are not realized in biological systems but could theoretically transfer the signal. We show how the double phosphorylation, which is exhibited by MAPK cascades, changes the properties of the transduced signal in a favorable way in comparison with single or triple phosphorylations. The analysis of signal transduction pathways often requires simplifying reductions, as noted for the MAPK cascade in section 4.4, although how these approximations change the system's behavior is a matter of ongoing discussion. Note that approximations can change the system's behavior both qualitatively and quantitatively (Millat et al., 2007).

Parameter sets were chosen for the HF model, as indicated in figures 4.4–4.6; the compatible parameters for the Hg and K models were calculated as in section 4.4. All three models display qualitatively similar behavior. Moreover, for some parameter combinations, they also have quantitatively similar behavior (see also figures 4.4–4.6). It can be seen that the behavior of the Hg model is more similar to that of the K model than the behavior of either is to that of the HF model. We suggest that the consequences of the approximations become obvious here.

In general, the average time until a signal is activated (τ) becomes slower with increasing number of phosphorylations and the average duration of signal activation (ϑ) becomes shorter. This means that multiple phosphorylations may have evolved to generate slow signals of short duration. A signal with a fast response may be of disadvantage in biological systems because the system may need a certain time to prove the validity of the stimulus. On the other hand, the proof of validity should be fast enough to realize protection from damage or intake of nutrients when exposed to high availabilities.

We also considered the product of τ and ϑ. If a minimum of this product exists, it shows an optimal trade-off between both. Indeed, as can be seen in figure 4.6, a minimum most often exists for structures with double phosphorylations. This feature matches the design of the MAPK cascade, which suggests that the MAPK cascade is favorably designed to trigger a fast response of short duration. For a few parameter combinations, such as $a = 4$, $d = 1$, $k = 2$, the product of τ and ϑ shows a minimum at triple phosphorylation. But in those cases, the minimal value of the triple phosphorylation is similar to that of a double phosphorylation, although there is a huge deviation between a structure with single and one with double or triple phosphorylations. Nevertheless, because a structure with two phosphorylations would imply lower complexity and require fewer processes than a structure with three, a double-phosphorylated structure could still be favored over a triple-phosphorylated one.

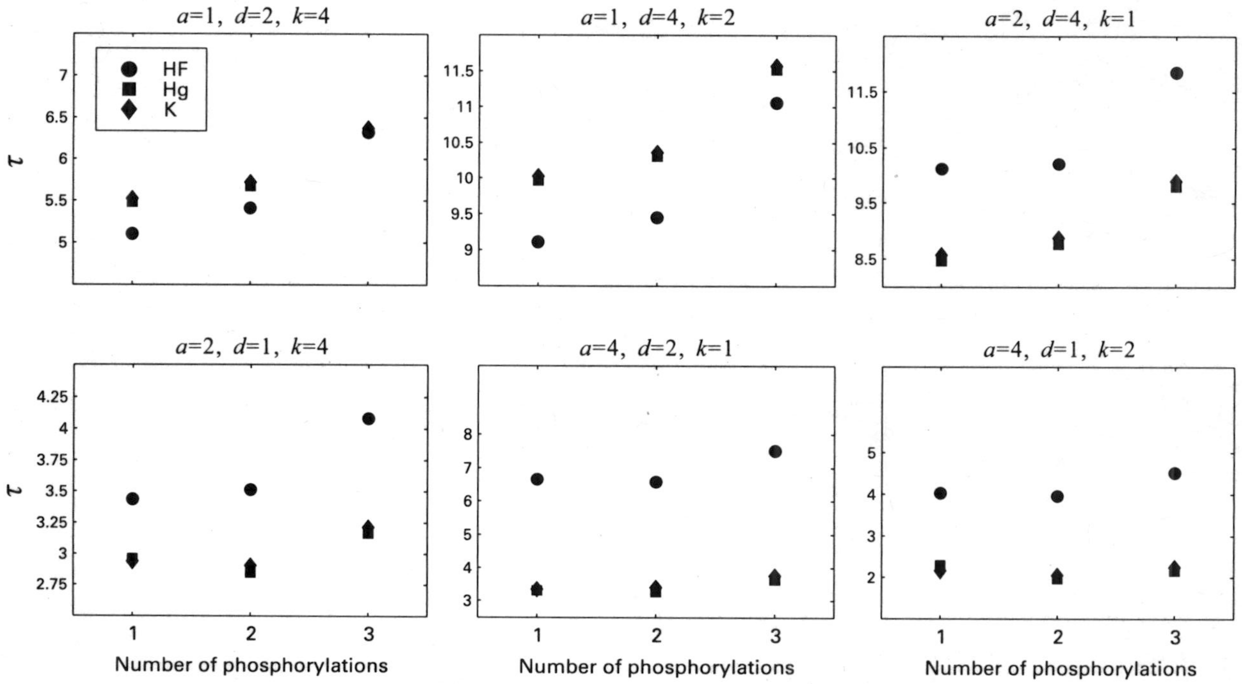

Figure 4.4
Comparison of the HF, Hg, and K models for a cascade of length three over a range of single, double, and triple phosphorylation based on signaling time τ. Parameters are indicated above each figure. Note that τ is increasing for all chosen parameter combinations indicating that a cascade with multiple phosphorylations might be favored for transmitting signals which can afford to be slow.

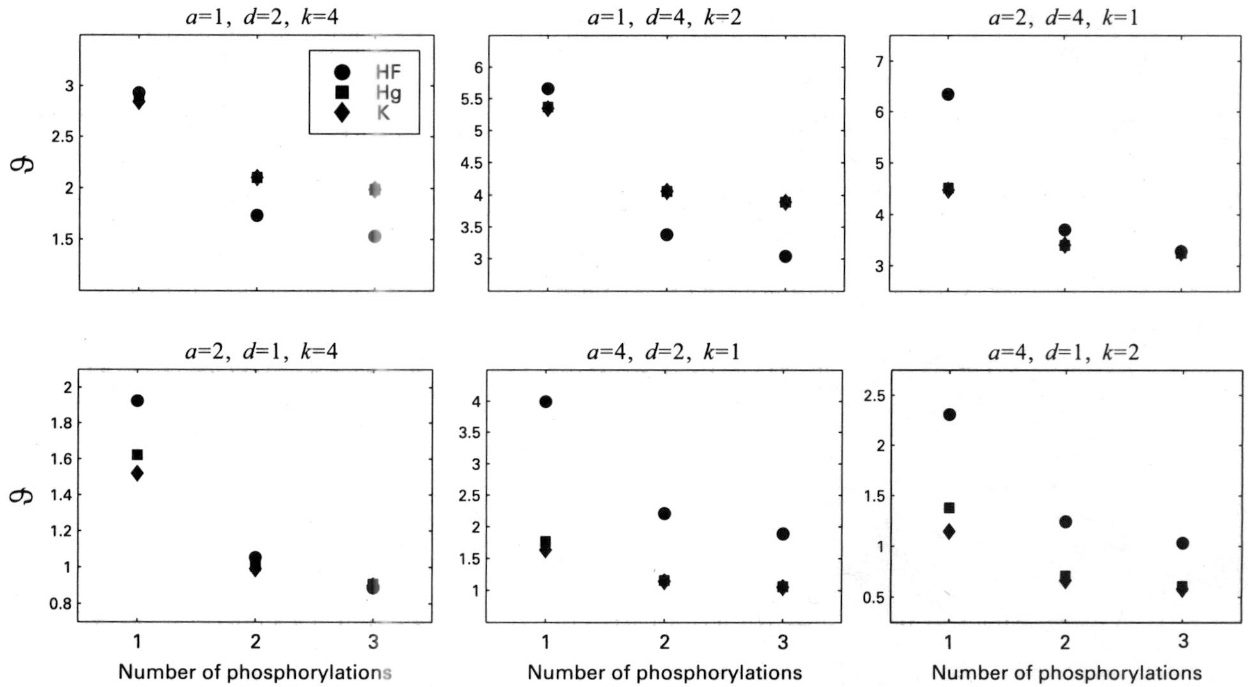

Figure 4.5
Comparison of the HF, Hg, and K models for a cascade of length three over a range of single, double, and triple phosphorylation based on signal duration ϑ. Parameters are indicated above each figure. Note that ϑ is decreasing for all chosen parameter combinations, indicating that a cascade with multiple phosphorylations might be favored for transmitting signals of short duration.

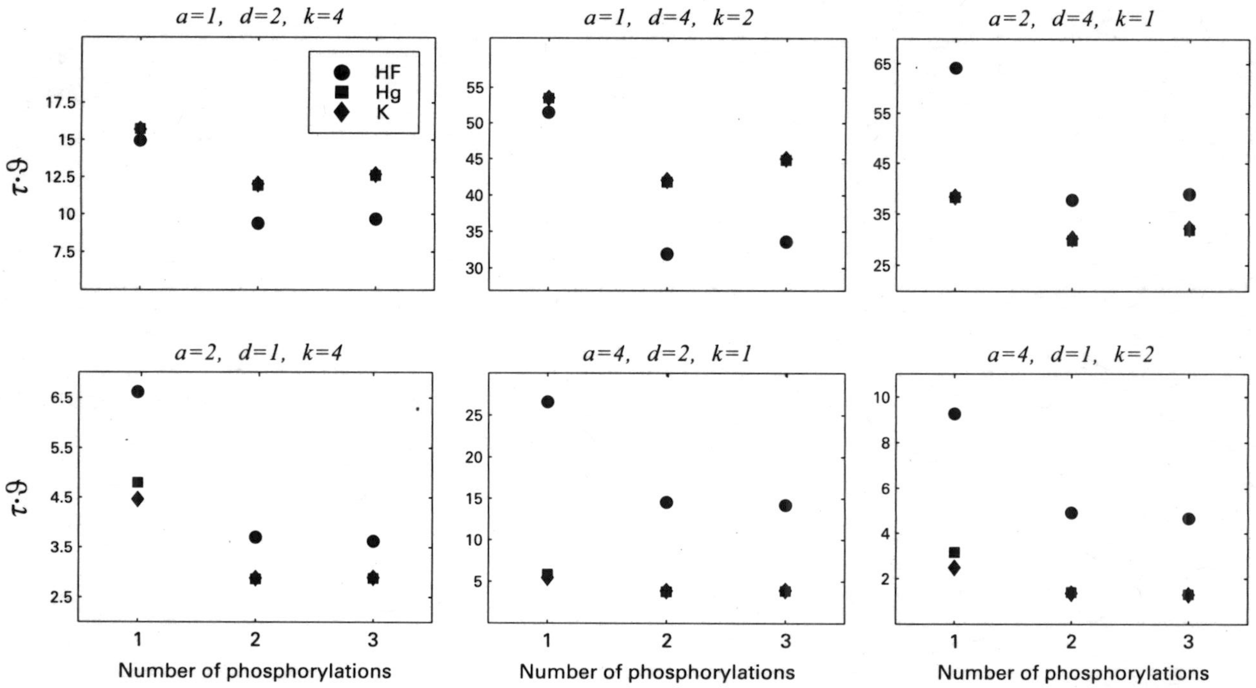

Figure 4.6
Comparison of the HF, Hg, and K models for a cascade of length three over a range of single, double, and triple phosphorylation based on product of signaling time and signal duration. Parameters are indicated above each figure. We consider the most downstream and phosphorylated protein of a cascade. This analysis of the product of τ and ϑ shows that a cascade of short duration and fast response most likely lies (as these parameter combinations indicate) in a double phosphorylated structure (for a cascade of length three).

4.6 Discussion and Outlook

Quantitative measures can be used to characterize the properties of signaling molecules and thus of signaling pathways. Thus our study compares the consequences of different degrees of phosphorylation in a pathway structure by varying from one to three phosphorylations while keeping a constant cascade length of three. Provided that the objective function mirrors the true purpose of the model structure, quantitative measures can be used to unravel design principles of biological systems. For the chosen sets of parameters, a minimum of the product of signaling time and signal duration exists in most of the cases for a double phosphorylated pathway structure. This suggests that the MAPK pathway has evolved into a structure that favors a fast response of short duration to a stimulus. Of course, taking the product of two components is only one approach of many to unravel the principle behind specific and perhaps even unique pathway structures. There are many more factors that contribute to the structure of the pathway as it is today. All possible factors need to be weighed against each other to determine a more precise criterion that might motivate the design of these structures.

A comparison of different model approximations describing the same (MAPK) pathway shows similar qualitative behavior among those models and exposes the deviations that arise from simplifying model assumptions. Among a set of candidate models, the aim is to choose the model closest to reality. The quantitative measures confirm that the assumptions of the Heinrich et al. (2002) model are more similar to those of the Kholodenko (2000) model than the assumptions of either are to those of the Huang and Ferrell (1996) model. Interestingly, it depends on the choice of parameters whether the HF model shows faster responses to a stimulus than the Hg or K model. This is a consequence of the quasi-steady-state approximation used in the Hg and K models. On the one hand, both Hg and K models neglect the time to establish the quasi–steady state of the complexes that is fully described in the HF model. On the other hand, if no quasi–steady state is established, the quasi-steady-state approximation used in Hg and K models is not valid and adds an artificial bottleneck in the dynamics of the system. In that case, the more detailed HF model becomes faster than the Hg and K models. As is known from investigations of enzyme kinetic reactions, the choice of parameters determines the validity of the quasi-steady-state assumption (Schnell and Mendoza, 1997).

The quantitative measures introduced in this chapter can also be used to analyze experimental data without any knowledge about the underlying network. That said, the quantitative measures of Heinrich et al. (2002) are defined for a special class of signals: they have to start at a finite value and eventually have to reach zero, which ensures that the integrals of equation (4.1) and equation (4.2) are finite. The challenge now is to develop quantitative measures capable of characterizing signals with an arbitrary final state.

5 Control Strategies in Times of Adversity: How Organisms Survive Stressful Conditions

Hana El-Samad

Life in the microbial world is characterized by fierce competition, nutritional hardship, and often dramatic changes of environmental parameters. Since a bacterial cell has limited chances and opportunities to choose and modify its environment actively, adaptive responses of a bacterium to its environment are a defining cornerstone of microbial life. These vital responses, collectively called stress responses, allow the organism to respond rapidly and effectively to restore stable intracellular homeostasis in the presence of a variety of environmental disturbances, such as perturbations in temperature, pH, oxygen concentration, nutrient availability, and osmolarity, that can threaten the integrity of the cell. This, however, requires the sensitive monitoring of numerous environmental parameters, followed by the orchestration of complex stress regulatory systems that maintain adequate cellular homeostasis in an intricate balancing act between costs and gains. It is not surprising as a result that stress-response systems are constructed using elaborate gene regulatory networks organized into hierarchies of interlocked feedback and feedforward loops. It also comes as no surprise that these stress-response systems, found embedded in the genomic blueprint of almost every bacterium studied to date, use a versatile battery of sophisticated control strategies to efficiently maintain the functionality and integrity of the bacterium under all challenging circumstances pertinent to its specific lifestyle.

5.1 Stress Responses: Universal Survival Strategies

During the last few decades, thorough experimental investigations of many bacterial stress-response pathways have generated a wealth of knowledge about many of their components. Impressive progress has also been made in uncovering the causality of the interactions between these components. Our current knowledge therefore encompasses extensive parts lists for many stress pathways, in addition to qualitative descriptions of their functionality and interconnections. Complex control systems, such as those implemented by stress-response systems, are nonetheless more than the sum of their parts. Their success hinges on their dynamic behavior, speed of response, robust performance, and adaptive efficiency as disturbances challenge the system. These are quantitative properties generated by specific architectures and the choice of specific parameters of the underlying regulatory networks. Deciphering how these properties arise and unraveling the trade-offs associated with them are therefore necessarily central to any meaningful understanding of stress responses. It can only be achieved through the systematic investigation of stress response systems as dynamic integrated systems rather than static collections of isolated parts. Such a system-level perspective, by placing complex cellular networks into a quantitative and predictive framework, holds the promise of generating unprecedented insights into the successes and failures of homeostatic cellular mechanisms.

This chapter presents a system-level perspective on an important stress-response pathway in bacteria: the cytoplasmic heat-shock response system. It discusses the salient organizational properties and design principles of the system in *Escherichia coli* and emphasizes the unique insights offered by using a multidisciplinary approach for probing the system. Finally, the chapter places the cytoplasmic heat-shock response into a broader cellular context by discussing its coordination with other stress-response pathways that act in the cytoplasm or other cellular compartments.

5.2 The Cytoplasmic Heat-Shock Response in *E. coli*: A Sophisticated Control System

5.2.1 The Biology of the Response System

The heat-shock response is a cellular reaction to elevated temperature. This physiological response is also the cellular answer to several other adverse conditions such as exposure to certain chemicals (solvents, certain antibiotics), hyperosmotic shock, and overproduction of foreign and the cell's own proteins. Such challenges frequently lead to misfolded, unfolded, or damaged proteins in the cell. These proteins may expose hydrophobic patches normally buried inside the folded protein; the exposed patches can in turn form aggregates, an event that constitutes a serious threat to the organization and functioning of cellular networks. To prevent this from occurring, the cell triggers the heat-shock response, resulting in the expression of specific genes that encode heat-shock proteins (HSPs; Gross, 1996). Many HSPs serve as molecular chaperones that help to refold denatured proteins; others are proteases that degrade and remove the denatured proteins. The bacterial cell must maintain a fine balance between the protective effect of the HSPs and the material and energy costs associated with overexpressing these proteins. In *E. coli*, this balance is achieved through an intricate architecture of feedback loops centered around the σ^{32}-factor that regulates the transcription of the heat-shock proteins under normal and stress conditions. The enzyme RNA polymerase (RNAP) bound to σ^{32} recognizes the heat-shock gene promoters and transcribes specific heat-shock genes, including molecular chaperones such as *GroEL*, *DnaK*, *DnaJ*, *GroES*, *GrpE*, and proteases such as *Lon*, and *FtsH*. This transcription process is controlled through the tight regulation of σ^{32}, whose synthesis, activity, and stability are modulated by intricate feedback and feedforward loops that incorporate information about the temperature and the folding state of the cell. The synthesis of σ^{32} is primarily regulated at the translational level. At low temperatures, the translation start site of *rpoH* mRNA (encoding σ^{32}) is occluded by base pairing with other regions of the σ^{32} mRNA. At elevated temperatures, this base pairing is destabilized, resulting in a "melting" of the secondary structure of σ^{32}, which enhances ribosome entry, therefore increasing translation efficiency (Morita et al., 1999). This mechanism implements both a temperature sensor and a feedforward control element that uses the temperature information to affect the production of heat-shock proteins independently of the folding state of the cellular proteins. The feedforward control thus enables the system to sense a disturbance (heat in this case) and react to it before its effects start to appear in the output of interest (unfolding of proteins).

The activity of σ^{32} is regulated through its interaction with chaperones such as *DnaK* and its cochaperone *DnaJ*. In addition to their role in protein folding, chaperones bind to σ^{32} and limit its capability to bind to core RNA polymerase. Raising the temperature produces an increase in the cellular levels of unfolded proteins, which titrate the chaperones away from σ^{32}, allowing it to bind to RNA polymerase and increasing transcription of heat-shock genes (Straus et al., 1990). The accumulation of high levels of HSPs leads to the efficient refolding of the denatured proteins thereby freeing up *DnaK/J* to sequester σ^{32} from RNA polymerase. This implements a negative feedback loop that is referred to as a sequestration feedback loop. The stability of σ^{32} is regulated through its interaction with proteases such as *FtsH*. During steady-state growth, σ^{32} is rapidly degraded ($t_{1/2} = 1$ minute), but is stabilized for the first five minutes after temperature upshift. One model postulated to explain this stabilization is based on the observation that *DnaK* and its cochaperone *DnaJ* seem to be required for the rapid in vivo degradation of σ^{32} by the protease (Morita et al., 2000). The sequestration of σ^{32} by chaperones is assumed to promote its degradation, either by presentation to the protease or through some other mechanism (Tatsuta et al., 2000). The titration of chaperones by high levels of unfolded proteins thus results in transient stabilization of σ^{32}, with the reverse effect upon a decrease in the level of unfolded proteins. This mechanism yields a feedback-regulated degradation of σ^{32} and is referred to as the *FtsH* degradation feedback loop. Alternatively, the direct titration of proteases by unfolded proteins may underlie σ^{32} stabilization. In either case, increased translation and stabilization lead to a transient 15–20-fold increase in the amount of σ^{32} at the peak of the heat-shock response, followed by a decrease to a new steady-state concentration dictated by the balance between the

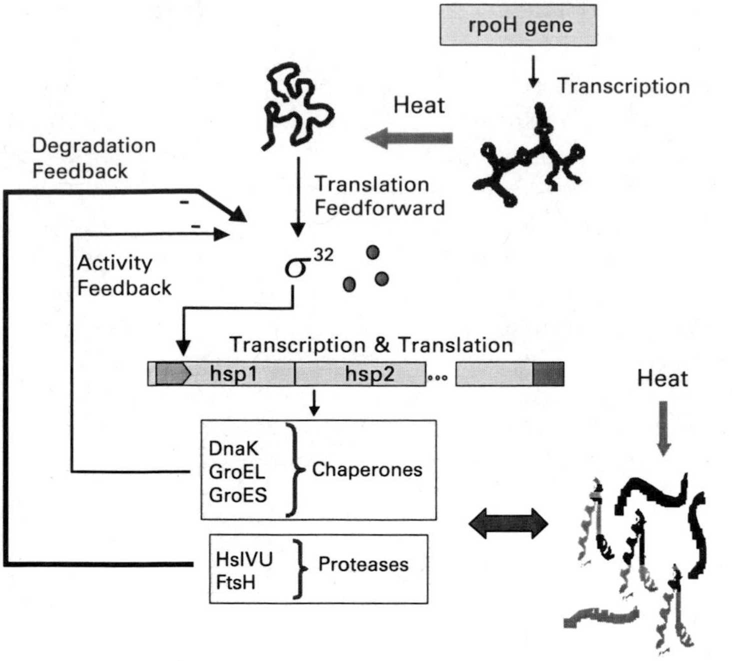

Figure 5.1
Synthesis, stability, and activity of σ^{32} are regulated in response to the temperature input.

temperature-dependent translation of the *rpoH* mRNA and the regulated degradation of σ^{32} (Morita et al., 1999).

The molecular interactions in the heat-shock system as described above are summarized in figure 5.1. The heat-shock response (HSR) has been thoroughly studied at the experimental level, yielding a wealth of information about the different components of the HSR system. Section 5.2 discusses how this qualitative biological information can be transformed into a quantitative description of the heat-shock system, which we then use to pose questions about the regulatory architecture of the system.

5.5.2 Functional Modules in the Response System: Building Cellular Block Diagrams

An important design feature that biological systems are thought to share with engineered systems is the propensity for *modular decompositions* (Hartwell et al., 1999), used extensively in control and dynamical systems theory to make modeling and model reduction of systems more tractable. The typical starting point of modular decompositions in control systems is isolating the process to be regulated, commonly referred to as a *plant*. The remaining components in the control system are then classified in terms of the function they accomplish to facilitate this regulation. For example, sensing and detection mechanisms constitute "sensing modules," while mechanisms responsible for making decisions based on information provided by sensor modules constitute "controller modules." The typical modular list of an engineering system also includes actuation modules. Actuation is necessary to transform the information-rich signal computed by the controller into a quantity of sufficient magnitude to drive the plant in the desired direction.

A similar modular decomposition can be carried in the heat-shock response if the protein-folding task is viewed as the process to be regulated. This plant is actuated by chaperones. The chaperone "signal" is produced by the high-gain transcription/translation machinery, which amplifies a modest σ^{32} input signal (few copies per cell) into a large chaperone output signal, much like an actuator. The σ^{32} control signal is the output of the "computational" or "controller" unit, which, based on the sensed temperature and folding state of the cell, modulates the number and activity of the σ molecules. The direct temperature measurement provided by the σ^{32} mRNA heat-induced melting and the indirect protein-folding information are assessed by the σ computational unit, producing an adequate control action by adjusting the synthesis, degradation, and the amount of sequestered σ. The conceptual modular decomposition of the HSR system is shown in figure 5.2.

Here a subtle difference between the modular structure of gene regulatory networks and that of engineering systems should be noted. A common practice in the

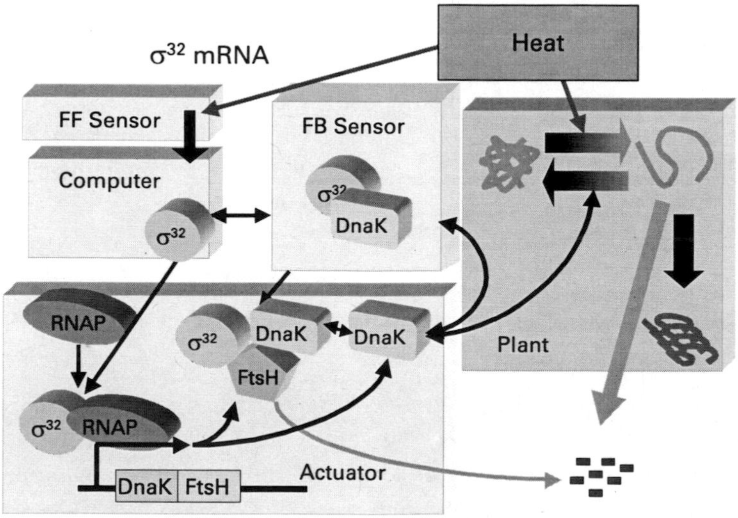

Figure 5.2
Functional modules of the heat-shock response (HSR) system, such as the plant, sensors, computational unit, and actuator, consist of various molecular species and their interactions.

design of engineering systems is to impose strict separation between the various functional modules. Sensors, controllers, and actuators are different physical entities coupled together through well-defined communication channels. Cross-talk between these entities is, by design, avoided. This design philosophy makes model analysis, debugging, and building more tractable. In gene networks, however, this strict separation between control, actuation, sensing, and signal propagation is often violated. Consider, for example, a signaling cascade whose function is to convey environmental information to a regulation mechanism. Although a sensor that receives various environmental cues, the cascade is also a controller, integrating these cues along the cascade and generating a decisive control signal before reaching the response-generating mechanism. This integration of various modules is also present in the heat-shock response system. For example, because it performs protein folding, the chaperone *DnaK* is part of the actuation module. *DnaK* is also part of the sensor module because it conveys temperature information through its binding to σ^{32}. Developing a modular decomposition framework from a control engineering perspective that takes into account the architectural differences inherent in gene regulatory networks is thus a challenging exercise.

5.2.3 Mathematical Modeling of the Response

The cytoplasmic heat-shock response under σ^{32} control consists of about 50–100 genes and includes many chaperones and proteases (Gross, 1996). My coworkers and I have developed a tractable mechanistic model of the σ^{32} HSR system (El-Samad et al., 2005), with *DnaK* as the chaperone representative and *FtsH* as the primary protease degrading σ^{32}. We assumed that *FtsH* action in degrading σ^{32} is mediated by interaction with the $[\sigma^{32}{:}DnaK]$ complex (Tatsuta et al., 2000). We also used the *HslVU* protease to account for the slow degradation of σ^{32} in *FtsH*-null mutants (Kanemori et al., 1997). Using first-order mass-action kinetics, our model consists of a set of 31 differential-algebraic equations (DAEs) with 27 kinetic parameters of the form

$$\frac{\mathrm{d}\boldsymbol{X}(t)}{\mathrm{d}t} = \boldsymbol{F}(t;\boldsymbol{X};\boldsymbol{Y}), \tag{5.1}$$

$$\boldsymbol{0} = \boldsymbol{G}(t;\boldsymbol{X};\boldsymbol{Y}), \tag{5.2}$$

where $\boldsymbol{X}$ is an 11-dimensional vector whose elements are differential (slow) variables and $\boldsymbol{Y}$ is a 20-dimensional vector whose elements are algebraic (fast) variables. Some kinetic rate parameters are picked from the relevant literature. The unavailable parameters are tuned to reproduce the steady-state levels of σ^{32} and chaperones at low temperatures. Those parameter sets reproducing steady-state behavior at high temperatures and transient response upon upshift are retained. Data from HSR mutants were used to discriminate among the remaining parameter sets, adopting the set that reproduces all the data without additional tuning (for details of the model equations, parameters, and data, along with time trajectories, see El-Samad et al., 2005). For numerical simulations of the deterministic models, we used the DAE solver DASSL; for sensitivity analysis, DASPK, a software for the sensitivity analysis of large-scale differential-algebraic systems (Li and Petzold, 2000). Stochastic simulations were implemented using the Gillespie stochastic simulation algorithm (Gillespie, 1976).

5.2.4 Control Analysis of the Response: Speed, Robustness, and Efficiency

In order to reverse-engineer the regulatory loops in the cytoplasmic heat-shock response system, we used the mathematical model to devise "virtual mutants" that lack some of these loops. When compared to the "wild type" heat shock, these mutants highlight the benefits of different aspects of the network architecture. To guarantee the equivalence of mutants with the wild type and to set a valid basis for comparison, we adjusted the levels of σ^{32}, chaperones, and unfolded proteins in the mutant designs to make them identical to those of the wild-type heat shock at low temperatures. We then investigated the behavior of the system after a temperature upshift. This procedure is necessary to reduce the accidental difference between the two models and ensure that any remaining dissimilarities in their dynamical behavior are the correct indicators of their actual difference (Alves and Savageau, 2000; Savageau, 1972).

Feedforward: A Dynamic Sensor

The feedforward component in the translation of σ^{32} is instrumental for many aspects of the HSR system's dynamic response. Indeed, the translation's switchlike behavior upon temperature upshift increases the efficiency of the protein folding by providing a sufficient number of σ^{32}, which then produces an adequate number of chaperones. To investigate the role of this feedforward loop, we devised a virtual mutant where this translation thermosensor is disabled by locking the translational switch in the off position upon temperature upshift, therefore imposing the same translational efficiency at both low and high temperatures. In this case, the accumulation of σ^{32} in the induction phase of the response is achieved through stabilization rather than the increase in synthesis rate of σ^{32}. This yields a more modest accumulation of σ^{32} at the peak of the heat-shock response, after which the level of σ^{32} recovers to a value slightly higher than that at the lower temperature. This, in turn, results in insufficient chaperone production and consequently, impaired and slightly delayed protein folding (figure 5.3). To address the question of whether any tuning of the feedback components themselves could compensate for the response deficiency in the absence of feedforward, we started by making the following observations.

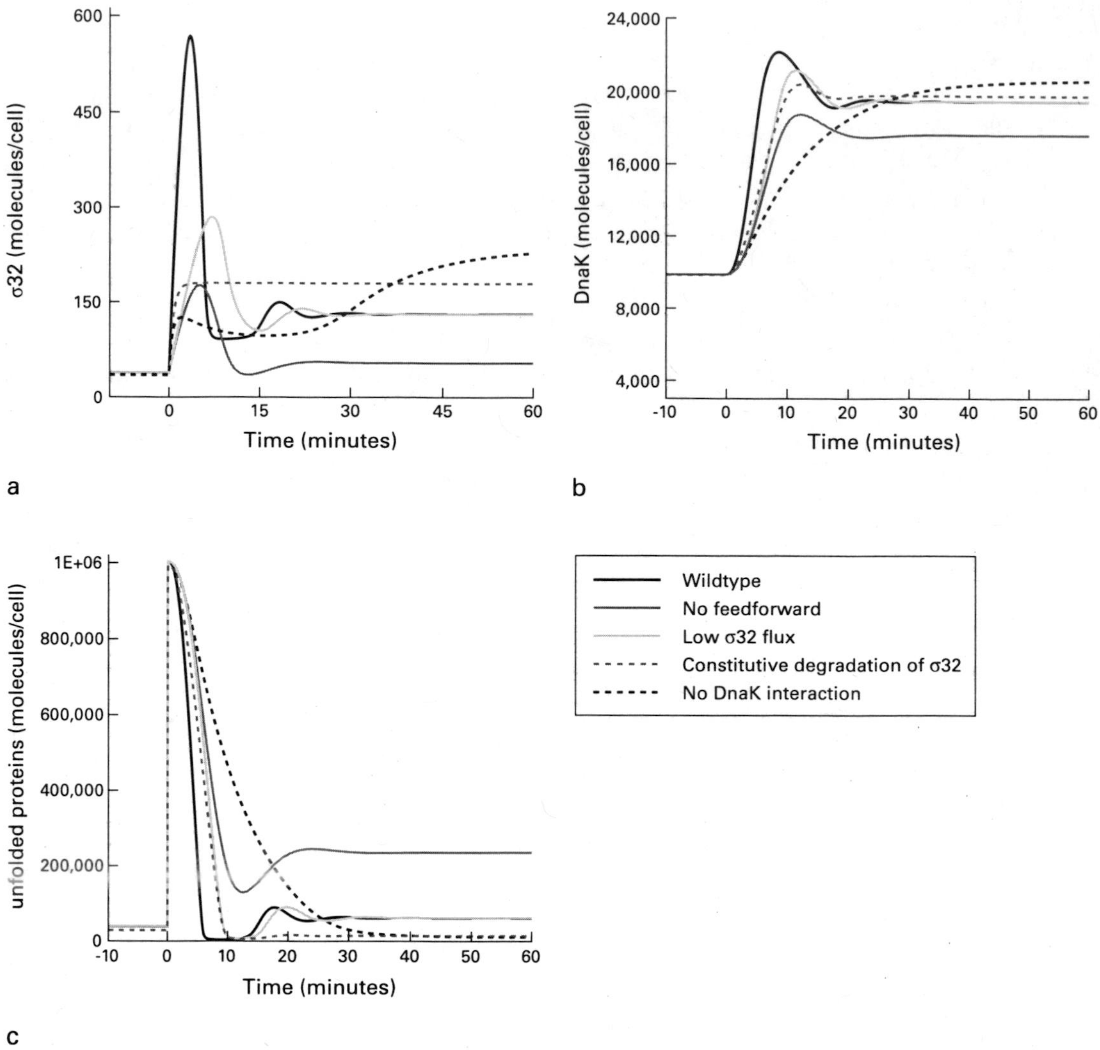

Figure 5.3
Levels of σ^{32} (a), *DnaK* (b), and unfolded proteins (c) for wild-type heat-shock response and its various virtual mutants. The wild type (*solid black line*) is taken as the canonical response. The model with no feed-forward term (*solid, dark gray line*) shows lower levels of σ^{32}, resulting in impaired and delayed protein folding. The model with low amplification flux (*solid, light gray line*) shows delayed proteins folding, achieved by a lower peak of σ^{32}. The model with constant degradation of σ^{32} (*dotted gray line*) shows delayed protein folding, and no peak of σ^{32}. The model with no *DnaK* interaction (*dotted black line*) shows a slightly higher performance in protein folding, but a dramatically delayed profile for this folding.

The lower level of σ^{32} at steady state in the feedforward mutant, as compared to its counterpart in the wild type, is the result of a new setpoint dictated by the balance between σ^{32}'s lower synthesis rate at high temperatures and its degradation rate. Because a certain level of folding is recovered in the adaptation phase of the response, and owing to the small number of σ^{32} as compared to chaperones, most σ^{32} return to the sequestered form, therefore limiting the number of new chaperones produced and stabilizing the number of unfolded proteins, although the process possesses higher thresholds when the synthesis of σ^{32} is larger in the presence of feedforward. To recover the threshold seen with feedforward control using feedback alone, one could decrease the degradation feedback gain to balance its reduced synthesis, for example, by making σ^{32} less susceptible to degradation by *FtsH*, although this would imply the presence of higher concentrations of σ^{32} even at low temperatures, hence an unneeded excess of heat-shock proteins. Moreover, the slower transient response of the feedforward mutant is attributed to the accumulation of σ^{32} through its feedback-mediated stabilization solely, a process possessing an inherent delay because it relies on protein unfolding rather than the direct sensing of temperature. Again, a possible mechanism that can compensate for this delay in the absence of the feedforward term is an increase in the turnover rate of σ^{32}. Although the steady state for σ^{32} (and subsequently chaperones and unfolded proteins) can be kept constant by simultaneously manipulating its synthesis and degradation rates, this results in modified response dynamics, which we investigate next.

The Amplification Flux: A Metabolically Costly Speeding Mechanism

To show the impact of the rate of production and destruction of σ^{32}, which we termed the *amplification flux*, on the transient response in the heat-shock system, we performed a simulation where we simultaneously decreased both the translation and the degradation of σ^{32} by 5-fold. The result of this experiment, shown in figure 5.3, indicates that the lower (or higher) turnover rate for σ^{32} necessarily results in a slower (or faster) response. Notice that, in this case, the translational switch is still operational in the sense that at high temperatures, σ^{32} translation is still being increased 5-fold relative to its value at low temperatures. Based on this analysis, one can postulate that a scenario where the synthesis and degradation of σ^{32}, tuned appropriately at low temperatures, can compensate for the delayed and impaired response in the absence of the feedforward loop at high temperatures. However, the by-product of such tuning is a high metabolic cost at low temperature due to the σ^{32} futile production/degradation cycle.

Feedback through Sequestration: An Efficient Mechanism to Achieve Robustness

Binding of σ^{32} to *DnaK* implements an important feedback loop in the HSR system. Feedback loops are commonly used in engineering applications to combat the effects of fluctuating parameters, thereby achieving robustness to parametric uncertainty.

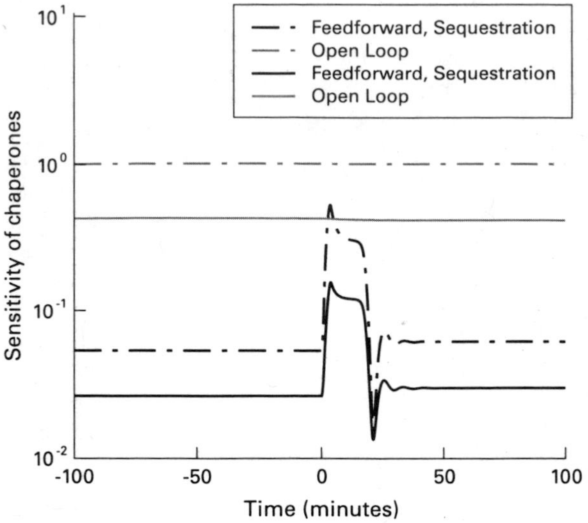

Figure 5.4
Small signal sensitivity of the level of chaperones to their synthesis rate (*dashed-dotted lines*) and to the binding between σ^{32} and the core RNAP (*solid lines*). The plots are shown for the wild-type heat-shock response and for the mutant model with no sequestration loop. Sensitivity is computed as the derivative of the chaperone level to the corresponding parameters along the trajectories of the system. The plots are in log space. Heat-shock begins at time 0 minutes.

The origins of uncertainty and perturbations in gene regulatory networks are numerous, including the intricate stages of the transcription and translation procedure. Following engineering terminology, we defined robustness as the relative insensitivity of the chaperone output to these perturbations. A systematic sensitivity analysis is possible through the computation of the derivative of the time solution of the model equations with respect to the parameters of interest. We again devised a mutant where the sequestration loop is missing and perform sensitivity computations for this case and the wild-type case. These computations confirmed the increased robustness in the presence of the sequestration loop. Figure 5.4 plots (dashed-dotted lines) the sensitivity of the total *DnaK* level to the transcription rate in the full model along with a plot of the same sensitivity in the case where the sequestration loop was eliminated. In the case when the sequestration of σ^{32} by *DnaK* is absent, the sensitivity values clearly increase, supporting the assertion that the sequestration loop is efficiently increasing the overall system robustness. A similar conclusion could also be drawn from the plot (solid lines) of the sensitivity of *DnaK* to the binding constant between σ^{32} and RNAP. Again the plotted sensitivities are for the full model and for the model where the sequestration of σ^{32} by *DnaK* is disabled, respectively.

Feedback through Degradation: Stability, Dynamic Response, and Noise Rejection
To investigate the properties of the regulated dynamic degradation of σ^{32} by *FtsH*, my coworkers and I (El-Samad et al., 2005) devised a model where this degradation is accomplished constitutively, that is, independently of the σ^{32} regulon. We performed simulation of the two models. The results shown in figure 5.3 (dotted gray line) indicate the impact of the regulated degradation of σ^{32} in implementing a faster transient response to a heat disturbance. Although very revealing, this outcome does not come as a real surprise considering that the regulated degradation scheme adjusts the stability of σ^{32} based on cellular needs. In comparison, the stability remains constant in the case of constitutive degradation. One feature of primary importance in cellular processes is their ability to attenuate undesirable noise. Environmental (extrinsic) and biochemical (intrinsic) sources of noise induce fluctuations in the concentration of cellular molecular species. The magnitude and nature of the fluctuations is thought to be dependent on the structure of the molecular networks, the concentration of molecules that populate this structure, and the rates of the underlying biochemical reactions. At the same time, molecular networks are expected to function reliably and robustly in the presence of this noise. It has been suggested that this robust operation is in part the outcome of feedback regulatory loops (Thattai and van Oudenaarden, 2001). To verify this prediction in the context of the heat shock response, we investigated the noise rejection merits of the *FtsH* degradation loop by comparing the stochastic performance of the virtual "mutant" where the degradation of σ^{32} is constitutive to that of the wild type, where the stability of σ^{32} is dynamically regulated. We used the Gillespie stochastic simulation algorithm (Gillespie, 1976), which reproduces the exact stochastic behavior of a system, thus permitting the rejection/amplification properties of a system to be assessed by observing the excursions of a quantity around its ensemble average. Figure 5.5 clearly shows that stochastic fluctuations are much more pronounced in the constitutive degradation case. Probability density functions compiled for the two cases also show a noticeably wider distribution for the constitutive than for the regulated degradation case. Although it has been experimentally established that the degradation of σ^{32} by *FtsH* depends on *DnaK* (Tatsuta et al., 2000), the advantages of such a dependence have been unknown. We looked for a justification for this mechanism through the use of our model. We removed this dependence and allowed σ^{32} to be degraded directly by *FstH* (i.e., without mediation by *DnaK*). The response shown in figure 5.3 (black line) indicates that, although the dependence of this degradation on the chaperones does not seem to be essential for the qualitative behavior of the heat-shock response, in its absence, the HSR is delayed by almost 20 minutes. This result can be qualitatively explained as follows. As temperature increases and the chaperones are sequestered by unfolded proteins, the stability of σ^{32} is enhanced because the interaction of σ^{32} with *DnaK* is necessary for its degradation by *FtsH*. This stabilization of σ^{32}, in

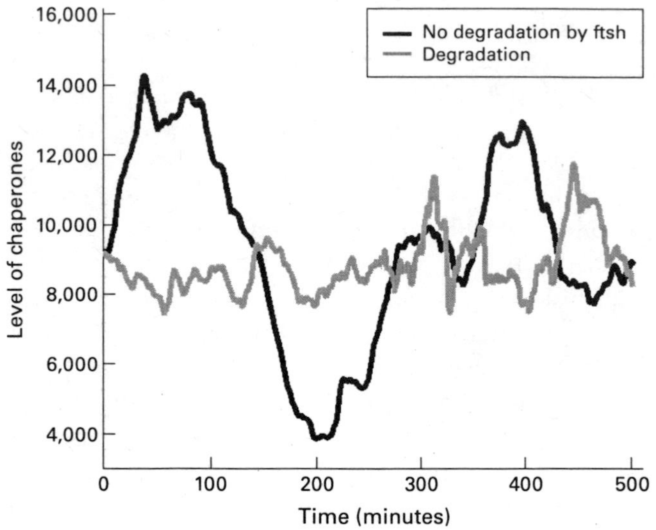

Figure 5.5
Stochastic simulation of the detailed heat-shock response model. The gray line corresponds to the wild-type heat-shock response; the black line corresponds to the mutant model, where σ^{32} is constitutively degraded.

addition to its increased translation, results in a sharp increase in its level. This in turn results in a fast response through the rapid induction of chaperone synthesis. Obviously, this enhanced stabilization effect is absent in the case of the direct degradation of σ^{32} by proteases.

5.3 Structural Organization of the Heat-Shock Response System: A Necessary Complexity

The previous sections have argued that the architecture of the heat-shock response exhibits a level of complexity necessary to satisfy various dynamical performance constraints; that it cannot be attributed solely to basic functionality demanded from an operational heat-shock response system. Indeed, a simple and operational HSR system would simply consist of a temperature sensor and transcriptional/translational apparatus that responds appropriately to temperature changes. This could be achieved in a hypothetical design where the melting of σ^{32} mRNA implements a temperature sensor, coupled with the synthesis of heat-shock proteins being dictated by the number of σ^{32} (figure 5.6a). Any number of σ^{32} and heat-shock proteins is achievable through the careful tuning of the synthesis rates of these proteins in an open-loop design. This, however, would result in a fragile and metabolically ineffi-

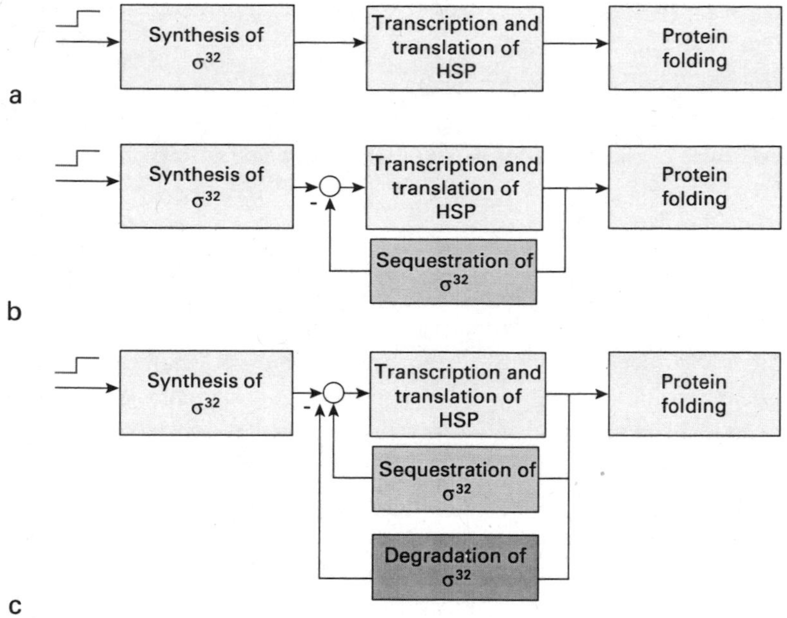

Figure 5.6
Hypothetical design models for the heat-shock response system. (a) Simple open-loop design. The feedforward element achieves temperature sensing. (b) Closed-loop design with feedforward and sequestration loop to regulate the activity of σ^{32}. (c) Closed-loop design with feedforward, sequestration, and degradation of σ^{32} loop, which corresponds to the wild-type heat-shock response system.

cient system that would quickly be at a disadvantage in the noisy environment of the cell. Adding the sequestration loop would help remedy these aspects (figure 5.6b), but would still lack sufficient responsiveness. Adding the degradation loop (figure 5.6c) finally generates the desired robustness and responsiveness. Considering the limited cellular energies and materials, these performance objectives often form a contradictory set of constraints and bring about various trade-offs. For example, a large turnover rate for σ^{32} yields necessarily a fast response but comes at the expense of fast production and degradation, which are mechanisms that require a continuous supply of cellular materials and energy. High feedback gains also contribute to improving the transient response while attenuating harmful cellular noise. Implementing these gains requires, among other things, increased binding specificities that result in highly complex or specialized proteins. This, again, is not without similarities to trade-offs considered in engineering systems where typical design requirements include the simultaneous minimization of the deviation from some desired operation and of the control effort needed to accomplish that behavior, obviously competing objectives. Drawing a parallel to the heat-shock response, this would correspond to the objective of limiting the deviation from an optimal number of unfolded proteins using a minimal number of chaperones and proteases. Adding transient performance and noise rejection considerations requires additional architectural complexity.

5.4 The Cytoplasmic Heat-Shock Response is One of Many Stress Responses in Bacteria

An array of coordinated stress response systems has evolved to tackle the multiple challenges a bacterium constantly faces. These stress response systems react to the damage induced by various environmental perturbations, and coordinate (cross-talk) with each other in order to mount an appropriate response that restores the necessary homeostasis in different cellular compartments. The heat-shock response, discussed above, uses the sigma factor σ^{32} to increase the expression of chaperones and proteases in response to misfolded proteins exclusively in the cytoplasm. In addition, the protein health of the periplasmic compartment of an *E. coli* cell is also monitored and regulated. Indeed, the *envelope stress response* (ESR) controlled by the sigma factor σ^E has recently moved to the spotlight as the counterpart of the σ^{32} system for alleviating the protein-folding problem in the periplasmic space and maintaining the integrity of the cell envelope (Grigorova et al., 2004). Reacting to heat-induced damage, however, is not the only type of stress that bacteria can tackle. Bacterial cells have, for example, the ability to trigger cold-shock responses. These responses govern the expression of RNA chaperones and ribosomal factors, ensuring accurate translation at low temperatures. Bacteria further sense, among other things, the nutritional content of their environment and trigger stringent responses to reduce the cellular protein synthesis capacity and control further global responses upon nutritional downshift.

Similar to cytoplasmic stress responses, all bacterial stress systems are implemented through intricate gene regulatory networks that endow them with the dynamical properties that are appropriate for their operation. Moreover, although these networks can make autonomous decisions, they also coordinate intimately with other stress networks to mount the appropriate integrated responses. For example, the periplasm and cytoplasm communicate their folding status through interdependence of their σ-factors, one aspect of which is the control of σ^{32} synthesis by σ^E. Thus, even though much has been learned in recent years about the mechanisms of action of single components in various stress responses, the main challenge for the future is to understand the interactions of these components and systems under different physiological conditions. It is becoming increasingly clear that the most promising vehicle for such an understanding is an approach based on the investigation of stress responses as integrated control systems. This approach will help delineate the dynamical properties supported by particular circuit topologies, feedback loops, and parameter regimes of individual systems and investigate their coupling to other systems. Ultimately, this integrated systems approach will be invaluable for the understanding of pathogenic bacteria and their persistence in the hostile environment of host cells. It also holds the promise of generating novel nontrivial predictions about parts of bacterial stress circuits that need to be targeted or rewired for therapeutic purposes.

6 Synthetic Biology: A Systems Engineering Perspective

Domitilla Del Vecchio and Eduardo D. Sontag

This chapter reviews some of the design challenges found in biomolecular systems from a systems engineering perspective, in particular, the problem of modularity. If components behave modularly, that is, if their behavior does not change upon interconnection, then one can predict the behavior of a circuit directly from the behavior of the composing units. In two instances of oscillating synthetic biomolecular systems, we demonstrate that, because of loading effects called "retroactivity" at interconnections, modularity does not necessarily hold in biomolecular systems. We propose a framework for quantifying retroactivity at interconnections between transcriptional circuits and present a mechanism, inspired by the design of electronic noninverting amplifiers, to counteract retroactivity.[1]

6.1 Background

Although biologists have long employed phenomenological and qualitative models to help discover the components of living systems and to describe their behaviors, the analysis of the dynamical properties of complex biomolecular reaction networks requires a more quantitative and systems-level approach. Thus, in recent years, the field of systems biology has emerged, whose focus is the quantitative analysis of cell behavior, with the goal of explicating the basic dynamic processes, feedback control loops, and signal-processing mechanisms underlying life. Complementary to systems biology is the new engineering discipline of *synthetic biology*, whose goal is to extend or modify the behavior of organisms, and to induce them to perform new tasks (Andrianantoandro et al., 2006; Endy, 2005). Through the de novo construction of simple elements and circuits, the field aims to foster an engineering framework for obtaining new cell behaviors in a predictable and reliable fashion. The ultimate goal is to develop synthetic biomolecular circuitry for a wide variety of applications from targeted drug delivery to the construction of biomolecular computers. In the process, synthetic biology helps us improve our quantitative and qualitative understanding of biological systems through designing and constructing instances of these systems in accordance with hypothesized models. Discrepancies between expected behavior and observed behavior highlight research issues that need more study, gaps in our knowledge, or inaccurate assumptions in models. One of the fundamental building blocks employed in synthetic biology is the process of transcriptional regulation. A transcriptional network is composed of a number of genes that express proteins that then act as transcription factors for other genes. The rate at which a gene is transcribed is controlled by the *promoter*, a regulatory region of DNA that lies upstream of the coding region of the gene. RNA polymerase binds a defined site (a specific DNA sequence) on the promoter. The quality of this site specifies the transcription rate of the gene (the DNA sequence at the site determines its chemical affinity for RNA polymerase). Although RNA polymerase acts on all of the genes, each transcription factor modulates only a particular set of target genes. Transcription factors affect the transcription rate by binding specific sites on the promoter region of the regulated genes. When bound, they change the probability per unit time that RNA polymerase binds the promoter region. Transcription factors thus affect the rate at which RNA polymerase initiates transcription. A transcription factor can act as a *repressor* when it prevents RNA polymerase from binding to the promoter site. A transcription factor acts as an *activator* if it facilitates the binding of RNA polymerase to the promoter. Such interactions can be generally represented as nodes connected by directed edges. Synthetic biomolecular circuits are typically fabricated in *Escherichia coli*, by cutting and pasting together coding regions and promotors (natural and engineered) according to designed structures. Because the expression of a gene is under the control of the upstream promoter region, this technique allows the production of any desired circuit of activation and repression interactions among genes. Early examples of such circuits include an activator-repressor system that can display toggle switch or clock behavior (Atkinson et al., 2003), a loop oscillator called the "repressilator" obtained by connecting three inverters in a ring topology (Elowitz and Leibler, 2000), a toggle switch obtained by connecting two inverters in a ring fashion (Gardner et al., 2000), and an autorepressed circuit (Becskei and Serrano, 2000; figure 6.1).

Several scientific and technological developments over the past four decades have set the stage for the design and fabrication of early synthetic biomolecular circuits. An early milestone in the history of synthetic biology was the discovery in 1961 of mathematical logic in gene regulation (Jacob and Monod, 1961). Only a few years later, special enzymes that can cut double-stranded DNA at specific recognition sites, known as restriction sites, were discovered (Arber and Linn, 1969). These enzymes, called restriction enzymes, were a major enabler of recombinant DNA technology. One of the most celebrated products of such technology is the large-scale production of insulin by *E. coli* bacteria, which serve as cellular factories (Villa-Komaroff et al.,

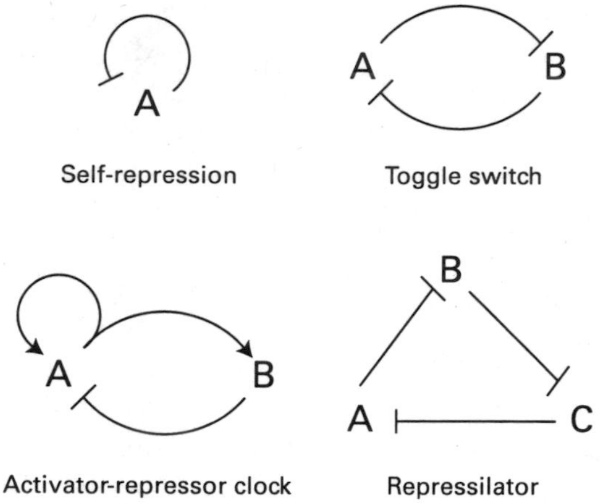

Figure 6.1
Early transcriptional circuits that have been fabricated in the bacterium *E. coli*: the self-repression circuit (Becskei and Serrano, 2000), the toggle switch (Gardner et al., 2000), the activator-repressor clock (Atkinson et al., 2003), and the repressilator (Elowitz and Leibler, 2000). Each node represents a gene and each arrow from node Z to node X indicates that the transcription factor encoded in z, denoted Z, regulates gene x. If z represses the expression of x, the interaction is represented by Z ⊣ X. If z activates the expression of x, the interaction is represented by Z → X (Alon, 2007)

1978). The development of recombinant DNA technology along with the demonstration in 1970 that genes can be artificially synthesized, provided the ability to cut and paste natural or synthetic promoters and genes in almost any fashion on plasmids of compatible size through the cloning process (Alberts et al., 1989). The polymerase chain reaction (PCR), devised in the 1980s, allows the exponential amplification of small amounts of DNA into amounts large enough to be used for transfection and transformation in living cells (Alberts et al., 1989). Today, commercial synthesis of DNA sequences and genes has become cheaper and faster, often costing less than $1 per base pair (Baker et al., 2006). Fluorescent proteins, such as GFP and its genetic variations, allow the in vivo measurement of the amount of protein produced by any target gene, providing a readout of gene circuit behavior. Circuit design also allows for external inputs in the form of inducers, used to probe the system. Inducers act by disabling repressor proteins, thus modulating the levels of transcription.

One of the current directions of the field is to create circuitry with more complex functionalities by assembling simpler circuits, such as those in figure 6.1. This tendency reflects the history of electronics after the bipolar junction transistor (BJT) was invented in 1947. In particular, a major breakthrough occurred in 1964 with the invention of the first operational amplifier (OPAMP), which led the way to standardized modular and integrated circuit design. By comparison, synthetic biology may be moving toward a similar development of modular and integrated circuit design. This is witnessed by several recent efforts toward formally characterizing between-module interconnection mechanisms, loading (or *impedance*) effects, and OPAMP-like devices to counteract loading problems (Del Vecchio et al., 2008b; Hartwell et al., 1999; Rubertis and Davies, 2003; Saez-Rodriguez et al., 2004, 2005; Sauro and Ingalls, 2007; Sauro and Kholodenko, 2004).

Section 6.2 describes the fundamental modeling assumption made for circuit analysis and design: *modularity*, which guarantees that building blocks maintain their behavior unchanged after interconnection. This property is fundamental for predicting the behavior of a complex system by the behavior of the composing units. Section 6.3 shows how the two synthetic oscillators of figure 6.1 can be designed assuming modular composition of their building blocks. Section 6.4 shows that modularity does not necessarily hold in transcriptional circuitry in the same manner as it occurs in many other engineering systems (Willems, 1999). Here we introduce the concept of *retroactivity* to characterize any change in the dynamics of a building block due to interconnection. We describe a procedure for quantifying retroactivity and thus for designing an interconnection so as to have low retroactivity, when possible. In section 6.5, we propose the concept of an insulation device as a system that enforces modularity by working as a buffer between a component that sends a signal and one that receives the signal. Large amplification and feedback gains are the key mechanisms for the design of insulation devices in many other engineering systems. We show that simple cycles, involving phosphorylation and dephosphorylation, which are ubiquitous in natural signal transduction systems, enjoy intrinsic insulation properties, and thus have the potential to serve as synthetic biomolecular insulation devices.

6.2 The Modularity Assumption

Each node y of a transcriptional network is usually modeled as an input-output module taking as input the concentrations of transcription factors that regulate gene y and giving as output the concentration of protein expressed by gene y, denoted Y. The transcription factors regulating y appear as inputs of the transcriptional module through their association/dissociation with the promoter site of gene y. We will denote by X the protein, by X the average protein concentration, and by x the gene expressing protein X. The internal dynamics of the transcriptional module are determined by the processes of transcription and translation, which are much slower than the dynamics of transcription factor binding (Alon, 2007). The binding of transcription factors to the promoter site reaches equilibrium in seconds, whereas transcription and translation of the target gene take minutes to hours. This time-scale separation, a key feature of transcriptional circuits, leads to the following central modeling simplification.

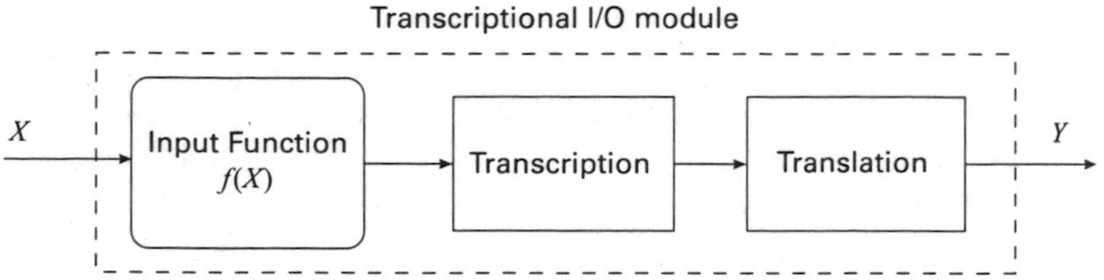

Figure 6.2
Transcriptional module modeled as an input-output system with input function given by the transcription regulation function $f(X)$ and with internal dynamics established by the transcription and translation processes.

According to the *modularity assumption*, the dynamics of transcription factor/DNA binding are considered at equilibrium, and each transcription factor concentration enters the input-output transcriptional module through a *static* input function that drives the transcription and translation dynamics (figure 6.2).

In the simplest case of one input acting as repressor or activator, the transcription regulation functions $f(X)$ take the Hill function form. When the transcriptional component takes several transcription factors as inputs, more complicated forms can be constructed from first principles (Alon, 2007).

Consider a transcriptional module with input function $f(X_1, \ldots, X_n)$. The internal dynamics of the transcriptional module usually model mRNA and protein dynamics through the processes of transcription and translation. Protein production is balanced both by decay, through *degradation*, which occurs when the protein is destroyed by specialized proteins in the cell that, for example, recognize a specific part of the protein and destroy it, and by *dilution*, which is caused by the reduction in concentration of the protein due to the increase of cell volume during growth. In a similar way, mRNA production is also balanced by degradation and dilution. Thus the dynamics of a transcriptional module are often well captured by the following ordinary differential equations:

$$\frac{\mathrm{d}r_Y(t)}{\mathrm{d}t} = f(X_1(t), \ldots, X_n(t)) - \alpha_1 r_Y(t), \tag{6.1a}$$

$$\frac{\mathrm{d}Y(t)}{\mathrm{d}t} = \gamma r_Y(t) - \alpha_2 Y(t), \tag{6.1b}$$

in which r_Y denotes the concentration of mRNA translated from gene Y, the constants α_1 and α_2 indicate the mRNA and protein decay rates, and γ is a constant that establishes the rate at which the mRNA is translated.

To engineer a system with prescribed behavior, one must be able to change the physical features so as to change the values of the parameters of the model. This is often possible. For example, the binding affinity of a transcription factor to its site on the promoter can be affected by single or multiple base pair substitutions. The protein decay rate (constant α_2 in equation [6.1b]) can be increased by adding degradation tags at the end of the gene expressing protein Y. Tags are genetic additions to the end of a sequence that modify expressed proteins in different ways such as marking the protein for faster degradation. Combinatorial promoters, which can accept multiple input transcription factors to implement regulation functions that take multiple inputs, can be realized by combining the operator sites of several simple promoters (Cox et al., 2007).

6.3 Design of Genetic Circuits under the Modularity Assumption

Based on the modeling assumptions outlined in the previous section, a number of synthetic genetic circuits have been designed and fabricated by composing transcriptional modules through input-output connection (figure 6.1). With such a procedure, one seeks to predict the behavior of a circuit from that of its composing units, once these have been well characterized in isolation. This approach is standard also in the design and fabrication of electronic circuitry.

6.3.1 The Repressilator

Elowitz and Leibler (2000) constructed the first operational oscillatory genetic circuit, which consists of three repressors arranged in ring fashion, and called it the "repressilator" (figure 6.1). The repressilator exhibits sinusoidal, limit cycle oscillations in periods of hours, slower than the period of the *E. coli* cell-division cycle. The state of the oscillator is transmitted between generations from mother to daughter cells. In the repressilator, the protein lifetimes are shortened to approximately two minutes (close to mRNA lifetimes). A dynamical model of the repressilator can be obtained by composing three transcriptional repression modules in a loop fashion:

$$\frac{\mathrm{d}r_A(t)}{\mathrm{d}t} = -\delta r_A(t) + f_1(C(t)), \qquad \frac{\mathrm{d}A(t)}{\mathrm{d}t} = r_A(t) - \delta A(t), \tag{6.2a}$$

$$\frac{\mathrm{d}r_B(t)}{\mathrm{d}t} = -\delta r_B(t) + f_2(A(t)), \qquad \frac{\mathrm{d}B(t)}{\mathrm{d}t} = r_B(t) - \delta B(t), \tag{6.2b}$$

$$\frac{\mathrm{d}r_C(t)}{\mathrm{d}t} = -\delta r_C(t) + f_3(B(t)), \qquad \frac{\mathrm{d}C(t)}{\mathrm{d}t} = r_C(t) - \delta C(t). \tag{6.2c}$$

We will consider two different cases for the input functions f_i: the symmetric case, with three identical repressions, and the nonsymmetric case, with two identical activations and one repression. For the symmetric case, we assume that

$$f_1(p) = f_2(p) = f_3(p) = \frac{\alpha^2}{1 + p^n}.$$

As the regulatory functions all have negative slope, and there is an odd number of them in the loop, there is only one equilibrium. One can then invoke well-known theorems of Mallet-Paret and Smith (1990) or of Hastings et al. (1977) to conclude that, if the equilibrium point is unstable, the system admits a nonconstant periodic orbit. Thus, to obtain periodic behavior, one can search for parameter values to guarantee the instability of the equilibrium point, a procedure followed in the design of the repressilator (Elowitz and Leibler, 2000). In particular, one can show that the symmetric repressilator in equation (6.2) has a periodic solution if the parameters α, δ, and n satisfy

$$\alpha^2/\delta^2 > \sqrt[n]{\frac{4/3}{n - 4/3}}\left(1 + \frac{4/3}{n - 4/3}\right).$$

This relationship is plotted in figure 6.3a. When n increases, the existence of an unstable equilibrium point is guaranteed for larger ranges of the other parameter values. Equivalently, for fixed values of α and δ, as n increases, the robustness of the circuit oscillatory behavior to parametric variations in the values of α and δ also increases. Of course, this "behavioral" robustness does not guarantee that other

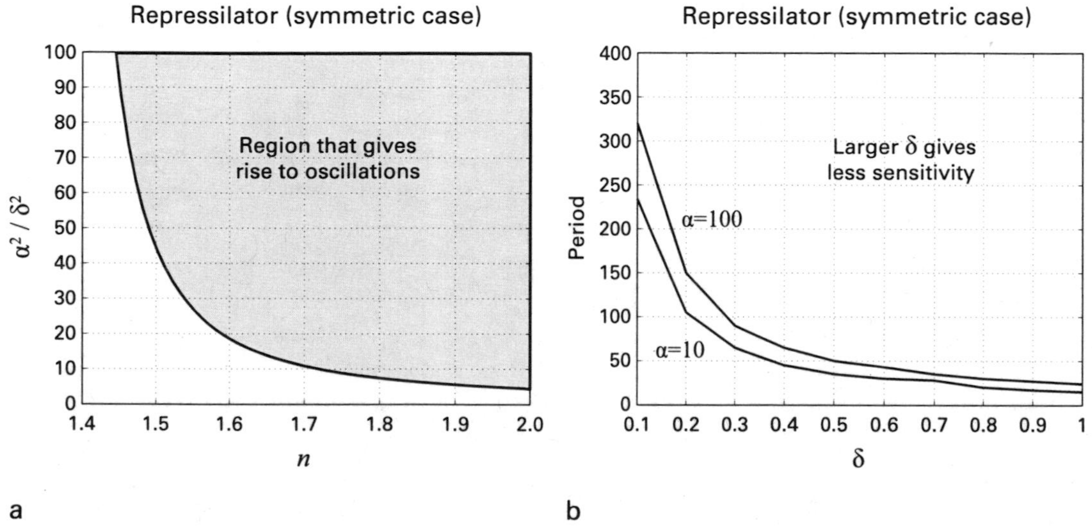

Figure 6.3
Repressilator (symmetric case). (a) Space of parameters that give rise to oscillations for the repressilator in equation (6.2). (b) Period as a function of δ and α.

important features of the oscillator, such as the period value, are insensitive to parameter variation. Numerical studies indicate that the period T approximately follows $T \propto 1/\delta$, and varies only slightly with α (figure 6.3b). From the figure, we can note that, as the value of δ increases, the sensitivity of the period to the variation of δ itself decreases. However, increasing δ would necessitate an increase of the cooperativity n. This analysis indicates a potential trade-off that should be taken into account in the design process in order to balance system complexity and the robustness of the oscillations.

A similar result for the existence of a periodic solution can be obtained for the nonsymmetric case, in which the input functions of the three transcriptional modules are modified to

$$f_1(p) = \frac{\alpha_3^2}{1+p^n}, \quad f_2(p) = \frac{\alpha^2 p^n}{1+p^n} \quad \text{and} \quad f_3(p) = \frac{\alpha^2 p^n}{1+p^n};$$

that is, two interactions are activations and one is a repression. One can verify that there is only one equilibrium point and again invoke the theorems of Mallet-Paret and Smith (1990) or Hastings et al. (1977) to conclude that if the equilibrium point is unstable, the system admits a nonconstant periodic solution. We can thus obtain the condition for oscillations again by establishing conditions on the parameters that guarantee an unstable equilibrium. These conditions are reported in figure 6.4. One can conclude that it is possible to design the circuit to be in the region of parameter space that gives rise to oscillations. It is also possible to show that, as the number of elements in the oscillatory loop increases, the value of n sufficient for oscillatory behavior decreases. The design criteria for obtaining oscillatory behavior are summarized in figures 6.3 and 6.4.

6.3.2 The Activator-Repressor Clock

Consider the activator-repressor clock diagram shown in figure 6.1, which is an example of a *relaxation oscillator*. The transcriptional module for A has an input function that takes two arguments: the activator concentration A and the repressor concentration B. The transcriptional module B has an input function that takes only the activator concentration A as its input. Let r_A and r_B represent the concentration of mRNA of activator and repressor, respectively. We consider the following four-dimensional model describing the rate of change of the species concentrations:

$$\frac{dr_A(t)}{dt} = -\delta_1/\epsilon r_A(t) + F_1(A(t), B(t)), \qquad \frac{dA(t)}{dt} = \nu(-\delta_A A(t) + k_1/\epsilon r_A(t)),$$

$$\frac{dr_B(t)}{dt} = -\delta_2/\epsilon r_B(t) + F_2(A(t)), \qquad \frac{dB(t)}{dt} = -\delta_B B(t) + k_2/\epsilon r_B(t),$$

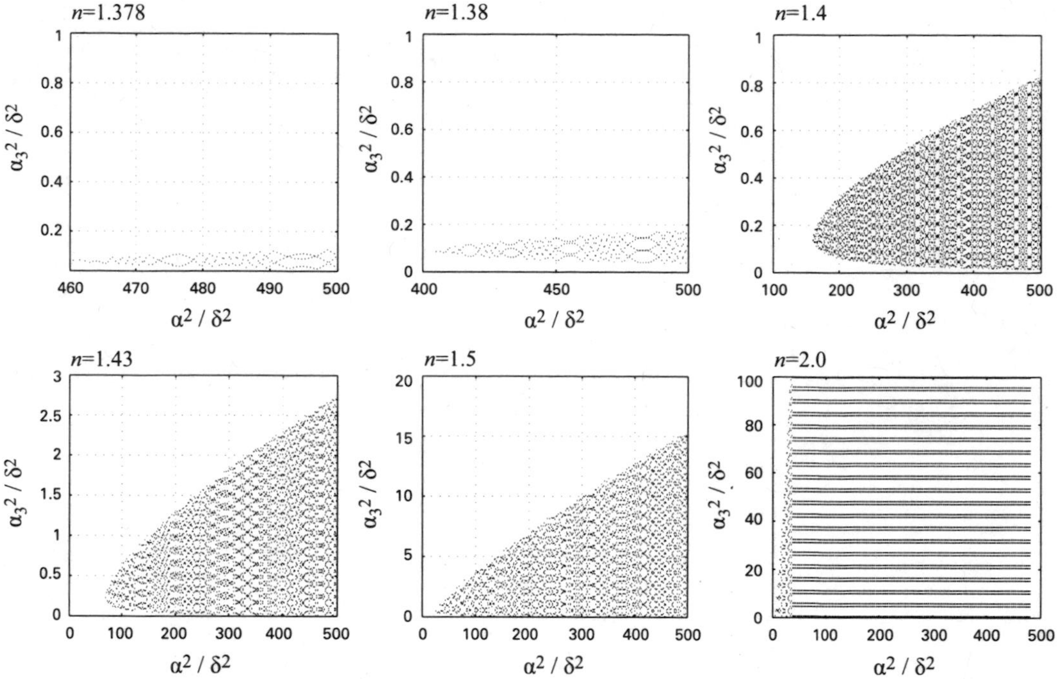

Figure 6.4
Space of parameters that give rise to oscillations for the repressilator (nonsymmetric case; El Samad et al., 2005).

in which the parameter ν regulates the difference of time scales between the repressor and the activator dynamics, and ϵ is a parameter that regulates the difference of time scales between the mRNA and the protein dynamics. The input functions F_1 and F_2 are given by

$$F_1(A,B) = \frac{K_1 A^n + K_{A0}}{1 + \gamma_1 A^n + \gamma_2 B^n} \quad \text{and} \quad F_2(A) = \frac{K_2 A^n + K_{B0}}{1 + \gamma_3 A^n},$$

in which K_1 and K_2 are the maximal activated transcription rates, while K_{A0} and K_{B0} are the basal transcription rates when no activator is present. The parameters $1/\gamma_i$ are the activation coefficients and are related to the affinity of the protein to the promoter site. The Hill coefficient is chosen to be $n = 2$.

The number of equilibria of the system is not influenced by the values of ϵ and of ν but is dependent on the other model parameters. The set of values of K_i, k_i, δ_i, γ_i, δ_A, and δ_B that result in the existence of a unique equilibrium can be determined by employing graphical techniques. In particular, one can plot the curves corresponding to the sets of A, B values for which $dr_B/dt = 0$ and $dB/dt = 0$ and the set of A, B

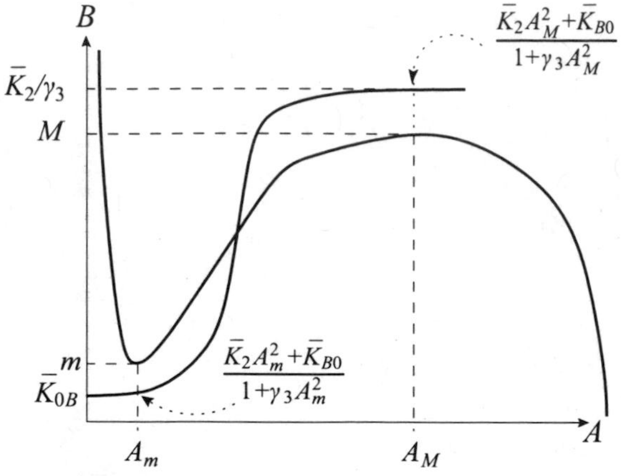

Figure 6.5
Shape of the curves in the A, B plane corresponding to $\mathrm{d}r_B/\mathrm{d}t = 0$, $\mathrm{d}B/\mathrm{d}t = 0$, and to $\mathrm{d}r_A/\mathrm{d}t = 0$, $\mathrm{d}A/\mathrm{d}t = 0$ as function of the parameters.

values for which $\mathrm{d}r_A/\mathrm{d}t = 0$ and $\mathrm{d}A/\mathrm{d}t = 0$ as in figure 6.5. The intersection of these two curves provides the equilibria of the system and conditions on the parameters can be determined that guarantee the existence of one equilibrium only.

We introduce scaled parameters:

$$\bar{K}_1 = \frac{K_1}{\delta_1/\epsilon}, \quad \bar{K}_{A0} = \frac{K_{A0}}{\delta_1/\epsilon}, \quad \bar{K}_2 = \frac{K_2}{\delta_2/\epsilon}, \quad \text{and} \quad \bar{K}_{B0} = \frac{K_{B0}}{\delta_2/\epsilon}.$$

In particular, we require that the basal activator transcription rate when B is not present, which is proportional to $\bar{K}_{A0}$, is sufficiently smaller than the maximal transcription rate of the activator, which is proportional to $\bar{K}_1$. Also, $\bar{K}_{A0}$ must be nonzero. In the case that $\bar{K}_1 \gg \bar{K}_{A0}$, one can verify that $A_M \approx \bar{K}_1/2\gamma_1$ and thus $M \approx \bar{K}_1/2\sqrt{\gamma_1\gamma_2}$. As a consequence, if $\bar{K}_1/\gamma_1$ increases, then so must $\bar{K}_2/\gamma_3$. This implies that the maximal transcription rate of the repressor divided by its protein and mRNA decay rates must be larger than the maximal transcription rate of the activator divided by its protein and mRNA decay rates. Finally, $A_m \approx 0$, and $m \approx \sqrt{\bar{K}_{A0}/\gamma_2 A_m}$. As a consequence, the smaller $\bar{K}_{A0}$ becomes, the smaller $\bar{K}_{B0}$ must be (see Del Vecchio, 2007, for details).

Given that the values of K_i, k_i, δ_i, γ_i, δ_A, and δ_B have been chosen so that there is a unique equilibrium, we numerically study the occurrence of periodic solutions as the difference in time scales between protein and mRNA, ϵ, and the difference in time scales between activator and repressor, ν, are changed. In particular, we perform bifurcation analysis with ϵ and ν, the two bifurcation parameters. These bifurcation results are summarized by figure 6.6 (see Del Vecchio, 2007, for the details of the numerical analysis).

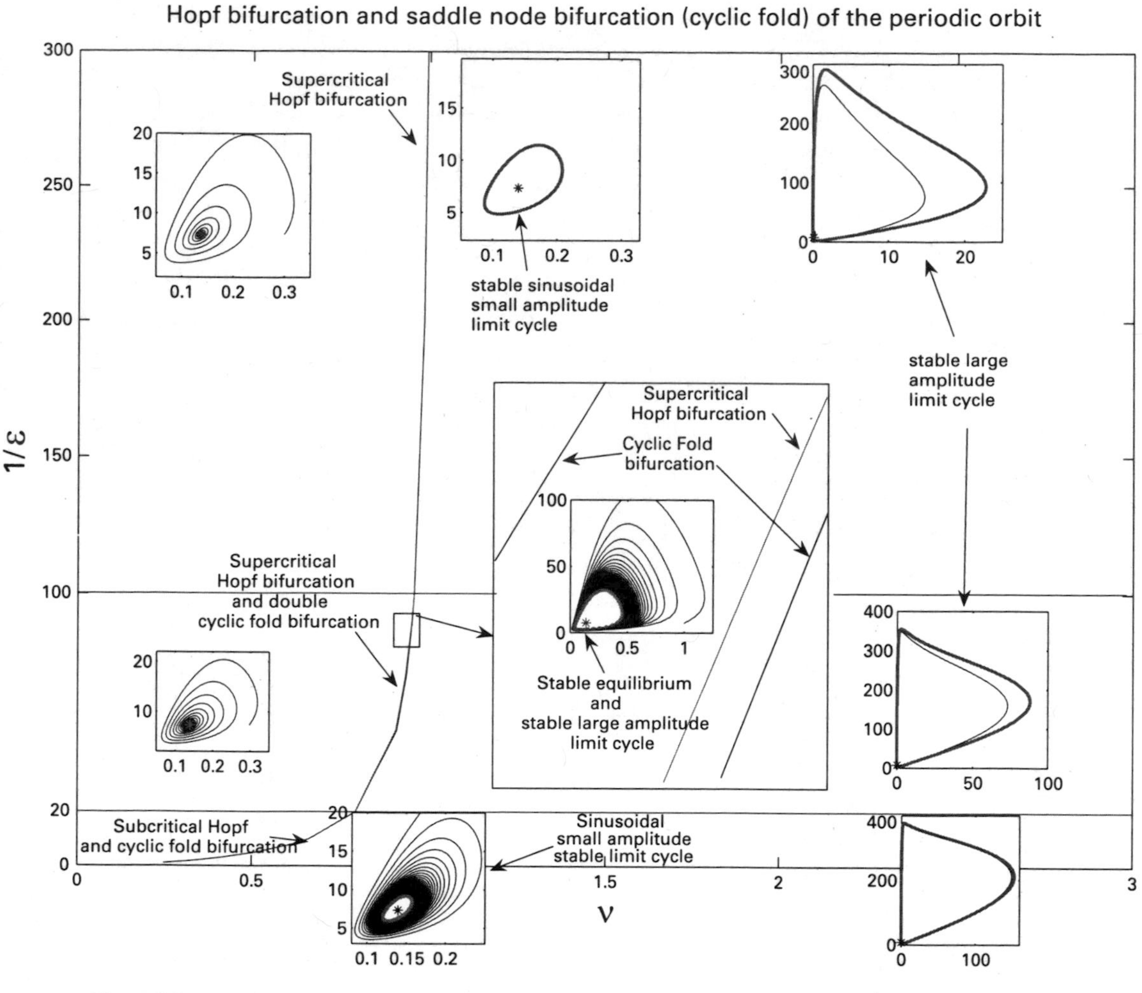

Figure 6.6
Design chart for the relaxation oscillator: For values of ν sufficiently large, one obtains sustained oscillations beyond the Hopf bifurcation, independently of the difference of time scales between the protein and the mRNA dynamics. Note also that there are values of ν for which a stable point and a stable orbit coexist and values of ν for which two stable orbits coexist. The interval of ν values for which two stable orbits coexist is too small to be able to set ν numerically in such an interval. Thus this interval is not practically relevant. The values of ν for which a stable equilibrium and a stable periodic orbit coexist, known as hard excitation, is instead relevant.

The situation described in figure 6.6 corresponds to the *hard excitation* condition (Leloup and Goldbeter, 2001) and occurs for realistic values of the separation of time scales between protein and mRNA dynamics. This simple oscillator motif described by a four-dimensional model can thus capture the features that lead to the long-term suppression of the rhythm by external inputs. *Birhythmicity* (Goldbeter, 1996) is also possible even if practically not relevant due to the numerical difficulty of moving the system to one of the two periodic orbits (see Del Vecchio, 2007; Conrad et al., 2008, for details). In terms of the ϵ and ν parameters, it is thus possible to design the system to achive oscillatory behavior: as the time scale of the activator dynamics increases with respect to the repressor dynamics, the system parameters move across a Hopf bifurcation and stable oscillations will arise. From a fabrication point of view, this can be achieved by adding suitable degradation tags to the activator protein. The region of the parameter space in which the system exhibits almost sinusoidal damped oscillations is on the left-hand side of the curve corresponding to the Hopf bifurcation. As the data of Atkinson et al. (2003) exhibit almost sinusoidal damped oscillations, it is possible that the clock is operating in a region of parameter space on the "left" of the curve corresponding to the Hopf bifurcation. If this were the case, increasing the separation of time scales between the activator and the repressor, ν, may lead to a stable limit cycle.

6.4 Beyond the Modularity Assumption: Retroactivity

The circuit design process outlined thus far relies deeply on the modularity assumption, by virtue of which the behavior of the circuit topology can be directly predicted by the properties of the composing units. For example, the monotonicity of the input functions of the transcriptional modules composing the repressilator was a key to formally showing the existence of periodic solutions. The form of the input functions in the activator-repressor clock design enabled easy predictions of the location and number of equilibria as the parameters were changed. The modularity assumption implies that, when two modules are connected to one another, their behavior does not change upon interconnection. However, a fundamental systems-engineering issue that arises when interconnecting subsystems is how the process of transmitting a signal to a downstream component affects the dynamic state of the sending component. Indeed, after designing, testing, and characterizing the input-output behavior of an individual component in isolation, it is certainly desirable if its characteristics do not change when another component is connected to its output channel. This problem, the effect of loads on the output of a system, is well understood in many fields of engineering, for example, in electrical circuit design. It has often been pointed out that similar issues arise for biological systems. Modules should have special features

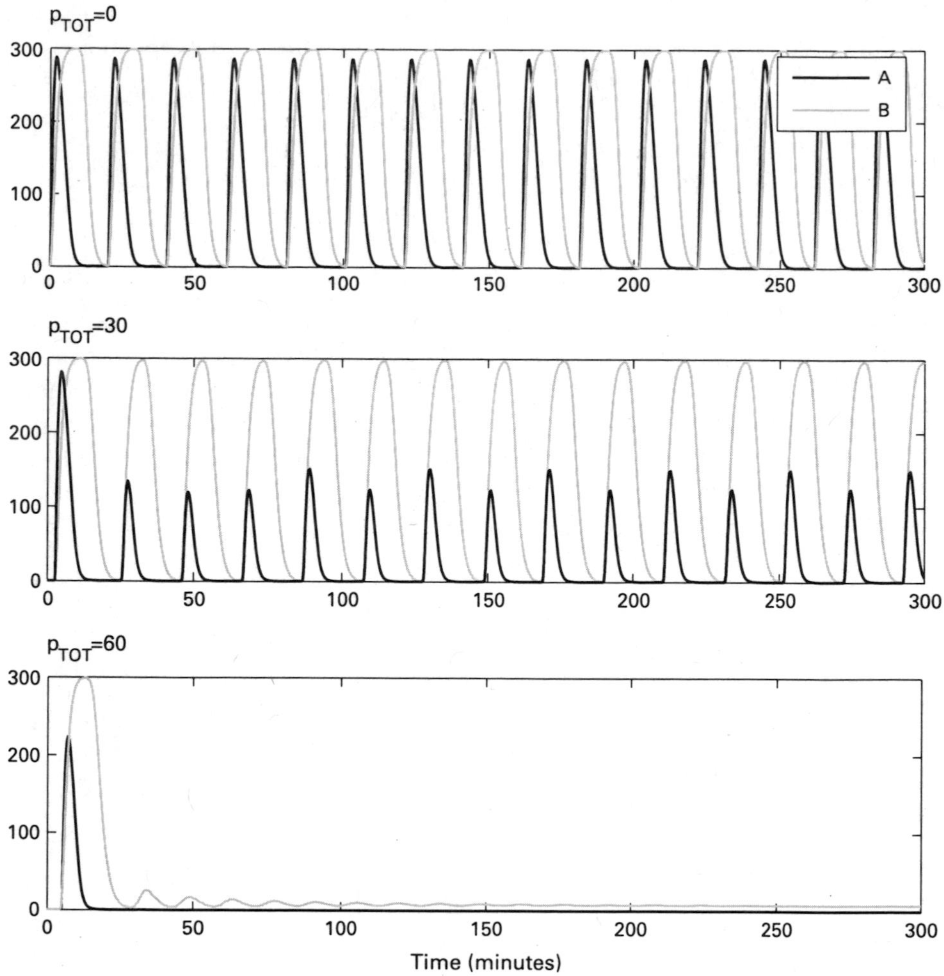

Figure 6.7
Activator-repressor clock behavior can be disrupted by a load on the activator A. As the number of downstream binding sites for A, p_{TOT}, is increased in the load, the activator and repressor dynamics lose their synchronization, and ultimately the oscillations disappear.

that allow them to be easily embedded in any system, such as zero output impedance and infinite input impedance. An extensive review on problems of loads and modularity in signaling networks can be found in Sauro (2004); Sauro and Ingalls (2007); Sauro and Kholodenko (2004), where the authors propose concrete analogies with similar problems arising in electrical circuits.

These questions are especially delicate in *synthetic* biology. For example, consider the activator-repressor clock of figure 6.1. Assume we want to employ this clock (upstream system) to drive one or more components (downstream systems), by using as its *output* signal the oscillating concentration $A(t)$ of the activator. From a systems/signals point of view, $A(t)$ becomes an *input* to the second system. The terms *upstream* and *downstream* reflect the direction in which we think of signals as traveling, *from* the clock *to* the systems being synchronized. This is only an idealization, however, because the binding and unbinding of A to promoter sites in a downstream system competes with the biochemical interactions that constitute the upstream block (retroactivity) and may therefore disrupt the operation of the clock itself (figure 6.7). One possible approach to avoid disrupting the behavior of the clock, motivated by the approach used with reporters such as GFP, is to introduce a gene coding for a new protein X, placed under the control of the same promoter as the gene for A, and to use the concentration of X, which presumably mirrors that of A, to drive the downstream system. This approach, however, has still the problem that the behavior of the X concentration in time may be altered and even disrupted by the addition of downstream systems that drain X. The net result is still that the downstream systems are not properly timed.

6.4.1 Modeling Retroactivity

We broadly call *retroactivity* the phenomenon by which the behavior of an upstream system is changed upon interconnection to a downstream system. As a simple example, consider a transcriptional component whose output is connected to downstream processes, which can be, for example, other transcriptional components (figure 6.8).

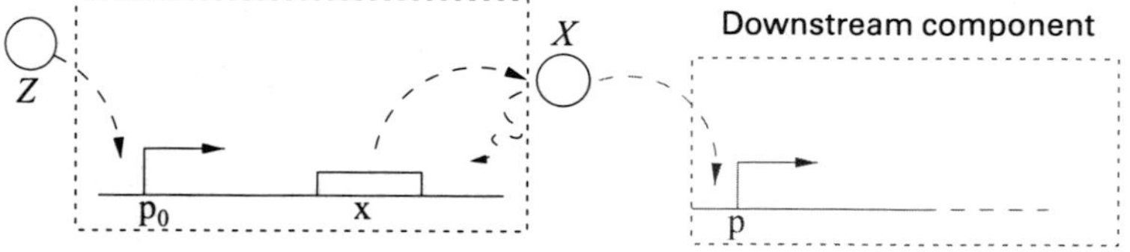

Figure 6.8
The transcriptional component takes as input u protein concentration Z and gives as output y protein concentration X.

The activity of the promoter controlling gene x depends on the amount of Z bound to the promoter. If $Z = Z(t)$, such an activity changes with time. We denote it by $k(t)$. By neglecting the mRNA dynamics, which are not relevant for the current discussion, we can write the dynamics of X as

$$\frac{dX(t)}{dt} = k(t) - \delta X(t), \tag{6.3}$$

in which δ is the decay rate of the protein. We refer to equation (6.3) as the isolated system dynamics. Now assume that X drives a downstream transcriptional module by binding to a promoter p with concentration p (figure 6.8). The reversible binding of X with p is then described by

$$\mathrm{X} + \mathrm{p} \underset{k_{\mathrm{off}}}{\overset{k_{\mathrm{on}}}{\rightleftharpoons}} \mathrm{C},$$

in which C is the protein-promoter complex and k_{on} and k_{off} are the association and dissociation rates. Because the promoter is not subject to decay, its total concentration p_{TOT} is conserved so that we can write $p + C = p_{\mathrm{TOT}}$. The new dynamics of X are therefore governed by the equations

$$\frac{dX(t)}{dt} = k(t) - \delta X(t) + \boxed{k_{\mathrm{off}}C(t) - k_{\mathrm{on}}(p_{\mathrm{TOT}} - C(t))X(t)}, \tag{6.4a}$$

$$\frac{dC(t)}{dt} = -k_{\mathrm{off}}C(t) + k_{\mathrm{on}}(p_{\mathrm{TOT}} - C(t))X(t), \tag{6.4b}$$

in which the terms in the box represent the signal

$$s(t) = k_{\mathrm{off}}C(t) - k_{\mathrm{on}}(p_{\mathrm{TOT}} - C(t))X(t), \tag{6.4c}$$

that is, the retroactivity to the output. We can interpret s as a mass flow between the upstream and the downstream system. When $s = 0$, equation (6.4a) reduces to the dynamics of the isolated system given in equation (6.3).

The effect of the retroactivity s on the behavior of X can be very large (figure 6.9). This is undesirable in a number of situations where we would like an upstream system to "drive" a downstream one, as is the case, for example, when a biological oscillator has to time a number of downstream processes. If, due to the retroactivity, the output signal of the upstream process becomes too low or out of phase with the output signal of the isolated system (as in figure 6.9), the coordination between the oscillator and the downstream processes will be lost. Here we focus on the retroactivity to the output s.

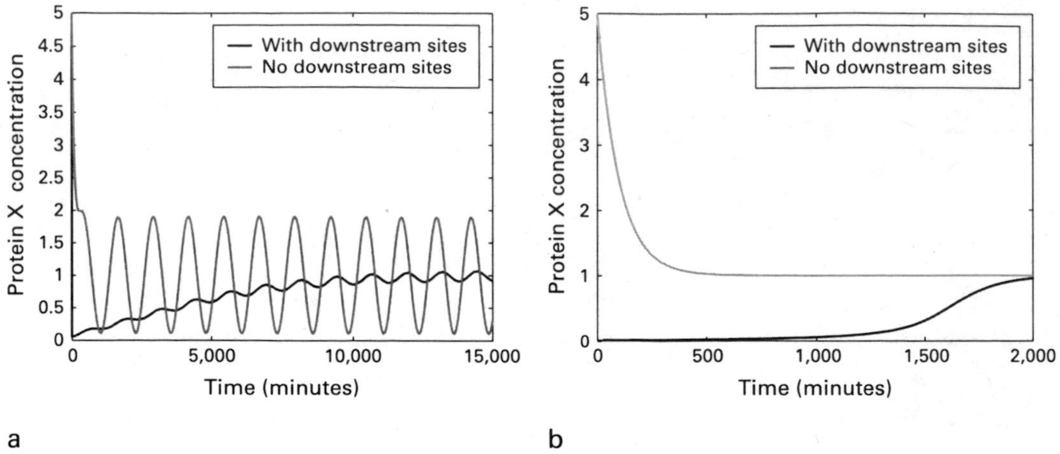

Figure 6.9
Dramatic effect of interconnection. Simulation results for the system in equation (6.4). The lighter line represents $X(t)$ as described by equations (6.3) (no retroactivity), while the darker line represents $X(t)$ obtained by equation (6.4) (with retroactivity). Both transient and longtime behaviors are different. Here $k(t) = 0.01(1 + \sin(\omega t))$ with $\omega = 0.005$ in panel a, and $\omega = 0$ in panel b, $k_{\mathrm{on}} = 10$, $k_{\mathrm{off}} = 10$, $\delta = 0.01$, $p_{\mathrm{TOT}} = 100$ and $X(0) = 5$. The choice of protein decay rate (in min^{-1}) corresponds to a half-life of about one hour. The oscillations are chosen to have a period of about 12 times the protein half-life in accordance to what is experimentally observed in the synthetic clock of Atkinson et al. (2003).

In general, we will model retroactivity by a signal that travels from downstream to upstream. We thus model a system by adding an additional input, s, to model any change in the system dynamics that may occur upon interconnection with a downstream system. Similarly, we add to a system a signal r as another output to model the fact that, when such a system is connected downstream of another system, it will send upstream a signal that will alter the dynamics of the upstream system. More generally, we define a system S to have internal state x, two types of inputs (I), and two types of outputs (O): an input u (I), an output y (O), a *retroactivity to the input* r (O), and a *retroactivity to the output* s (I) (figure 6.10). We will thus represent a system S by the equations

$$\frac{\mathrm{d}x(t)}{\mathrm{d}t} = f(x(t), u(t), s(t)), \tag{6.5a}$$

$$y(t) = Y(x(t), u(t), s(t)), \tag{6.5b}$$

$$r(t) = R(x(t), u(t), s(t)), \tag{6.5c}$$

in which the functions f, Y, and R can take any form and the signals x, u, s, r, and y may be scalars or vectors. In such a formalism, the input-output model of the isolated system is recovered from equations (6.5) by eliminating r and setting $s = 0$.

Let system S_i have inputs u_i and s_i and outputs y_i and r_i. Let S_1 and S_2 be two systems with disjoint sets of internal states. We define the interconnection of an upstream system S_1 with a downstream system S_2 by setting $y_1 = u_2$ and $s_1 = r_2$. We will only consider the interconnections of systems that do not have internal states in common.

6.4.2 Quantification of the Retroactivity to the Output

An operative quantification of the retroactivity to the output can be obtained by exploiting the difference of time scales between the dynamics of the output of the upstream module and the dynamics of the input stage of the downstream module. This separation of time scales is always encountered in transcriptional circuits as discussed in section 6.3. We quantify the difference between the dynamics of X in the isolated system (6.3) and the dynamics of X in the connected system (6.4) by establishing conditions on the biological parameters under which the two systems exhibit similar behavior. This is achieved by exploiting the difference of time scales between the protein production and decay processes and the binding and unbinding process at the promoter p. By virtue of this separation of time scales, we can approximate system (6.4) by a one-dimensional system describing the evolution of X on the slow manifold (Kokotović et al., 1999). This reduced system takes the form

$$\frac{\mathrm{d}\bar{X}(t)}{\mathrm{d}t} = k(t) - \delta\bar{X}(t) + \bar{s}(t),$$

where $\bar{X}$ is an approximation of X and $\bar{s}$ is an approximation of s, which can be written as $\bar{s} = -\mathcal{R}(\bar{X})(k(t) - \delta\bar{X})$ with

$$\mathcal{R}(\bar{X}) = \frac{1}{1 + \dfrac{(1 + \bar{X}/k_d)^2}{p_{\mathrm{TOT}}/k_d}}, \tag{6.6}$$

where k_d is the dissociation constant for transcription factor binding, $k_d = k_{\mathrm{off}}/k_{\mathrm{on}}$ (see Del Vecchio et al., 2008b, for details). The expression $\mathcal{R}(\bar{X})$ quantifies the retroactivity to the output on the dynamics of X after a fast transient, when we approximate X with $\bar{X}$ in the limit where $\delta/k_{\mathrm{off}} \approx 0$. The retroactivity measure is thus low if the affinity of the binding sites p is small (k_d large) or if the signal $X(t)$ is large enough compared to p_{TOT}. Thus the form of $\mathcal{R}(\bar{X})$ provides an operative quantification of the retroactivity: such an expression can in fact be evaluated once the association and dissociation constants of X to p, the concentration of the binding sites p_{TOT}, and the range of operation of the signal $\bar{X}(t)$ that travels across the interconnection are all known. Thus the modularity assumption introduced in section 6.2 holds if the value of $\mathcal{R}(\bar{X})$ is low enough.

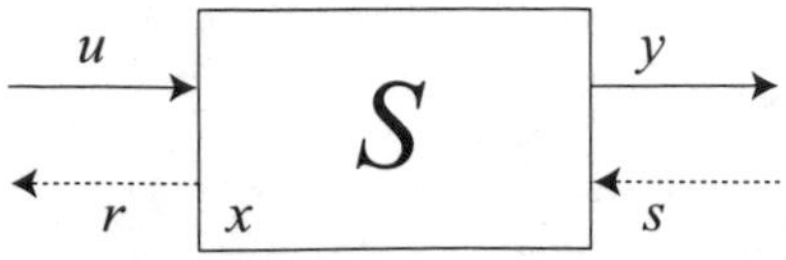

Figure 6.10
System S input and output signals. The dotted arrows denote signals originating by retroactivity upon interconnection.

6.5 Insulation Devices to Enforce Modularity

Of course, it is not always possible to design an interconnection such that the retroactivity is low. This is, for example, the case of an oscillator that has to time a downstream load: in general, the load cannot be included in the design as the oscillator must perform well in the face of unknown and possibly variable loads. That said, as with electrical circuits, one can design a device, to be placed between the oscillator and the load, such that the output of the device is unaffected by the load and the device itself does not affect the behavior of the upstream oscillator. Specifically, consider a system S as shown in figure 6.10 that takes u as input and gives y as output, designed in such a way that

1. the retroactivity r to the input is very small;
2. the effect of the retroactivity s to the output on the internal dynamics of the system is very small;
3. its input-output relationship is about linear.

A system like this is said to enjoy the *insulation* property and will be called an insulation device. Indeed, it will not affect an upstream system because $r \approx 0$ and it will keep the same output signal y *independently* of any connected downstream system. Other researchers have considered the insulation from external perturbations and robustness properties of amplifiers in the context of biochemical networks (Sauro and Ingalls, 2007; Sauro and Kholodenko, 2004). Here we revisit the amplifier mechanism in the context of gene transcriptional networks with the objective of mathematically and computationally demonstrating how suitable biochemical realizations of such a mechanism can attain properties 1, 2, and 3.

In electrical circuits, the standard insulation device is the operational amplifier, or OPAMP. In electronic amplifiers, r is very small because the input stage of an OPAMP absorbs almost no current. This way, there is no voltage drop across the output impedance of an upstream voltage source. Equation (6.6) quantifies the effect of retroactivity on the dynamics of X as a function of biochemical parameters that characterize the interconnection mechanism with a downstream system. These parameters are the affinity of the binding site $1/k_d$, the total concentration of the

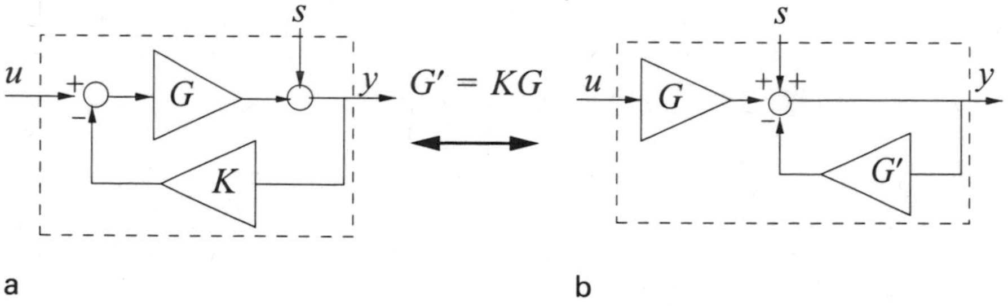

Figure 6.11
(a) Basic feedback/amplification mechanism by which amplifiers attenuate the effect of the retroactivity to the output s. (b) Alternative representation of the same mechanism, which will be employed to design biological insulation devices.

promoter, p_{TOT}, and the level of the signal $X(t)$. To reduce retroactivity, we can choose k_d large (low affinity) and p_{TOT} small, for example. Having a small value for p_{TOT}, low affinity, or both implies that there is a small "flow" of protein X toward its target sites. Thus we can say that a low retroactivity to the input is obtained when the "input flow" to the system is small. This interpretation establishes a nice analogy to the electrical case, in which low retroactivity to the input is obtained by a low input current.

In an electronic amplifier, the effect of the retroactivity to the output s on the amplifier behavior is reduced to almost zero by virtue of a large (theoretically infinite) input amplification gain and a negative output feedback. Such a mechanism can be illustrated in its simplest form by figure 6.11a, which is very well known to control engineers. For simplicity, we have assumed in such a diagram that the retroactivity s is just an additive disturbance. That the effect of the retroactivity s to the output is negligible for large gains G can be verified through the following simple computation. The output y is given by $y = G(u - Ky) + s$, which leads to

$$y = u\frac{G}{1 + KG} + \frac{s}{1 + KG}.$$

As G grows, y tends to u/K, which is independent of the retroactivity s. To attenuate the retroactivity effect at the output of a component one (1) amplifies the input of the component through a large gain and (2) applies a large negative output feedback (figure 6.11b). We next illustrate this idea in the context of the transcriptional example. Consider the approximated dynamics of X. Let us assume that we can apply a gain G to the input $k(t)$ and a negative feedback gain G' to X with $G' = KG$. This leads to the new differential equation for the connected system given by

$$\frac{\mathrm{d}X(t)}{\mathrm{d}t} = (Gk(t) - (G' + \delta)X(t))(1 - \mathcal{R}(X(t))). \tag{6.7}$$

It can be shown (see Del Vecchio et al., 2008a, for details) that, as G and thus as G' grows, the signal $X(t)$ generated by the connected system given by equation (6.7) becomes close to the solution $X(t)$ of the isolated system

$$\frac{dX(t)}{dt} = Gk(t) - (G' + \delta)X(t); \tag{6.8}$$

that is, the presence of the disturbance term $\mathcal{R}(X)$ will not significantly affect the time behavior of $X(t)$. A key question arises: How can we obtain a large amplification gain G and a large negative feedback G' in a biological insulation component? This question is addressed in the following section, in which we show that a simple phosphorylation/dephosphorylation cycle has remarkable insulation properties (for additional designs of biomolecular insulation devices, see Del Vecchio et al., 2008b).

6.5.1 A Biomolecular Realization of an Insulation Device through Protein Phosphorylation

In this design, we propose to obtain input amplification through a fast phosphorylation reaction and negative feedback through a fast dephosphorylation reaction. In particular, this is realized by having the input Z activate the phosphorylation of a protein X, which is available in the system in abundance. That is, Z is a kinase for a protein X. The phosphorylated form of X, X_p, binds to the downstream sites, whereas X does not. A negative feedback on X_p is obtained by having a phosphatase Y activate the dephosphorylation of protein X_p. Protein Y is also available in abundance in the system. This mechanism is depicted in figure 6.12. A similar design has been proposed by Sauro and Kholodenko (2004) and Sauro and Ingalls (2007), in which a MAPK cascade plus a negative feedback loop that spans the length of the MAPK cascade is considered as a feedback amplifier. Our design is much simpler, involving only one phosphorylation cycle and requiring no additional feedback loop.

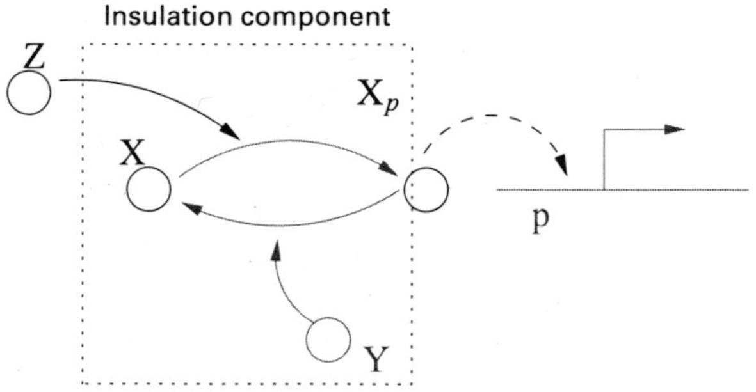

Figure 6.12
Insulation device (*dotted box*).

To convey the idea of how this device realizes the insulation function, we consider the one-step reaction model for the phosphorylation reactions analyzed by Heinrich et al. (2002):

$$\mathrm{Z} + \mathrm{X} \xrightarrow{k_1} \mathrm{Z} + \mathrm{X}_p \quad \text{and} \quad \mathrm{Y} + \mathrm{X}_p \xrightarrow{k_2} \mathrm{Y} + \mathrm{X}.$$

We assume that there is plenty of protein X and of phosphatase Y in the system and that these quantities are conserved. The conservation of X gives $X + X_p + C = X_{\mathrm{TOT}}$, in which X is the inactive protein, X_p is the phosphorylated protein that binds to the downstream sites p, and C is the complex of the phosphorylated protein X_p bound to the promoter p. The X_p dynamics can be described by the first equation in the following model:

$$\frac{\mathrm{d}X_p(t)}{\mathrm{d}t} = k_1 X_{\mathrm{TOT}} Z(t) \left(1 - \frac{X_p(t)}{X_{\mathrm{TOT}}} - \boxed{\frac{C(t)}{X_{\mathrm{TOT}}}}\right) - k_2 Y(t) X_p(t) + \boxed{k_{\mathrm{off}} C(t) - k_{\mathrm{on}} X_p(t)(p_{\mathrm{TOT}} - C(t))}, \tag{6.9}$$

$$\frac{\mathrm{d}C(t)}{\mathrm{d}t} = -k_{\mathrm{off}} C(t) + k_{\mathrm{on}} X_p(t)(p_{\mathrm{TOT}} - C(t)). \tag{6.10}$$

The boxed terms represent the retroactivity s to the output of the insulation system of figure 6.12. For a weakly activated pathway, $X_p \ll X_{\mathrm{TOT}}$ (Heinrich et al., 2002). Also, if we assume that the total concentration of X is large compared to the concentration of the downstream binding sites, that is, $X_{\mathrm{TOT}} \gg p_{\mathrm{TOT}}$, equation (6.9) is approximated by

$$\frac{\mathrm{d}X_p(t)}{\mathrm{d}t} = k_1 X_{\mathrm{TOT}} Z(t) - k_2 Y(t) X_p(t) + k_{\mathrm{off}} C(t) - k_{\mathrm{on}} X_p(t)(p_{\mathrm{TOT}} - C(t)).$$

If we denote $G = k_1 X_{\mathrm{TOT}}$ and $G' = k_2 Y$ and again exploit the difference of time scales between the X_p dynamics and the C dynamics after a fast initial transient, we can approximate the dynamics of X_p accurately by

$$\frac{\mathrm{d}X_p(t)}{\mathrm{d}t} = (G(t)Z(t) - G'(t)X_p(t))(1 - \mathcal{R}(X_p(t))), \tag{6.11}$$

in which $\mathcal{R}(X_p)$ is the measure of the retroactivity s to the output after a short transient. Therefore, for G and G' large enough, $X_p(t)$ is well described by the isolated system $\mathrm{d}X_p(t)/\mathrm{d}t = G(t)Z(t) - G'(t)X_p(t)$. As a consequence, the effect of the retroactivity to the output s is attenuated by increasing $k_1 X_{\mathrm{TOT}}$ and $k_2 Y$. That is, to obtain large input and feedback gains, one should have large phosphorylation/dephosphorylation rates, large amounts of protein X and phosphatase Y in the system, or both.

To highlight the roles of the various parameters for attaining the insulation properties, we can consider a more complex model for the phosphorylation and dephosphorylation reactions and perform parametric analysis. In particular, let us consider a two-step reaction model such as that of Huang and Ferrell (1996). According to this model, we have the following two reactions for phosphorylation and dephosphorylation, respectively:

$$\mathrm{X} + \mathrm{Z} \underset{\beta_2}{\overset{\beta_1}{\rightleftharpoons}} \mathrm{C}_1 \overset{k_1}{\rightarrow} \mathrm{X}_p + \mathrm{Z}, \tag{6.12}$$

and

$$\mathrm{Y} + \mathrm{X}_p \underset{\alpha_2}{\overset{\alpha_1}{\rightleftharpoons}} \mathrm{C}_2 \overset{k_2}{\rightarrow} \mathrm{X} + \mathrm{Y}, \tag{6.13}$$

in which C_1 is the complex of protein X with kinase Z and C_2 is the complex of phosphatase Y and protein X_p. Additionally, we have the conservations $Y_{\mathrm{TOT}} = Y + C_2$, $X_{\mathrm{TOT}} = X + X_p + C_1 + C_2 + C$, because proteins X and Y are not degraded. The differential equations modeling the insulation system of figure 6.12 thus become

$$\frac{\mathrm{d}Z(t)}{\mathrm{d}t} = k(t) - \delta Z(t)$$

$$\boxed{-\beta_1 Z(t) X_{\mathrm{TOT}}\left(1 - \frac{X_p(t)}{X_{\mathrm{TOT}}} - \frac{C_1(t)}{X_{\mathrm{TOT}}} - \frac{C_2(t)}{X_{\mathrm{TOT}}} - \boxed{\frac{C(t)}{X_{\mathrm{TOT}}}}\right) + (\beta_2 + k_1) C_1(t)}, \tag{6.14}$$

$$\frac{\mathrm{d}C_1(t)}{\mathrm{d}t} = -(\beta_2 + k_1) C_1(t)$$

$$+ \beta_1 Z(t) X_{\mathrm{TOT}}\left(1 - \frac{X_p(t)}{X_{\mathrm{TOT}}} - \frac{C_1(t)}{X_{\mathrm{TOT}}} - \frac{C_2(t)}{X_{\mathrm{TOT}}} - \boxed{\frac{C(t)}{X_{\mathrm{TOT}}}}\right), \tag{6.15}$$

$$\frac{\mathrm{d}C_2(t)}{\mathrm{d}t} = -(k_2 + \alpha_2) C_2(t) + \alpha_1 Y_{\mathrm{TOT}} X_p(t)\left(1 - \frac{C_2(t)}{Y_{\mathrm{TOT}}}\right), \tag{6.16}$$

$$\frac{\mathrm{d}X_p(t)}{\mathrm{d}t} = k_1 C_1(t) + \alpha_2 C_2(t) - \alpha_1 Y_{\mathrm{TOT}} X_p(t)\left(1 - \frac{C_2(t)}{Y_{\mathrm{TOT}}}\right)$$

$$+ \boxed{k_{\mathrm{off}} C(t) - k_{\mathrm{on}} X_p(t)(p_{\mathrm{TOT}} - C(t))}, \tag{6.17}$$

$$\frac{\mathrm{d}C(t)}{\mathrm{d}t} = -k_{\mathrm{off}} C(t) + k_{\mathrm{on}} X_p(t)(p_{\mathrm{TOT}} - C(t)), \tag{6.18}$$

in which C is the complex of X_p with the downstream promotor p, and the expression of gene z is controlled by a promoter with activity $k(t)$. The terms in the large box in equation (6.14) represent the retroactivity r to the input, while the terms in the small box in equation (6.14) and in the boxes of equations (6.15) and (6.17) represent the retroactivity s to the output. A detailed analysis of the system in equations (6.14–6.18) also provides analytical relationships among the parameters that indicate how to obtain small retroactivity to the input r and linear input-output behavior (see Del Vecchio et al., 2008b, for details). As for the simplified model, we have again that $G \propto k_1 X_{TOT}$ and $G' \propto k_2 Y_{TOT}$.

The system in equations (6.14–6.18) was simulated with and without the downstream binding sites p, that is, with and without the terms in the small box of equation (6.14) and in the boxes of equations (6.17) and (6.15). This analysis highlights the effect of the retroactivity to the output s on the dynamics of X_p. The simulations

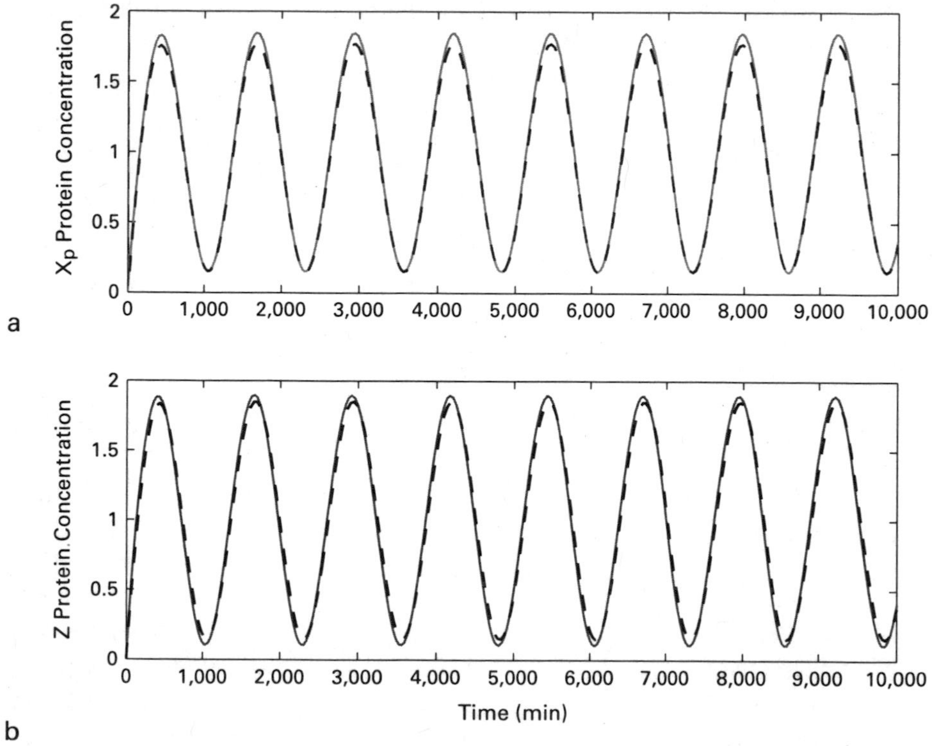

Figure 6.13
Simulation results for system described by equations (6.14–6.18). In all plots, $p_{TOT} = 100$, $k_{off} = k_{on} = 10$, $\delta = 0.01$, $k(t) = 0.01(1 + \sin(\omega t))$, and $\omega = 0.005$. In subplots A and B, $k_1 = k_2 = 50$, $\alpha_1 = \beta_1 = 0.01$, $\beta_2 = \alpha_2 = 10$, and $X_{TOT} = Y_{TOT} = 1{,}500$. Panel a shows the signal $X_p(t)$ without the downstream binding sites p (*solid line*) and the same signal with the downstream binding sites p (*dashed line*). The small error shows that the effect of the retroactivity to the output s is attenuated very well. Panel b shows the signal $Z(t)$ without X, to which Z binds (*solid line*), and the same signal $Z(t)$ with X present in the system (*dashed line*). The small error confirms a small retroactivity to the input.

validate our theoretical study indicating that, when $X_{TOT} \gg p_{TOT}$ and the time scales of phosphorylation/dephosphorylation are much faster than the time scale of decay and production of the protein Z, the retroactivity to the output s is very well attenuated (figure 6.13a). Similarly, the time behavior of Z was simulated with and without the terms in the large box of equation (6.14), that is, with and without X to which Z binds, to verify whether the insulation component exhibits retroactivity to the input r. In particular, the accordance of the behaviors of $Z(t)$ with and without its downstream binding sites on X (figure 6.13b) indicates that there is no substantial retroactivity to the input r generated by the insulation device.

6.6 Conclusions and Future Challenges

We have reviewed some design methods employed in synthetic biology that rely on the modularity assumption according to which modules maintain their dynamic behavior unchanged upon interconnection. By virtue of this assumption, one can first characterize a module in isolation and then predict the behavior of a circuit directly by the behavior of its composing modules. This is a powerful approach to circuit design, which is employed also in other engineering areas, such as electronics. We pointed out, however, that, just as in several other engineering systems, because of loading effects at interconnections, modularity does not necessarily hold in biomolecular systems. As with historical developments in electronics, where researchers focused on characterizing impedance (loading) effects and on counteracting them with the aid of operational amplifiers, similar efforts are currently taking place in the area of biomolecular circuit design.

Biomolecular circuit design presents many challenges to systems and control engineers. We need to address problems of loading effects at interconnections between general biomolecular systems, such as signaling systems, as opposed to purely transcriptional systems. These include not only the effects of loads but also impedance-matching problems and frequency sensitivity analysis. We need to consider energetic constraints in the design of active devices such as insulation devices, which could possibly impose tough requirements on the cell, as well as possible trade-offs between the need for large gains in the design of insulation devices and the effects of such large gains on biological noise. Finally, to these engineering challenges must be added the scientific challenge of uncovering design principles that are already employed by natural systems for coping with similar problems.

Note

1. A significant portion of the contents of this chapter appeared in Del Vecchio et al., 2008b.

7 Graphs and the Dynamics of Biochemical Networks

David Angeli and Eduardo D. Sontag

Because biochemical networks may give rise to complex dynamical behaviors, their analysis is often difficult. Several tools based on testing graph-theoretical properties of these networks are useful in that context. This chapter discusses two approaches, one based on bipartite graphs and Petri net concepts, and another based on decompositions into order-preserving subsystems.

7.1 Dynamical Behavior

Models of biochemical networks studied in the literature are often asymptotically "well behaved": upon external stimulation—such as presentation of a ligand to a receptor—and after transient activity, variables approach simple steady-state behaviors, and seldom enter "chaotic" regimes. One may thus ask, what is special about such networks when viewed in the context of more general dynamical systems? On the one hand, precise *quantitative* information (parameters, exact forms of reactions) is often unavailable; on the other, *qualitative* information (network structures showing how chemical species interact) is much more abundant. This motivates the development of analysis tools that effectively combine sparse quantitative with more readily available qualitative information, in the form of graphs. In the present chapter, we review two such tools. The presentation will be kept informal so as to emphasize intuition; references will be provided for the precise mathematical results. For definiteness, we will restrict attention to deterministic systems described by ordinary differential equations (ODEs), but similar approaches can be developed for systems described by partial differential equations, such as those discussed in chapter 3. Thus we will consider systems described by sets of simultaneous ODEs of the following general form:

$$\dot{x}_1(t) = \frac{\mathrm{d}x_1(t)}{\mathrm{d}t} = f_1(x_1(t), \ldots, x_n(t), u_1(t), \ldots, u_m(t)),$$

$$\vdots$$

$$\dot{x}_n(t) = \frac{\mathrm{d}x_n(t)}{\mathrm{d}t} = f_n(x_1(t), \ldots, x_n(t), u_1(t), \ldots, u_m(t)),$$

$$\boldsymbol{y}(t) = \boldsymbol{h}(x_1(t), \ldots, x_n(t)),$$

or, in vector form, $\dot{\boldsymbol{x}} = \boldsymbol{f}(\boldsymbol{x}, \boldsymbol{u})$ and $\boldsymbol{y} = \boldsymbol{h}(\boldsymbol{x})$ (the argument t is often omitted from now on), where the variables $x_i(t)$ represent the concentrations of chemical species (proteins, mRNA, metabolites), the variables $u_i(t)$ are inputs (stimuli, external signals, controls, forcing functions), and the coordinates of the vector $\boldsymbol{y}(t)$ represent outputs (responses, measurements, readouts, reporter variables), at any given time t.

The problem of combining qualitative and quantitative information can be approached in several ways. We single out two of them: (1) species-reaction (s/r) representations and (2) input-output (i/o) decompositions.

The *species-reaction approach* is based on the representation of chemical reaction networks by bipartite *species-reaction graphs*. For systems that can be modeled by ordinary differential equations, such a representation translates algebraically into the factorization of the state evolution equations $\dot{\boldsymbol{x}} = \boldsymbol{f}(\boldsymbol{x}, \boldsymbol{u})$ in the form

$$\dot{\boldsymbol{x}} = \boldsymbol{\Gamma} \boldsymbol{R}(\boldsymbol{x}, \boldsymbol{u}),$$

where $\boldsymbol{\Gamma}$ denotes the stoichiometry matrix of the reaction, $\boldsymbol{R}(\boldsymbol{x}, \boldsymbol{u})$ is the vector of reaction rates (which may depend on external inputs $\boldsymbol{u}$), and $\boldsymbol{x}$ is a vector that describes the concentrations of the various species that take part in the reaction, at each time t. Feinberg-Horn-Jackson deficiency theory (Feinberg, 1987; Horn, 1974), and the closely related work of Craciun and Feinberg (2005, 2006), as well as methods based on Petri net formalisms (Angeli et al., 2007) all rely on the species-reaction formalism. In this chapter, as an illustration, we will briefly survey some mathematical results based on Petri net methods.

The *input-output component* or modular approach is based on viewing larger networks as made up of several subnetworks ("subsystems" in control theory, or modules), interconnected through input and output variables, as illustrated in figure 7.1.

This approach may be particularly useful when the component subsystems have low dynamical complexity, for example, when they are monostable (discussed below), in which case all the complexity of the overall system arises from interconnections, especially if conclusions about global behavior can be drawn using only simple input-output "black box" information on components (also discussed below). Two examples are (1) methods based on the notion of passivity (Angeli, 2006; Arcak and Sontag, 2006, 2008; Sontag, 2006); and (2) methods based on order-preserving

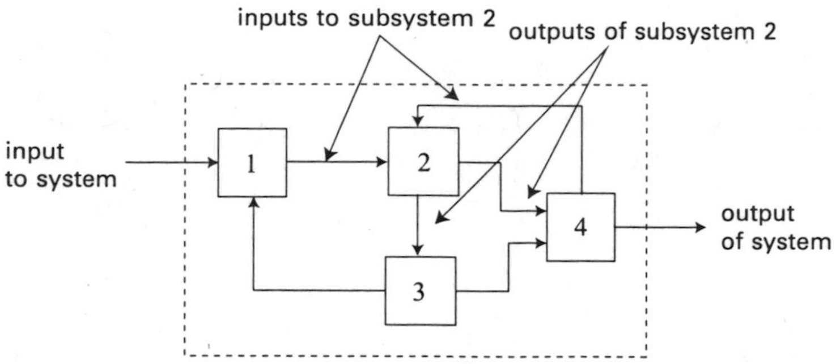

Figure 7.1
Larger system made up of interconnected input-output subsystems.

("monotone") flows (Angeli and Sontag 2003, 2004; Enciso and Sontag, 2006). In this chapter, as an illustration, we will briefly survey some mathematical results based on monotone methods.

An important caveat when employing input-output interconnection approaches to analyze biochemical networks (or, for that matter, any physical system) is that the input-output formalism ignores the possible "impedance" or "retroactivity" effects that arise from the actual interconnections (for details of this most important issue, see chapter 6 and Del Vecchio et al., 2008b).

Each of the methods discussed in this chapter relies on a different graphical representation of chemical reactions. In the case of species-reaction methods, and the analysis based upon Petri nets in particular, we make use of a bipartite graph in which there are two types of nodes, one representing reactions and the other representing species. The definition of monotone systems, or to be more precise *orthant-monotone* systems, relies on a graph that shows the interactions (positive or negative) among the different species.

This chapter is organized as follows. Section 7.2 reviews the basic formalism used for modeling biochemical networks. Section 7.3 reviews some results that use bipartite graphs, and specifically Petri nets. As a motivation for the consideration of methods based upon monotone input-output systems, section 7.4 postulates a quasi-steady-state reduction principle (QSSRP) and asks the question, for what types of components is the QSSRP a valid mathematical tool? Section 7.5 provides a (partial) answer to this question, after introducing monotone systems.

7.2 Chemical Network Formalism

Let us first review a formalism that allows us to write up differential equations associated with chemical reactions easily. In general, we consider a collection of chemical reactions that involves a set of n_s species:

$S_j, \quad j \in \{1, 2, \ldots n_s\}.$

Although these species may be ions, atoms, or molecules (even large molecules, such as proteins), we will call them "molecules," for simplicity. In general, a *chemical reaction network* (CRN) is a set of chemical reactions $\mathcal{R}_i$, $i \in \{1, 2, \ldots, n_r\}$:

$$\mathcal{R}_i : \quad \sum_{j=1}^{n_s} \alpha_{ij} S_j \to \sum_{j=1}^{n_s} \beta_{ij} S_j, \tag{7.1}$$

where the α_{ij} and β_{ij}, as nonnegative integers, are the *stoichiometry coefficients*.

The species with nonzero coefficients on the left-hand side of the equation are usually referred to as the *reactants*, and the ones on the right-hand side are called the *products*, of the respective reaction. (Zero coefficients are not shown in diagrams.) The interpretation is that, in reaction 1, α_{11} molecules of species S_1 combine with α_{12} molecules of species S_2, and so forth, to produce β_{11} molecules of species S_1, β_{12} molecules of species S_2, and so forth, and similarly for each of the other $n_r - 1$ reactions. The forward arrow means that the transformation of reactants into products only happens in the direction of the arrow.

It is convenient to arrange the stoichiometry coefficients into an $n_s \times n_r$ matrix, called the *stoichiometry matrix*, $\mathbf{\Gamma} = \Gamma_{ij}$, defined as follows:

$$\Gamma_{ji} = \beta_{ij} - \alpha_{ij}, \quad i = 1, \ldots, n_r, \quad j = 1, \ldots, n_s. \tag{7.2}$$

(Note the reversal of indices.) The matrix $\mathbf{\Gamma}$ has as many columns as there are reactions. Each column shows, for all species (ordered according to their index i), the net "produced–consumed." The graphical information given by reaction diagrams is summarized by the matrix $\mathbf{\Gamma}$.

We now describe how the state of the network evolves over time, for a given chemical reaction network. We need to find a rule for the evolution of the column vector:

$$\boldsymbol{x}(t) = \begin{bmatrix} x_1(t) \\ \vdots \\ x_{n_s}(t) \end{bmatrix},$$

where $x_i(t)$ represents the concentration of the species S_i at time t. The variables x_i take only nonnegative values.

Another ingredient that we require is a formula for the actual rate at which the individual reactions take place. We denote by $R_i(\boldsymbol{x})$ the algebraic form of the jth reaction rate as a function of species concentrations. The most common assumption is that of *mass-action kinetics*, where

$$R_i(\boldsymbol{x}) = k_i \prod_{j=1}^{n_s} x_j^{\alpha_{ij}}, \qquad \text{for all } i = 1, \ldots, n_r,$$

which says simply that the reaction rate is proportional to the product of the concentrations of the reactants, with higher exponents when more than one molecule is needed. The coefficients k_i are called "reaction constants"; they usually label the arrows in diagrams. We introduce a column vector of reactions

$$\boldsymbol{R}(\boldsymbol{x}) = \begin{bmatrix} R_1(\boldsymbol{x}) \\ \vdots \\ R_{n_r}(\boldsymbol{x}) \end{bmatrix}.$$

With these conventions, the system of differential equations associated to the chemical reaction network is given as follows:

$$\frac{\mathrm{d}\boldsymbol{x}}{\mathrm{d}t} = \boldsymbol{\Gamma}\boldsymbol{R}(\boldsymbol{x}). \tag{7.3}$$

This formalism assumes that none of the species x_i are inputs to the system. On the other hand, in some experimental situations, there may be species whose concentration is affected by inflows from outside the reaction vessel, or whose values may be assumed to be clamped externally. To model such external input signals, one needs to extend the formalism. This can be done by omitting the differential equation for the species in question, and considering this concentration instead as a forcing function, that is, as part of the input channels $\boldsymbol{u}$ in a system $\dot{\boldsymbol{x}} = \boldsymbol{f}(\boldsymbol{x}, \boldsymbol{u})$. For simplicity, we will discuss only reactions having no such external inputs.

By way of illustration, let us consider the following set of chemical reactions:

$$E + S_0 \underset{k_{-1}}{\overset{k_1}{\rightleftharpoons}} ES_0 \overset{k_2}{\rightarrow} E + S_1, \tag{7.4a}$$

$$F + S_1 \underset{k_{-3}}{\overset{k_3}{\rightleftharpoons}} FS_1 \overset{k_4}{\rightarrow} F + S_0, \tag{7.4b}$$

which may be thought of as a model of the activation of a protein substrate S_0 by an enzyme (a kinase denoted by E); ES_0 is an intermediate complex, which dissociates either back into the original components or into a product (activated protein) S_1 and the enzyme. The second reaction transforms S_1 back into S_0, and is catalyzed by another enzyme (a phosphatase denoted by F). A system of reactions of this type is sometimes called a "futile cycle," and reactions of this type are ubiquitous in cell biology (Samoilov et al., 2005). The mass-action kinetics model is obtained as follows. Denoting concentrations with the same letters as the species themselves, we have the following vector of species, stoichiometry matrix $\mathbf{\Gamma}$ and vector of reaction rates $\boldsymbol{R}(\boldsymbol{x})$:

$$\boldsymbol{x} = \begin{bmatrix} S_0 \\ S_1 \\ E \\ F \\ ES_0 \\ FS_1 \end{bmatrix}, \quad \mathbf{\Gamma} = \begin{bmatrix} -1 & 1 & 0 & 0 & 0 & 1 \\ 0 & 0 & 1 & -1 & 1 & 0 \\ -1 & 1 & 1 & 0 & 0 & 0 \\ 0 & 0 & 0 & -1 & 1 & 1 \\ 1 & -1 & -1 & 0 & 0 & 0 \\ 0 & 0 & 0 & 1 & -1 & -1 \end{bmatrix}, \quad \boldsymbol{R}(\boldsymbol{x}) = \begin{bmatrix} k_1 E \times S_0 \\ k_{-1} ES_0 \\ k_2 ES_0 \\ k_3 F \times S_1 \\ k_{-3} FS_1 \\ k_4 FS_1 \end{bmatrix}.$$

From here, we can write the equations (7.3). For example,

$$\frac{\mathrm{d}S_0}{\mathrm{d}t} = (-1)(k_1 E \times S_0) + (1)(k_{-1} ES_0) + (1)(k_4 FS_1)$$

$$= k_4 FS_1 - k_1 E \times S_0 + k_{-1} ES_0.$$

Conservation Laws

Let us consider the set of row vectors $\boldsymbol{c}$ such that $\boldsymbol{c}\mathbf{\Gamma} = \mathbf{0}$. Any such vector gives rise to a (linear) *conservation law* in the sense that

$$\frac{\mathrm{d}(\boldsymbol{c}\boldsymbol{x})}{\mathrm{d}t} = \boldsymbol{c}\frac{\mathrm{d}\boldsymbol{x}}{\mathrm{d}t} = \boldsymbol{c}\mathbf{\Gamma}\boldsymbol{R}(\boldsymbol{x}) = 0,$$

for all t, and therefore

$$\boldsymbol{c}\boldsymbol{x}(t) = \text{constant}$$

along all solutions (a first integral of the motion). The set of such vectors forms a linear subspace Δ (of the vector space consisting of all row vectors of size n_s). The existence of conservation laws is important in the analysis of dynamics since affine subspaces orthogonal to Δ are invariant for motions.

Thus, in the example cited above, we have that, along all solutions,

$$S_0(t) + S_1(t) + ES_0(t) + FS_1(t) \equiv \text{constant}$$

because $[1, 1, 0, 0, 1, 1]\mathbf{\Gamma} = \mathbf{0}$. Likewise, because we have two more linearly independent conservation laws, namely, $[0, 0, 1, 0, 1, 0]$ and $[0, 0, 0, 1, 0, 1]$,

$$E(t) + ES_0(t) \quad \text{and} \quad F(t) + FS_1(t)$$

are also constant along all trajectories. Since $\mathbf{\Gamma}$ has rank three (easy to check) and has six rows, its left-nullspace has dimension three. Thus, a basis of the set of conservation laws is given by the three that we have found.

7.3 Petri Nets and Persistence Analysis

As an example of the use of species-reaction representations and associated graph-theoretic concepts in biochemical systems analysis, we will briefly review recent work employing Petri net notions (since no interconnections are studied, we will not consider input and output variables here). *Petri nets*, also called place/transition nets, are a popular mathematical and graphical modeling tool, typically used for representing processes in which data (tokens) are concurrently handled by asynchronous and distributed processors. They are widely employed in fields as diverse as reliability engineering, work-flow management, and software design. Although most Petri net models are discrete, continuous Petri nets are usually studied also. The modeling of chemical reaction networks using Petri net formalism was pioneered by Reddy et al. (1993); see also the more recent work of Zevedei-Oancea and Schuster (2003).

We associate to a chemical reaction network a bipartite directed graph (i.e., a directed graph with two types of nodes) with weighted edges, called the *species-reaction Petri net* (SR net). Formally, this is a quadruple (V_S, V_R, E, W), where V_S is a finite set of nodes, each one associated to a species, V_R is a finite set of nodes (disjoint from V_S), each one corresponding to a reaction, and E is a set of edges as described below. (We write S or V_S interchangeably, or R instead of V_R, by identifying species or reactions with their respective indices.) The set of all nodes is $V := V_R \cup V_S$. The edge set $E \subset V \times V$ is defined as follows. For every reaction R_i:

$$\sum_{j \in S} \alpha_{ij} S_j \rightarrow \sum_{j \in S} \beta_{ij} S_j,$$

we draw an edge from $S_j \in V_S$ to $R_i \in V_R$ for each species S_j such that $\alpha_{ij} > 0$. That is, $(S_j, R_i) \in E$ exactly when $\alpha_{ij} > 0$, and we say in this case that R_i is an *output reaction for* S_j. Likewise, $(R_i, S_j) \in E$ whenever $\beta_{ij} > 0$, and we say that R_i is an *input reaction for* S_j. The SR net is a bipartite graph: edges only connect species to reactions and vice versa; they never connect two species or two reactions to each other. The notion of an SR net is very closely related to that of an SR graph (Craciun and Feinberg, 2005, 2006). The only difference is that an SR net is a directed graph, whereas an SR graph is not and that reversible reactions in an SR net are represented by two distinct reaction nodes, whereas only one reaction node appears in the SR graph for a reversible reaction. The function $W : E \rightarrow \mathbf{N}$ associates to each edge a positive integer according to the rule: $W(S_j, R_i) = \alpha_{ij}$ and $W(R_i, S_j) = \beta_{ij}$.

For a vector $\boldsymbol{v}$, we write $\boldsymbol{v} \geq 0$ if each entry of $\boldsymbol{v}$ is nonnegative, $\boldsymbol{v} > \mathbf{0}$ if $\boldsymbol{v} \geq \mathbf{0}$ and $\boldsymbol{v} \neq \mathbf{0}$, and $\boldsymbol{v} \gg \mathbf{0}$ if $v_i > 0$ for all i. In the Petri net literature, a conservation law, that is, a row vector $\boldsymbol{c} > \mathbf{0}$ such that $\boldsymbol{c}\boldsymbol{\Gamma} = \mathbf{0}$, is called a *P-semiflow*. (The terminology is unfortunate because these vectors do not correspond to fluxes in the system.) The *support* of $\boldsymbol{c}$ is the set of indices $\{i \in V_S : c_i > 0\}$. A Petri net is said to be *conservative* if there exists a P-semiflow $\boldsymbol{c} \gg \mathbf{0}$. (Petri net theory views Petri nets as "token-passing" systems, and, in that context, P-semiflows, also called place invariants, amount to conservation relations for the "place markings" of the network, that show how many tokens there are in each "place," the nodes associated to species in SR nets. We do not make use of that interpretation here.) The net is said to be *consistent* if there exists a $\boldsymbol{v} \gg \mathbf{0}$ (a T-semiflow) such that $\boldsymbol{\Gamma v} = \mathbf{0}$. The vector $\boldsymbol{v}$ may be viewed as a set of fluxes that is in equilibrium (Zevedei-Oancea and Schuster, 2003). A nonempty set $\Sigma \subset V_S$ is called a *siphon* if each input reaction associated to Σ is also an output reaction associated to Σ. A siphon is said to be *minimal* if it does not contain (strictly) any other siphons.

Persistence

The *persistence property* for differential equations defined on nonnegative variables is the requirement that solutions starting in the positive orthant do not approach the boundary of the orthant. For chemical reactions and population models, this translates into the nonextinction property: provided that every species is present at the start of the reaction, no species will tend to be eliminated in the course of the reaction. Mathematically, this property can be equivalently expressed as the requirement that the ω-limit set of any trajectory which starts in the interior of the positive orthant (all concentrations positive) does not intersect the boundary of the positive orthant: $\omega(\boldsymbol{x}_0) \cap \partial\mathcal{O}_+^{n_s} = \emptyset$ for each $\boldsymbol{x}_0 \in \operatorname{int}(\mathcal{O}_+^{n_s})$.

Angeli et al. (2007) provide checkable conditions for persistence of chemical species in reaction networks, using concepts and tools from Petri net theory, and verify these conditions on various systems which arise in the modeling of cell signaling pathways. Besides its applied interest, persistence is a key enabling theoretical property because it may be used in conjunction with other techniques in order to guarantee convergence of solutions to equilibria. For example, if a strictly decreasing Lyapunov function exists on the interior of the positive orthant (see, for example, Feinberg and Horn, 1974; Horn, 1974; Sontag, 2001, for classes of networks where this can be guaranteed), persistence allows such a conclusion. For complex networks, determining persistence, or lack thereof, is in general an extremely difficult mathematical problem. In fact, the study of persistence is a classical one in the (mathematically) related field of population biology, where species correspond to individuals of different types instead of chemical units (see, for example, Butler and Waltman, 1986; Gard, 1980).

The main persistence theorems from Angeli et al., 2007, are as follows:

1. A necessary condition: *A conservative and persistent chemical reaction network has a consistent Petri net.*

2. A sufficient condition: *If its associated Petri net is conservative, and each siphon contains the support of a P-semiflow, then the chemical reaction network is persistent.*

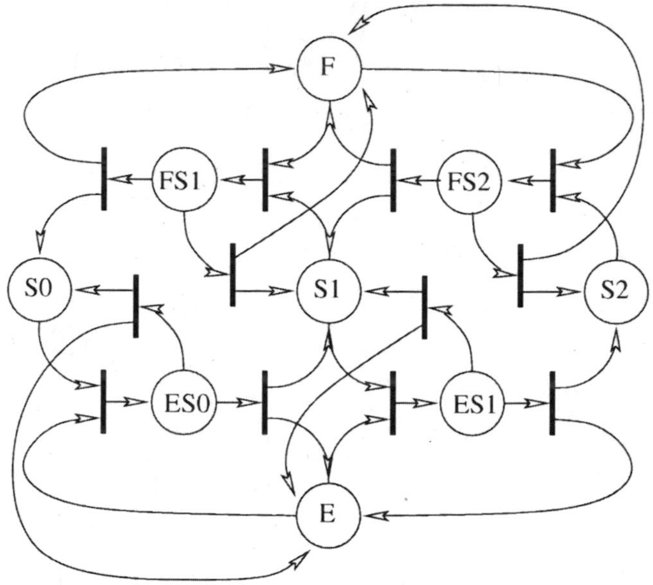

Figure 7.2
Associated Petri net.

As an example of application of the sufficiency condition, let us consider the following set of reactions:

$$E + S_0 \leftrightarrow ES_0 \rightarrow E + S_1 \leftrightarrow ES_1 \rightarrow E + S_2,$$

$$F + S_2 \leftrightarrow FS_2 \rightarrow F + S_1 \leftrightarrow FS_1 \rightarrow F + S_0.$$

These model a double futile cycle, similar to the one discussed in section 7.2, except that now two rather than one enzymatic modifications are produced; ES_0 represents the complex consisting of E bound to S_0 and so forth. We denote reversible reactions by a "$\leftrightarrow$" in order to avoid having to write them twice. The network comprises nine distinct species, labeled S_0, S_1, S_2, E, F, ES_0, ES_1, FS_2, and FS_1. Its associated Petri net is shown in figure 7.2. This net is indeed consistent: to see this, we order the species and reactions by the obvious order obtained when reading the equations from left to right and from top to bottom (e.g., S_1 is the fourth species, and the reaction $E + S_1 \rightarrow ES_1$ is the fourth reaction), introducing $\mathbf{\Gamma}$ and then verifying that

$$\mathbf{\Gamma} v = \mathbf{0}, \quad \text{when} \quad v = [2\ 1\ 1\ 2\ 1\ 1\ 2\ 1\ 1\ 2\ 1\ 1]^T.$$

Also, there are three minimal siphons, $\{E, ES_0, ES_1\}$, $\{F, FS_1, FS_2\}$, and $\{S_0, S_1, S_2, ES_0, ES_1, FS_2, FS_1\}$, each of which contains the support of a P-semiflow, which arise from the following three independent conservation laws: $E + ES_0 + ES_1 = \text{const}_1$,

$F + FS_2 + FS_1 = \mathrm{const}_2$, and $S_0 + S_1 + S_2 + ES_0 + ES_1 + FS_2 + FS_1 = \mathrm{const}_3$. Since the sum of these three conservation laws is also a conservation law, the network is conservative and, by the cited sufficiency theorem, also persistent.

7.4 A Quasi-Steady-State Reduction Principle

We next turn to input-output decompositions, and specifically decompositions into monotone subsystems. As discussed earlier, we will impose the requirement that components be "dynamically simple."

Specifically, we will assume that each subsystem is monostable, in the following sense: a system $\dot{\boldsymbol{x}} = \boldsymbol{f}(\boldsymbol{x}, \boldsymbol{u})$ with inputs $\boldsymbol{u}$ and outputs $\boldsymbol{y} = \boldsymbol{h}(\boldsymbol{x})$ has a well-defined steady-state response to step inputs if, for each step input $\boldsymbol{u}(t) \equiv \boldsymbol{u}$ there is a (necessarily unique) globally asymptotically stable steady state $\boldsymbol{x}_{\boldsymbol{u}}$ of the system (see figure 7.3; consult Angeli and Sontag, 2003, 2004; Enciso and Sontag, 2006, for precise definitions). The map $\boldsymbol{k}(\boldsymbol{u}) = \boldsymbol{h}(\boldsymbol{x}_{\boldsymbol{u}})$ will be called the *input-output characteristic* of the system. Often, input-output characteristics may be obtained from experimental data by presenting systems with constant inputs, letting them relax to steady state, and then measuring the value of the reporter variable (or more generally variables, if $\boldsymbol{y}$ is a vector whose components indicate the measured quantities). Characteristics are also called, depending on the context, "nonlinear DC gains," "dose-response curves," "receptor activity plots," and so forth.

The only "quantitative" information required by the results to be discussed below is the plot of the characteristics of the individual subsystems. The requirement of monostability can be weakened to some extent (see, for example, Enciso and Sontag, 2008).

A Reduction Principle

Suppose that two single-input, single-output systems, having respective characteristics k and g, are placed in a feedback loop as shown in figure 7.4. Let us perform the following thought experiment, ignoring dynamics. First, we suppose that a step signal u with constant value $u(t) \equiv u_1$ is applied to the system with characteristic k, letting this system relax to steady state, and taking note of its steady-state output

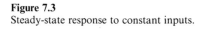

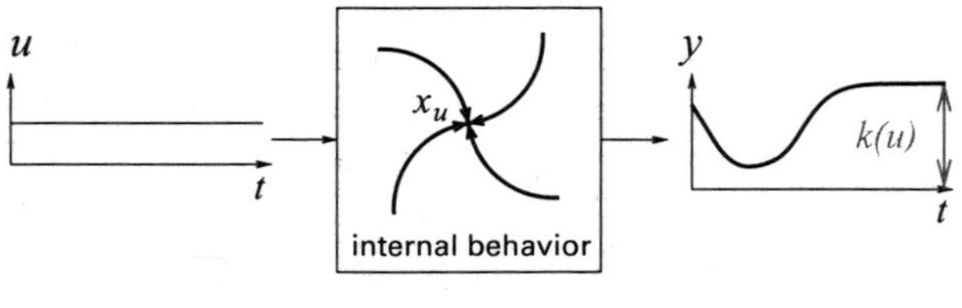

Figure 7.3
Steady-state response to constant inputs.

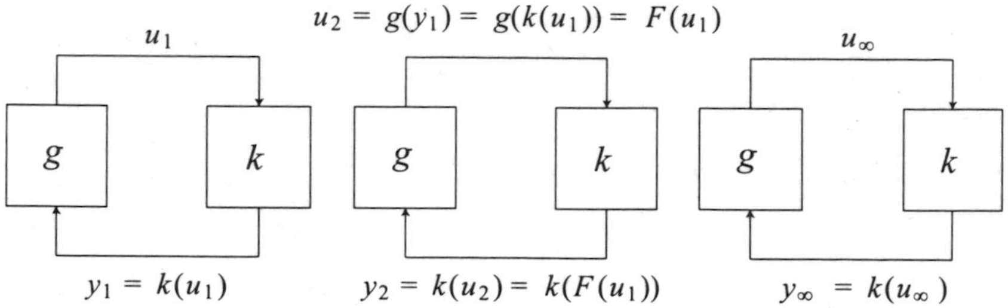

Figure 7.4
Iteration of characteristics, ignoring dynamics.

$y_1 = k(u_1)$ (leftmost panel in figure 7.4). Next, we apply the constant input y_1 to the system with characteristic g, and take note of its steady-state output $u_2 = g(y_1) = g(k(u_1))$, which we write as $F(u_1)$, where F is the *return map* for the loop. Likewise, we apply the resulting u_2 as an input to the k-system (middle panel in figure 7.4). Iterating, let us suppose that we converge to a pair of values, u_∞, $y_\infty = k(u_\infty)$.

This process would make physical sense if there were a theoretically infinite time-scale separation between the speed of response of the individual systems and the rate of change of the external signals, which is clearly unrealistic. Nonetheless, we may still ask the following question, is it possible to find all true asymptotic behaviors in this fashion? More precisely, we ask whether the following quasi-steady-state reduction principle (QSSRP) holds: suppose that generic trajectories of the discrete iteration $u \mapsto F(u)$ converge to one of $k \geq 1$ stable points $u^1, \dots, u^k$; is it then true that, generically, bounded trajectories of the closed-loop system, obtained as the feedback interconnection of the two subsystems, globally converge to one of k possible steady states, corresponding in 1-1 fashion to the $u^1, \dots, u^k$?

Provided that the QSSRP holds for a class of feedback structures, the only information needed to characterize the stability of equilibria is what is encapsulated in the graphs of the characteristics of the individual systems, and this is true even for systems involving large numbers of variables (chemical species). When the input and output signals u and y are scalar, this type of analysis is especially simple. Of course, one must have access to the graphs of the characteristics k and g, to begin with, and this may be difficult in particular instances. On the other hand, such steady-state response information can often be approximated on the basis of experimental data, and such data on steady-state responses is often far easier to obtain than data on the internal parameters (e.g., kinetic constants) that describe the component subsystems. Thus, it is most interesting to ask for what types of systems the QSSRP is valid.

To explore this question, we consider a simple example, in which each system is linear and one-dimensional, with respective equations as follows:

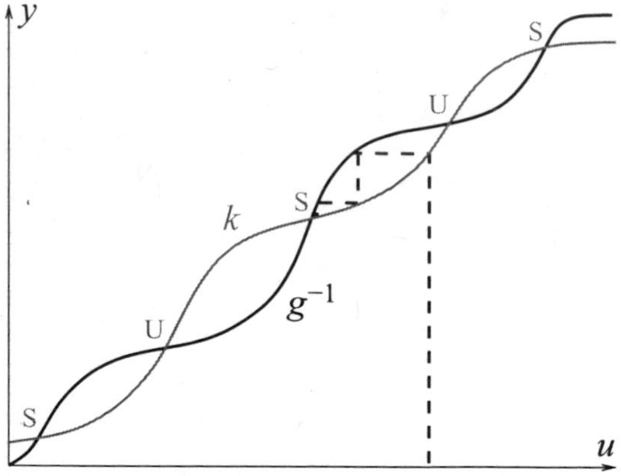

Figure 7.5
Two characteristics for the example.

$$\dot{y} = -y + k(u),$$

$$\dot{u} = -u + g(y),$$

where both functions k and g are increasing (positive feedback). It is clear that both systems admit characteristics, and these are precisely k and g, respectively. To find steady states of the interconnection, we plot the graphs of both k and g on the (u, y)-plane (to be more precise, the graph of g^{-1}), as shown in figure 7.5. Observe that the discrete iterations of $F = g \circ k$ converge to those points where the graphs intersect, and the slope of k is less than the slope of g^{-1}, marked "S," and unstable points of the iteration are marked "U," as is easy to verify through a standard "cobwebbing" argument, as illustrated in figure 7.5. (In this informal presentation, we ignore delicate technical issues that arise at those points where the intersections are not transversal, that is, points where the two curves meet tangentially; see the cited papers for details.)

In this example, the QSSRP is valid. Indeed, local stability of the closed-loop steady state $(u, k(u))$ holds when $k'(u) < (g^{-1})'(u)$ (the trace of the Jacobian is always negative, and the determinant is positive when this condition holds). Moreover, drawing nullclines and directions of flow confirms global stability, as sketched in figure 7.6. Of course, one cannot expect the QSSRP to hold for arbitrary systems. A counterexample is as follows. Consider the following two-species system:

$$\dot{x} = x(-x + y),$$

$$\dot{y} = 3y\left(-x + c + \frac{bu^4}{K + u^4}\right),$$

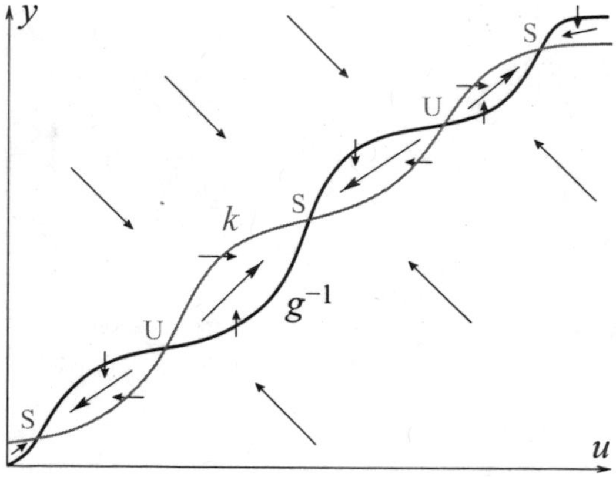

Figure 7.6
Phase plane for the example.

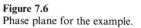

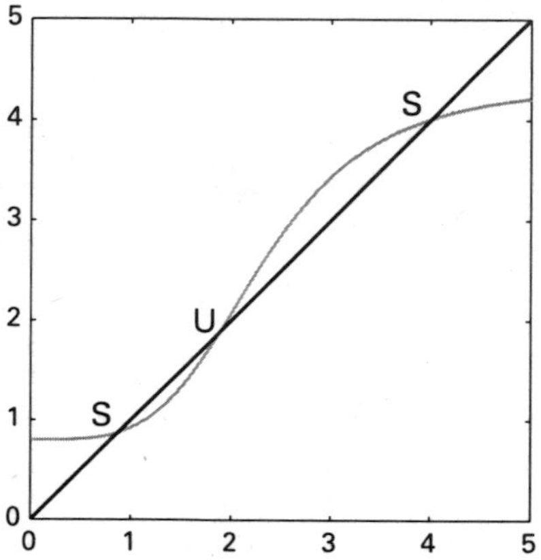

Figure 7.7
Characteristics for the counterexample.

which is monostable, with the following characteristic:

$$k(u) = c + \frac{bu^4}{K + u^4}.$$

(To be precise, there are also steady states with $x = 0$ or $y = 0$. One may, however, restrict the state space of the system to the interior of the positive orthant, $x > 0$, $y > 0$, or one may consider a slight perturbation of the system, replacing the x in the first equation by $x + \varepsilon$, and y in the second by $y + \varepsilon$, for $\varepsilon > 0$ sufficiently small.) For the feedback system, we pick a memoryless unity feedback $u = g(y) = y$. See

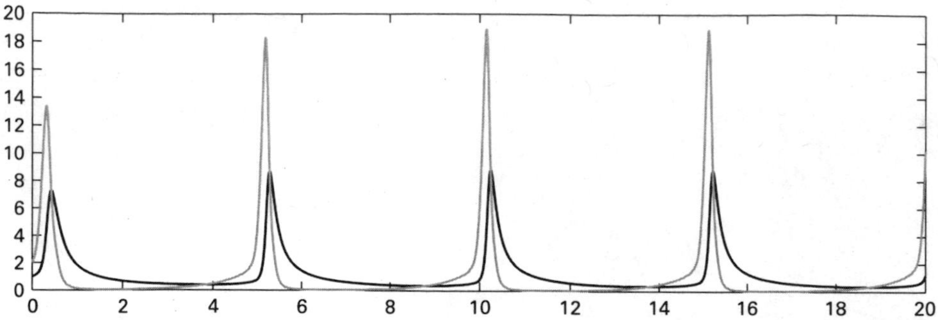

Figure 7.8
Trajectories for the counterexample.

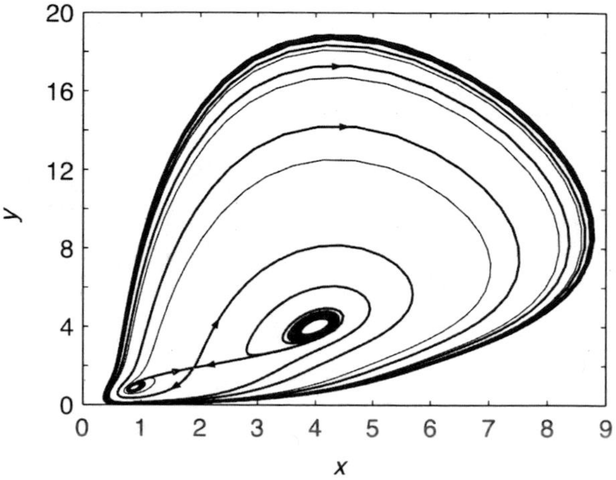

Figure 7.9
Phase plane for the counterexample.

figure 7.7 for the superimposed plots of k and g^{-1}. If the quasi-steady-state reduction principle were true for this interconnection, then one would predict that generic solutions of the closed-loop system

$$\dot{x} = x(-x + y),$$

$$\dot{y} = 3y\left(-x + c + \frac{by^4}{K + y^4}\right),$$

must globally converge to one of two stable states (and there is also a saddle-point unstable steady state). This is not true, however. For example, picking these parameters: $c = 0.8$, $b = 50/14$, $K = 405/14$, one sees that generic trajectories are relaxation-like oscillations (see figure 7.8 for a plot with initial conditions $x(0) = 1$, $y(0) = 2$ and figure 7.9 for the phase plane associated to this system, which shows two unstable spirals in heteroclinic connection with a saddle, as well as a limit cycle). The counterexample illustrates that finding conditions for validity of the QSSRP is nontrivial. This leads us to the study of monotone systems (see also Angeli, 2007, for yet another class of systems to which QSSRP applies).

7.5 Monotone Input-Output Components

Monotone input-output systems generalize monotone dynamical systems (with no external inputs nor outputs) as introduced by Morris Hirsch (1983) and further developed by many, notably Hal Smith (Hirsch and Smith, 2005; Smith, 1995). A *monotone input-output system* is one for which trajectories preserve partial orders on states, inputs, and outputs. Given partial orders in spaces of input and output values and states (species), a monotone system satisfies the following axiom: for any two input signals $\boldsymbol{u}$ and $\boldsymbol{v}$ for which $\boldsymbol{u}(t) \leq \boldsymbol{v}(t)$ for all times t, and for any two states $\boldsymbol{x}$, $\boldsymbol{z}$ such that $\boldsymbol{x} \leq \boldsymbol{z}$, it follows that $\varphi(t, \boldsymbol{x}, \boldsymbol{u}(\cdot)) \leq \varphi(t, \boldsymbol{z}, \boldsymbol{v}(\cdot))$ for all t, where $\varphi(t, \boldsymbol{x}, \boldsymbol{u}(\cdot))$ is the solution at time t if the initial state is $\boldsymbol{x}$ and the input is $\boldsymbol{u}$, and similarly for $\varphi(t, \boldsymbol{z}, \boldsymbol{v}(\cdot))$. (The "$\leq$" signs must be interpreted as referring to the respective orders.) The output mapping $\boldsymbol{h}$ should likewise preserve orders. For example, suppose that the ordering picked for states is the "northeast" (NE) order, in which $\boldsymbol{x} = (x_1, x_2) \leq \boldsymbol{z} = (z_1, z_2)$ is defined by the requirement that both $x_1 \leq z_1$ and $x_2 \leq z_2$, as shown in figure 7.10. Monotonicity strongly constrains dynamics. As an extremely simple illustration of these constraints, let us show that no periodic orbits can exist in a system that is two-dimensional ($n = 2$), under the above NE order. (There are no inputs $\boldsymbol{u}$ in this example.) Our proof by contradiction is as follows. Suppose that there would exist some counterclockwise trajectory (the argument is similar in the clockwise case). Suppose that, on this trajectory, two initial conditions $\boldsymbol{x}(0) < \boldsymbol{z}(0)$ are chosen, as shown in figure 7.11. There is some time $T > 0$ such that $\boldsymbol{x}(T) =$

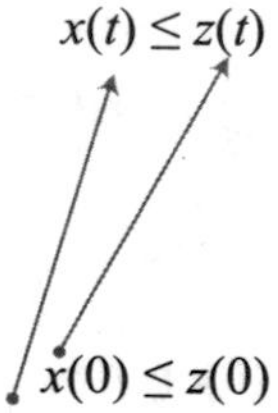

Figure 7.10
Monotonicity.

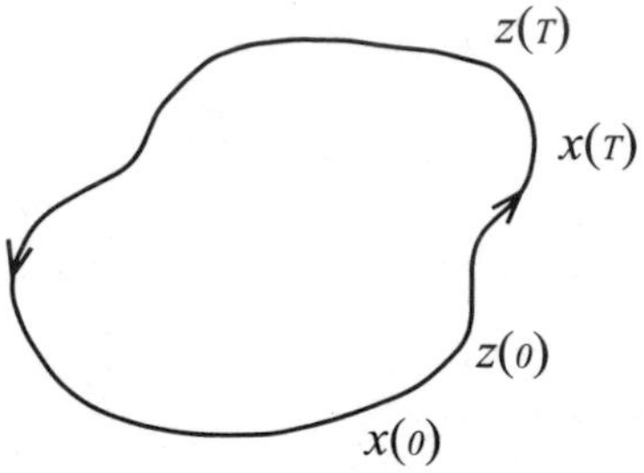

Figure 7.11
No periodic orbits.

$\varphi(t, \boldsymbol{x}(0))$ has a maximal x_1-coordinate, as shown. Since (under standard regularity assumptions) solutions may not cross, the state $z(T) = \varphi(T, z(0))$ fails to satisfy $\boldsymbol{x}(T) \leq z(T)$, contradicting the monotonicity assumption.

More generally (not merely for two-dimensional systems and the NE order), monotone systems have "low dynamical complexity" and, in that sense, constitute a good class of elementary components for a decomposition approach. They behave in many ways like one-dimensional systems, in that, for constant inputs, no "chaotic attractors" (or even stable oscillations) can occur and, generically, (bounded) solutions converge to steady states as $t \to \infty$. More precisely, these statements are true under an additional technical condition of irreducibility (strong monotonicity), which is often satisfied. A precise version is given by one of the fundamental results in the field, Hirsch's generic convergence theorem (Hirsch, 1983; Hirsch and Smith, 2005; Smith, 1995).

Systems that are in an appropriate sense "close" to monotone share some of these global properties. For example, if nonmonotone behavior occurs at a faster time scale, generic convergence to steady states is still valid (see the singular perturbation result in Wang and Sontag, 2008). In that sense, fast regulatory negative feedback loops do not affect regularity of behavior. Monotonicity (or even closeness to monotonicity) may be far too strong a requirement when analyzing large systems, but it is useful as a constraint on *components*, in an interconnection approach, because the quasi-steady-state reduction principle does indeed hold for several feedback configurations involving monotone systems, as discussed below.

An important subclass of monotone systems is that of *orthant-monotone systems*, defined mathematically as monotone with respect to a conic order, where the cone is an orthant. A far more concrete and transparent, but equivalent, definition is as follows. Associate to each system a *species graph* in which there are $n + m + p$ nodes v_i, one per species, input, and output variable. If v_i and v_j are vertices corresponding to state variables, we draw an edge from v_j to v_i (only when $i \neq j$) if $\partial f_i / \partial x_j(x, u) \not\equiv 0$. If v_j is associated to an input variable and v_i to a state variable, we draw an edge

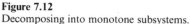

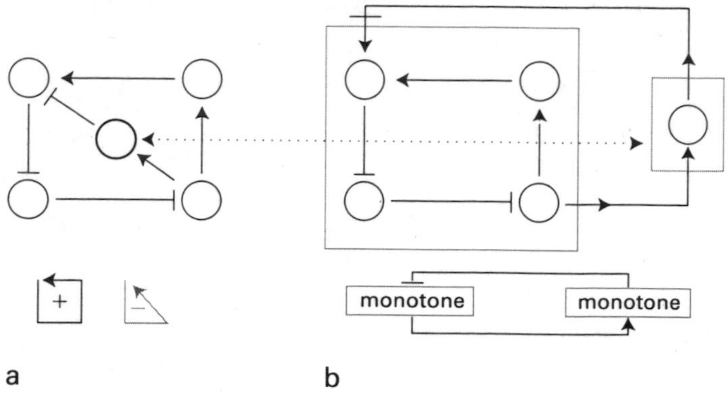

Figure 7.12
Decomposing into monotone subsystems.

from v_j to v_i if $\partial f_i/\partial u_j(x,u) \not\equiv 0$ (where u_j is the jth coordinate of the input). Finally, if v_i is associated to an output variable and v_j to a state variable, we draw an edge from v_j to v_i if $\partial h_i/\partial x_j(x,u) \not\equiv 0$. We label edges as positive or negative if $\partial f_i/\partial x_j(x,u) \geq 0$ for all (x,u) or $\partial f_i/\partial x_j(x,u) \leq 0$ for all (x,u), respectively, (and likewise for edges involving inputs or outputs). If the sign is ambiguous, that is, if $\partial f_i/\partial x_j(x,u) > 0$ for some (x,u) and also $\partial f_i/\partial x_j(x,u) < 0$ for some (x,u) (and similarly for edges involving inputs or outputs), we label the edge with an "ambiguous" sign. We will say that a system has "well-defined signs of interactions" if there are no ambiguous edges; this is often the case with biochemical models. An orthant-monotone system is one in which every undirected cycle (that is, every cycle in which the direction of arrows is ignored) has a net positive parity, meaning that there are no ambiguous labels in the path, and the product of the labels is positive.

It is easy to see that, if a system has well-defined signs of interactions, then it can be thought of as an interconnection of monotone components. This simple idea is illustrated in figure 7.12a, which shows a system that fails the positive loop test, for example, because the triangular path shown at the bottom has three negative edges. We may, however, remove the diagonal vertex and incident edges, and think of the system as an interconnection, using negative (inhibitory) feedback, of two monotone subsystems (figure 7.12b).

Decompositions into monotone components are particularly useful if one can decompose a system of interest into a small number of such components. The minimal such number is the solution of an integer programming problem associated to the system species graph (labeled "max-cut problem"), which is also related to the question of "balancing" in signed graphs and to the "degree of frustration" of Ising spin-glass models (see Sontag, 2007, for details). It is noteworthy that some gene regulatory networks can be decomposed into a smaller number of monotone components than would be expected from random graphs with the same characteristics (Sontag, 2007).

Two Quasi-Steady-State Reduction Principle Theorems

There are several theorems that validate the QSSRP for interconnections of monotone systems with well-defined characteristics. The basic theorem for positive feedback analyzes an interconnection of two systems

$$\dot{x}_1 = f_1(x_1, u_1), \qquad y_1 = h_1(x_1),$$

$$\dot{x}_2 = f_2(x_2, u_2), \qquad y_2 = h_2(x_2),$$

each of which has an increasing characteristic, denoted by k and by g, respectively. (A special case occurs when one of the systems is memoryless, for example, if there are no state variables x_1 and y_1 is simply a static function $y_1(t) = k(u_1(t))$.) The "positive feedback interconnection" of these two systems is formally defined by letting the output of each of them serve as the input of the other ($u_2 = y_1 = y$ and $u_1 = y_2 = u$), For simplicity of exposition, we restrict here to systems with scalar inputs and outputs (see Enciso and Sontag, 2005, for a generalization to vector inputs and outputs). As in the discussion of the QSSRP, we plot the graphs of k and g^{-1} together. It is quite obvious that there is a bijective correspondence between the steady states of the feedback system and the intersection points of the two graphs. Moreover, just as in the general discussion, let us attach labels to the intersection points between the two graphs as follows: a label "S" is placed at those points at which the slope of k is smaller than the slope of g^{-1}, and a label "U" if the slope of k is larger than the slope of g^{-1}. (We assume that the graphs do not intersect tangentially.) Under mild nondegeneracy technical conditions (transversality and notions related to controllability and observability in control theory), one can conclude that "almost all" (in a measure-theoretic or a Baire-category sense) bounded solutions of the feedback system must converge to one of the steady states corresponding to intersection points labeled with an S. This theorem, which instantiates the QSSRP for feedback loops of two monotone systems with scalar inputs and outputs, is proved by Angeli and Sontag (2004; see also Enciso and Sontag, 2008, for additional work, which weakens the assumed technical conditions and extends the result to vector input-output signals).

We now turn to negative feedback. One mathematical way to define negative feedback in the context of monotone systems is to say that the orders on inputs and outputs are "inverted" (for example, an inhibition term of the form $\frac{V}{K+y}$, as usual in biochemistry). Equivalently, and more conveniently, we may incorporate the inhibition into the output of the second system, which is then seen as an "antimonotone" input-output system, and this is how we proceed from here on. We emphasize that the closed-loop system that results is generally *not* monotone.

The basic theorem, proved by Angeli and Sontag (2003), is as follows, still assuming that inputs and outputs are scalar (Enciso and Sontag, 2006, generalizes these results). We once again plot together k and g^{-1}, and consider the discrete iteration $u_{i+1} = (g \circ k)(u_i)$. The theorem states that, provided that solutions of the closed-loop system are bounded, if this iteration has a globally attractive fixed point $\bar{u}$, then the feedback system has a globally attracting steady state. (An equivalent condition, as shown in Enciso and Sontag, 2006, is that the iteration have no nontrivial period-two orbits.) Note that, for negative feedback loops involving systems with scalar inputs and outputs, there is never more than one intersection of the plots, since k is increasing and g^{-1} is decreasing; thus the QSSRP has been shown to be valid in this case.

7.6 Discussion

Several mathematical examples of applications of the quasi-steady-state reduction principle theorems discussed can be found in published papers, and many of them are surveyed by Sontag (2007), including mathematical models of MAPK cascades, blood testosterone levels, and the *Lac* operon system. From an experimental perspective, the QSSRP has been recently validated using tools from synthetic biology: in the 2007 International Genetically Engineered Machines competition, Thattai's group project (Rai et al., 2008; Thattai, 2007) showed that one recovers the closed-loop behavior from the intersections of characteristics for a genetically engineered system constructed for that purpose. Many theoretical questions remain open, among them the formulation of precise theorems that instantiate the QSSRP for general networks (not only feedback loops).

8 A Control-Theoretic Interpretation of Metabolic Control Analysis

Brian P. Ingalls

In this chapter, the main results of metabolic control analysis (MCA) are reinterpreted from the point of view of engineering control theory. To begin, the standard model of metabolic systems is identified as redundant in both state dynamics and input effects. A key feature of these systems is that, whereas the dynamics are typically nonlinear, these redundancies appear linearly, through the stoichiometry matrix. This means that the effect of the input can be linearly decomposed into a component driving the state and a component driving the output. A statement of this separation principle is shown to be equivalent to the main theorems of MCA. Presenting a control-theoretic treatment of stoichiometric systems, the chapter arrives at an alternative derivation of some of the fundamental results in the theory of control of biochemical systems.

8.1 Background

Biochemical mechanisms for implementation of feedback control were first discovered in the biosynthetic pathways of metabolism (Pardee and Reddy, 2003), and it was within the study of metabolism that a quantitative theory of the control and regulation of biochemical networks was first developed. In the 1970s, researchers on both sides of the Atlantic, led by Michael Savageau in the United States and by Henrik Kacser and Reinhart Heinrich in Europe, elucidated theoretical frameworks for addressing issues of regulation in metabolic networks. A fundamental tool used by both groups was local parametric sensitivity analysis, applied primarily at steady state. The European camp, whose theory was dubbed metabolic control analysis (MCA), or sometimes metabolic control theory (MCT), made use of a standard linearization technique in addressing steady state behavior (Heinrich and Rapoport, 1974a,b; Kacser and Burns, 1973). Savageau's work, known as biochemical systems theory (BST), makes use of a more sophisticated log linearization that provides an improved approximation of nonlinear dynamics (Savageau, 1976). With respect to local parametric sensitivity analysis, the two approaches yield identical results.

The analysis in the present chapter follows the linearization method used in metabolic control analysis, which provides a direct connection between these biochemical studies and the general theory of local parametric sensitivity analysis. Moreover, linearization leaves intact the stoichiometric relationships that are exploited in studies of these networks. Indeed, as will be shown below, it is this stoichiometric nature that distinguishes the mathematics of metabolic control analysis from that of standard sensitivity analysis. As first shown by Reder (1988), an application of some basic linear algebra provides an extension of sensitivity analysis that captures the features of stoichiometry. Beyond these mathematical underpinnings, the field of metabolic control analysis deals with myriad intricacies of application to biochemical networks that demand careful interpretation of experimental and theoretical results (surveyed in Fell, 1992, 1997; Heinrich and Schuster, 1996).

Local parametric sensitivity analysis addresses the behavior of dynamical systems under small perturbations in system parameters. Such analysis plays an important role in control theory, and several texts on sensitivity analysis have been written with control applications in mind (see, for example, Frank, 1978; Rosenwasser and Yusupov, 2000; Tomović, 1963; and Varma et al., 1999).

The analysis in this chapter is based on the standard ordinary differential equation–based description of biochemical systems (chapter 1) in which the states are the concentrations of the chemical species involved in the network and the inputs are parameters influencing the reaction rates. In addressing metabolic systems, researchers commonly take enzyme activity as the parameter input. This choice of input channel typically results in an overactuated system—with more inputs than states. Additionally, the reaction rates are important outputs. Because they depend directly on the parameter inputs, these rates enjoy some autonomy from the state dynamics and can, to a degree, be manipulated separately.

The discussion that follows highlights a procedure for making explicit the separation between manipulating metabolite concentrations, on the one hand, and reaction rates, on the other, which complements investigations of metabolic "redesign" that have appeared in the literature (Dean and Dervakos, 1998; Hatzimanikatis et al., 1996; Torres and Voit, 2002). Within the metabolic control analysis community, a significant step in this direction was taken by Kacser and Acerenza (1993), who described a "universal method" for altering pathway flux. Later, the goal of increasing specific metabolite concentrations was taken up by Kacser and Small (1994). A local description of the combined problem was given by Westerhoff and Kell (1996). These results can all be seen as contained within the "metabolic design" approach described by Kholodenko et al. (1998, 2000). In the sections that follow, equivalent results are derived from a control-engineering viewpoint, culminating in a control-theoretic interpretation of the main results of metabolic control analysis: the summation and connectivity theorems.

8.2 Redundancy in Control Engineering

The results presented here are a consequence of redundancies that appear in stoichiometric systems. Before addressing these, let us briefly review the standard manner in which such redundancies are treated in control engineering.

8.2.1 State Redundancy: Nonminimal Realizations

Recall from chapter 1 the standard description of a linear, time-invariant system:

$$\frac{\mathrm{d}}{\mathrm{d}t}\boldsymbol{x}(t) = \boldsymbol{A}\boldsymbol{x}(t) + \boldsymbol{B}\boldsymbol{u}(t), \tag{8.1a}$$

$$\boldsymbol{y}(t) = \boldsymbol{C}\boldsymbol{x}(t) + \boldsymbol{D}\boldsymbol{u}(t), \tag{8.1b}$$

where $\boldsymbol{x} \in \mathbf{R}^{n_0}$, $\boldsymbol{u} \in \mathbf{R}^{m_0}$, $\boldsymbol{y} \in \mathbf{R}^{p_0}$ and $\boldsymbol{A}$, $\boldsymbol{B}$, $\boldsymbol{C}$ and $\boldsymbol{D}$ are constant matrices of the appropriate dimensions. In systems theory, one is often interested primarily in the input-output behavior associated with this system, characterized by the output trajectories that arise from various choices of the input $\boldsymbol{u}(\cdot)$ with initial condition $\boldsymbol{x}(0) = \mathbf{0}$. Given a particular system of the form (8.1), the associated input-output behavior can be equally generated from a whole class of systems of this form. That is, the representation, or *realization*, of these input-output behaviors is not unique.

A realization is said to be *minimal* if there are no alternative systems of smaller order that represent the same behavior. Nonminimal realizations exhibit redundancy (typically due to a symmetry or to decoupled behavior); they can be improved by removal of the redundant components. A simple instance of nonminimality is when there is a redundancy among the state variables, regardless of the input or output structure. Biochemical systems typically exhibit such simple redundancies, as will be seen in section 8.3.

8.2.2 Input Redundancy: Overactuation

In control engineering, much effort has gone into the analysis of system (8.1) in the *underactuated* case ($n_0 > m_0$), where one attempts to manipulate a system for which there are fewer input channels than degrees of freedom. In the case that the number of input channels equals the number of degrees of freedom ($n_0 = m_0$) the system is *fully actuated*, and much of that analysis is trivial. Finally, if $n_0 < m_0$, the system is *overactuated*, in which case a redundancy in the control inputs presents an embarrassment of riches to the control designer; the state dynamics can be controlled without completely specifying the input. The additional degrees of freedom in the input can then be used to meet further performance criteria (Härkegård and Glad, 2005).

In the overactuated case, system (8.1) can be treated as follows. For simplicity, take the case that $\boldsymbol{B}$ has rank n_0 (so there are exactly $n_0 - m_0$ redundancies among the inputs). Because $\boldsymbol{B}$ does not have full column rank, it can be factored as $\boldsymbol{B} = \boldsymbol{B}_0\boldsymbol{B}_1$, where $\boldsymbol{B}_0$ is $n_0 \times n_0$ and has full rank, while $\boldsymbol{B}_1$ is $n_0 \times m_0$ and has rank n_0. The control input $\boldsymbol{u}$ can them be mapped to a *virtual control input* $\tilde{\boldsymbol{u}} \in \mathbf{R}^{n_0}$ by $\tilde{\boldsymbol{u}} = \boldsymbol{B}_1\boldsymbol{u}$, resulting in the fully actuated system

$$\frac{\mathrm{d}}{\mathrm{d}t}\boldsymbol{x}(t) = \boldsymbol{A}\boldsymbol{x}(t) + \boldsymbol{B}_0\tilde{\boldsymbol{u}}(t),$$

where two different control inputs $\boldsymbol{u}_1$ and $\boldsymbol{u}_2$ whose difference lies in the nullspace of $\boldsymbol{B}_1$ (and hence of $\boldsymbol{B}$) have an identical effect on the state dynamics because they give rise to the same virtual input $\tilde{\boldsymbol{u}}$.

This redundancy can be made explicit by writing $\boldsymbol{u}$ as the sum of two terms that lie inside and outside of the nullspace of $\boldsymbol{B}$, respectively:

$$\boldsymbol{u}(t) = \boldsymbol{K}\boldsymbol{a}_1(t) + \boldsymbol{M}\boldsymbol{a}_2(t),$$

where the columns of matrix $\boldsymbol{K}$ form a basis for the nullspace of $\boldsymbol{B}$ and the columns of $\boldsymbol{M}$ are linearly independent of one another and of the columns of $\boldsymbol{K}$. Through this decomposition, the state dynamics can be manipulated by the choice of $\boldsymbol{a}_2(\cdot)$, while $\boldsymbol{a}_1(\cdot)$ can be chosen to satisfy other design criteria. In particular, if the system output involves a feedthrough term (that is, $\boldsymbol{D}$ in system (8.1) is nonzero) then the choice of $\boldsymbol{a}_1$ may reveal itself in the output. Stoichiometric systems, as defined in the next section, have this property, allowing the separate design of strategies for controlling state and output behavior.

8.3 Stoichiometric Systems

Consider n chemical species involved in m reactions in a fixed volume. The concentrations of the species make up the n-dimensional vector $\boldsymbol{s}$. The rates of the reactions are the elements of the m-vector $\boldsymbol{v}$. These rates depend on the species concentrations and on a set of parameter inputs that are collected into vector $\boldsymbol{p}$. The network topology is described by the $n \times m$ *stoichiometry matrix* $\boldsymbol{N}$, whose i, jth element indicates the net number of molecules of species i produced in reaction j (negative values indicate consumption).

The system dynamics are described by

$$\frac{\mathrm{d}}{\mathrm{d}t}\boldsymbol{s}(t) = \boldsymbol{N}\boldsymbol{v}(\boldsymbol{s}(t), \boldsymbol{p}(t)), \qquad \text{for all } t \geq 0. \tag{8.2}$$

In addition to the state, $\boldsymbol{s}(\cdot)$, and the input $\boldsymbol{p}(\cdot)$, the variables of primary interest in this system are the reaction rates $\boldsymbol{v}(\boldsymbol{s}, \boldsymbol{p})$. Thus, in interpreting (8.2) as a control system, we will choose the vector of reaction rates as the system output:

$$\boldsymbol{y}(\boldsymbol{s},\boldsymbol{p}) = \boldsymbol{v}(\boldsymbol{s},\boldsymbol{p}). \tag{8.3}$$

Systems of the form of equations (8.2) and (8.3) can be defined as *stoichiometric systems* precisely because the reaction rates $\boldsymbol{v}$ in (8.2) are outputs of interest.

As will be shown below, the structure of the stoichiometry matrix can be exploited to yield insights into the behavior of the concentration and reaction rate variables. The key to exploiting the stoichiometric structure of (8.2) is to describe how dependencies among the rows and columns of $\boldsymbol{N}$ have consequences for the input-output behavior of the system.

Linearly dependent rows within the stoichiometry matrix correspond to integrals of motion of the system: quantities that do not change with time. Each redundant row identifies a chemical species whose dynamics are completely determined by the behavior of other species in the system. Biochemically, such structural constraints most often appear as *conserved moieties*, where the concentration of some species is a function of the concentration of others due to a chemical conservation. (A simple example is a system that models the interconversion of two chemical species A and B, but does not incorporate the production or consumption of either species. In this case, the total concentration $[A] + [B]$ is conserved.) An extensive theory has been developed to determine preferred conservation relations from algebraic descriptions of the system network (section 3.1 in Heinrich and Schuster, 1996).

The consequences of linear dependence among the columns of $\boldsymbol{N}$ will be explored below. If the stoichiometry matrix has full column rank then steady state can only be attained when $\boldsymbol{v}(\boldsymbol{s},\boldsymbol{p}) = \boldsymbol{0}$. Biochemical systems typically admit steady states in which there is a nonzero flux through the network. These correspond to reaction rate vectors $\boldsymbol{v}$ that lie in the nullspace of $\boldsymbol{N}$. The dimension of this nullspace determines the number of degrees of freedom in these steady-state reaction profiles.

8.4 Rank Deficiencies

Networks that describe metabolic systems often have highly redundant stoichiometries. As an example, consider a metabolic map from *Escherichia coli* published by Reed et al. (2003) that has a 770×931 stoichiometry matrix of rank 733. Clearly, in attempting an analysis of such a system, it is worthwhile to begin with a reduction afforded by linear dependence.

8.4.1 Deficiencies in Row Rank

As mentioned, structural conservations in the reaction network reveal themselves as linear dependencies among the rows of the stoichiometry matrix $\boldsymbol{N}$. Let r denote the rank of $\boldsymbol{N}$. Following Reder (1988), we relabel the species so that the first r rows of $\boldsymbol{N}$ are independent. The species concentration vector can then be partitioned as

$$s = \begin{bmatrix} s_i \\ s_d \end{bmatrix},$$

where $s_i \in \mathbf{R}^r$ is the vector of *independent species* and $s_d \in \mathbf{R}^{n-r}$ contains the *dependent species*. Next, we partition N into two submatrices. Calling the first r rows N_R, we can write $N = LN_R$, where the matrix L, referred to as the *row link matrix*, has the form

$$L = \begin{bmatrix} I_r \\ L_0 \end{bmatrix}.$$

System (8.2) can then be written as

$$\frac{\mathrm{d}}{\mathrm{d}t}\begin{bmatrix} s_i(t) \\ s_d(t) \end{bmatrix} = \begin{bmatrix} I_r \\ L_0 \end{bmatrix} N_R v(s(t), p(t)).$$

It follows that

$$\frac{\mathrm{d}}{\mathrm{d}t} s_d(t) = L_0 \frac{\mathrm{d}}{\mathrm{d}t} s_i(t), \qquad \text{for all } t \geq 0.$$

Integrating gives $s_d(t) = L_0 s_i(t) + \tilde{T}$, for all time, where $\tilde{T} = s_d(0) - L_0 s_i(0)$. Finally, concatenating $\tilde{T}$ with $\mathbf{0}_r \in \mathbf{R}^r$, we define $T = [\mathbf{0}_r^T, \tilde{T}^T]^T$, and write

$$s(t) = L s_i(t) + T. \tag{8.4}$$

As a consequence of this decomposition, attention can be restricted to a reduced version of (8.2), namely,

$$\frac{\mathrm{d}}{\mathrm{d}t} s_i(t) = N_R v(L s_i(t) + T, p(t)). \tag{8.5}$$

It follows that the n-dimensional state enjoys only r degrees of freedom, because the $n - r$ dependent species are fixed by the behavior of the r independent species. From an input-output perspective, we conclude that, provided $r < n$, the original description in terms of n state variables is a nonminimal realization of the system's input-output behavior, regardless of the form of the reaction rates.

8.4.2 Deficiencies in Column Rank

Recalling that r denotes the rank of the stoichiometry matrix N, we relabel the reactions so that the first $m - r$ columns of N are linearly dependent on the remaining r. We partition the vector of reaction rates v correspondingly into $m - r$ independent (v_i) and r dependent (v_d) rates as

$$v = \begin{bmatrix} v_i \\ v_d \end{bmatrix}.$$

Following the procedure outlined above, one might hope to reach a reduced description of the system dynamics in which some of these reaction rates are eliminated, but this is an impossible task. Such an elimination could, for instance, decouple an input channel from the dynamics.

As with the construction of the row link matrix, we let $\boldsymbol{N}^C$ denote the submatrix of $\boldsymbol{N}$ consisting of the last r columns, from which $\boldsymbol{N}$ can be recovered as $\boldsymbol{N} = \boldsymbol{N}^C\boldsymbol{P}$, where the *column link matrix* $\boldsymbol{P}$ is of the form

$$\boldsymbol{P} = [\boldsymbol{P}_0 \quad \boldsymbol{I}_r].$$

The column link matrix can be determined by constructing a matrix of the form

$$\boldsymbol{K} = \begin{bmatrix} \boldsymbol{I}_{m-r} \\ -\boldsymbol{P}_0 \end{bmatrix},$$

whose columns span the nullspace of $\boldsymbol{N}$, forming a basis for the nullspace of $\boldsymbol{P}$, and hence of $\boldsymbol{N}$. To realize an alternative system description, we write

$$\begin{aligned} \frac{\mathrm{d}}{\mathrm{d}t}\boldsymbol{s}(t) &= \boldsymbol{N}\boldsymbol{v}(\boldsymbol{s}(t), \boldsymbol{p}(t)) \\ &= \boldsymbol{N}^C\boldsymbol{P}\boldsymbol{v}(\boldsymbol{s}(t), \boldsymbol{p}(t)) \\ &= \boldsymbol{N}^C[\boldsymbol{P}_0 \quad \boldsymbol{I}_r]\boldsymbol{v}(\boldsymbol{s}(t), \boldsymbol{p}(t)). \end{aligned} \tag{8.6}$$

At steady state, this factored description reveals a dependence among the reaction rates. Denoting the steady-state rate vector by $\boldsymbol{v}^{ss} = \boldsymbol{J}$ (for system flux), we have a partitioning of $\boldsymbol{J}$ into dependent and independent components:

$$\boldsymbol{J} = \begin{bmatrix} \boldsymbol{J}_i \\ \boldsymbol{J}_d \end{bmatrix}.$$

From equation (8.6), steady state occurs when $\boldsymbol{J}_d = -\boldsymbol{P}_0\boldsymbol{J}_i$. As described by, for example, Heinrich and Schuster (1996), this steady-state dependence can be written as

$$\boldsymbol{J} = \boldsymbol{K}\boldsymbol{J}_i = \begin{bmatrix} \boldsymbol{I}_{m-r} \\ -\boldsymbol{P}_0 \end{bmatrix}\boldsymbol{J}_i. \tag{8.7}$$

Note that Heinrich and Schuster (1996) refer to the submatrix $-\boldsymbol{P}_0$ as $\boldsymbol{K}_0$. The notation proposed here is dual to the notation used in addressing row redundancy.

The partitioning of reaction rates is nonunique, and the advantages of one choice over another are not addressed here. A straightforward procedure for choosing independent reaction rates as the "entry" and "exit" points from the network is outlined by Westerhoff et al. (1994).

8.4.3 Complete Reduction

The two types of dependence described above lead to complementary system decompositions. Reducing the system by eliminating redundancies in rows *and* columns leads to an alternative description of the dynamics:

$$\begin{aligned}\frac{d}{dt}\boldsymbol{s}(t) &= \boldsymbol{L}\boldsymbol{N}_{\boldsymbol{R}}^{\boldsymbol{C}}\boldsymbol{P}\boldsymbol{v}(\boldsymbol{s}(t),\boldsymbol{p}(t)) \\ &= \begin{bmatrix}\boldsymbol{I}_r \\ \boldsymbol{L}_0\end{bmatrix}\boldsymbol{N}_{\boldsymbol{R}}^{\boldsymbol{C}}[\boldsymbol{P}_0 \quad \boldsymbol{I}_r]\boldsymbol{v}(\boldsymbol{s}(t),\boldsymbol{p}(t)),\end{aligned} \tag{8.8}$$

where the factored form of the original $n \times m$ stoichiometry matrix involves the invertible $\boldsymbol{N}_{\boldsymbol{R}}^{\boldsymbol{C}}$, defined as the upper right $r \times r$ submatrix of $\boldsymbol{N}$.

8.5 Overactuation

We now consider the consequence of these linear dependencies on input-output behavior. To begin with, observe that, if the reaction rates were considered as *inputs* (that is, $\boldsymbol{u} = \boldsymbol{v}$) then, restricting to the nonredundant dynamics, system (8.5) would be an overactuated system of the form (8.1) (with $\boldsymbol{A} = \boldsymbol{0}$, referred to as a *driftless* system). Identifying $\boldsymbol{B}$ with $\boldsymbol{N}_{\boldsymbol{R}}$, $\boldsymbol{B}_0$ with $\boldsymbol{N}_{\boldsymbol{R}}^{\boldsymbol{C}}$ and $\boldsymbol{B}_1$ with $\boldsymbol{P}$, we could define the corresponding virtual input as $\tilde{\boldsymbol{u}} = \boldsymbol{P}\boldsymbol{v}$ and any input satisfying $\boldsymbol{P}\boldsymbol{v} = \boldsymbol{0}$ would have no effect on the state dynamics. Of course, because the reaction rates depend on the species concentrations, they cannot be treated directly as inputs. Nevertheless the behavior resulting from this supposition can be realized from both biochemical and control design viewpoints.

One is often interested in the case where the system inputs (to be manipulated by an experimenter or through inherent regulation) are the activity levels of the enzymes associated with the reactions in the network. In most kinetic models, each reaction rate varies linearly with the activity of the corresponding enzyme, and there is one specific enzyme associated with each reaction. In such cases, we may write for each reaction

$$v_k(\boldsymbol{s},\boldsymbol{p}) = p_k w_k(\boldsymbol{s}),$$

where the function w_k is referred to as the *turnover rate* for reaction k. In this framework, the parameter inputs can be identified directly with the reaction rates in two ways. If one is interested in the effect of *relative* changes in reaction rates, then changes in the input are equivalent to changes in the reaction rate, for example, a 1% change in p_k amounts to a 1% change in v_k. Alternatively, one can follow a standard procedure in control engineering known as *input redefinition* by setting

$$\tilde{u}_k(t) = p_k(t)w_k(s(t)),$$

so the system dynamics become simply

$$\frac{\mathrm{d}}{\mathrm{d}t}s(t) = \boldsymbol{N}\tilde{\boldsymbol{u}}(t).$$

The system overactuation can then be analyzed as follows. Any change in $\tilde{\boldsymbol{u}}$ that lies in the nullspace of the stoichiometry matrix $\boldsymbol{N}$, or equivalently of the column link matrix $\boldsymbol{P}$, will have no effect on state dynamics; the redefined input can be decomposed into a component that lies in the nullspace of $\boldsymbol{P}$ and another that does not, as discussed in section 8.2.2. Recall that the columns of $\boldsymbol{K}$ form a basis for the nullspace of $\boldsymbol{P}$. We take $\boldsymbol{M}$ to be an independent extension of the columns of $\boldsymbol{K}$ to a basis for $\mathbf{R}^m$. Then we can decompose

$$\tilde{\boldsymbol{u}}(t) = \boldsymbol{K}\boldsymbol{a}_v(t) + \boldsymbol{M}\boldsymbol{a}_s(t), \tag{8.9}$$

where $\boldsymbol{a}_v \in \mathbf{R}^{m-r}$ and $\boldsymbol{a}_s \in \mathbf{R}^r$, and where an input $\tilde{\boldsymbol{u}}$ has no effect on the state exactly when $\boldsymbol{a}_s = \mathbf{0}$. Because this holds regardless of the choice of $\boldsymbol{M}$, the question arises as to which form of $\boldsymbol{M}$ will make the decomposition most useful. We will consider two alternatives.

8.5.1 Input Decomposition: General Dynamics

We first take

$$\boldsymbol{M} = \overline{\boldsymbol{M}} = \begin{bmatrix} \mathbf{0}_{(m-r)\times r} \\ (\boldsymbol{N}_{\boldsymbol{R}}^{\boldsymbol{C}})^{-1} \end{bmatrix}.$$

With this decomposition in place, the independent state dynamics take the form

$$\begin{aligned} \frac{\mathrm{d}}{\mathrm{d}t}s_i(t) &= \boldsymbol{N}_{\boldsymbol{R}}\tilde{\boldsymbol{u}}(t) \\ &= \boldsymbol{N}_{\boldsymbol{R}}^{\boldsymbol{C}}\boldsymbol{P}(\boldsymbol{K}\boldsymbol{a}_v(t) + \overline{\boldsymbol{M}}\boldsymbol{a}_s(t)) \\ &= \boldsymbol{N}_{\boldsymbol{R}}^{\boldsymbol{C}}\boldsymbol{P}\overline{\boldsymbol{M}}\boldsymbol{a}_s(t) \end{aligned}$$

$$= N_R^C [P_0 \quad I_r] \begin{bmatrix} \mathbf{0}_{(m-r)\times r} \\ (N_R^C)^{-1} \end{bmatrix} a_s(t)$$

$$= N_R^C (N_R^C)^{-1} a_s(t)$$

$$= a_s(t), \tag{8.10}$$

indicating that concentration dynamics are manipulated directly by the choice of the coefficients of $a_s(\cdot)$.

The dynamics of the reaction rates, though also of interest, do not appear in such a simple form. The decomposition with $\overline{M}$ leads to

$$v(t) = \begin{bmatrix} v_i(t) \\ v_d(t) \end{bmatrix} = \begin{bmatrix} a_v(t) \\ -P_0 a_v(t) + (N_R^C)^{-1} a_s(t) \end{bmatrix}, \tag{8.11}$$

which provides a dynamic generalization of equation (8.7) and confirms that manipulation of the state variables has been decoupled from the *independent* reactions' rates, which can be manipulated directly through the coefficients of $a_v(\cdot)$.

Equations (8.10) and (8.11) indicate that, once outputs are taken into consideration, it is inappropriate to refer to the system as overactuated. Since the number of input channels (m) corresponds exactly to the number of degrees of freedom of the system (r for the independent species dynamics and $m - r$ for the independent reaction rates), the system can be interpreted as fully actuated. An equivalent conclusion can be reached when attention is restricted to local analysis, as we next consider.

8.5.2 Input Decomposition: Local Steady-State Analysis

Fixing a particular parameter input value p^0 and a corresponding steady state s^0, which is assumed asymptotically stable, we can describe the local effect of the input on concentrations and fluxes through a linearization around this steady state. The treatment of local input response is equivalent to a local parametric sensitivity analysis.

For an arbitrary input parameter vector p, if $\frac{\partial v}{\partial p}$ is invertible at the steady state, then changes in the parameter input can be identified with changes in the reaction rates by redefining the input as

$$\tilde{u} = \frac{\partial v}{\partial p}(p - p^0), \tag{8.12}$$

where the derivatives are evaluated at the steady state.

We require that $\frac{\partial v}{\partial p}$ be invertible so that we can recover p from $\tilde{u}$. This redefined input realizes the direct connection between rate and input in a local sense since

$$\frac{\partial \boldsymbol{v}}{\partial \tilde{\boldsymbol{u}}} = \frac{\partial \boldsymbol{v}}{\partial \boldsymbol{p}} \frac{d\boldsymbol{p}}{d\tilde{\boldsymbol{u}}} = \frac{\partial \boldsymbol{v}}{\partial \boldsymbol{p}} \left(\frac{\partial \boldsymbol{v}}{\partial \boldsymbol{p}} \right)^{-1} = \boldsymbol{I}_m.$$

Following the construction above, we decompose this redefined input as $\tilde{\boldsymbol{u}}(t) = \boldsymbol{K}\boldsymbol{a}_v(t) + \boldsymbol{M}\boldsymbol{a}_s(t)$ with

$$\boldsymbol{M} = -\frac{\partial \boldsymbol{v}}{\partial \boldsymbol{s}} \boldsymbol{L},$$

where the derivatives are evaluated at the nominal steady state $\boldsymbol{s}^0$. (The negative sign is chosen to follow convention.) The independence of $\boldsymbol{K}$ and $\boldsymbol{M}$ follows from the fact that $\boldsymbol{N_R} \frac{\partial \boldsymbol{v}}{\partial \boldsymbol{s}} \boldsymbol{L}$ is the Jacobian of system (8.5) at $\boldsymbol{s}^0$, which is invertible by the assumption of asymptotic stability.

Now, to consider the steady-state response of the system to changes in the redefined input $\tilde{\boldsymbol{u}}$, we note that, at steady state:

$$\boldsymbol{0} = \boldsymbol{N_R}\boldsymbol{v}(\boldsymbol{L}\boldsymbol{s}_i + \boldsymbol{T}, \boldsymbol{p}). \tag{8.13}$$

Under the assumption of asymptotic stability, equation (8.13) yields a local implicit description of $\boldsymbol{s}_i$ as a function of $\boldsymbol{p}$, with $\boldsymbol{s}_i(\boldsymbol{p}^0) = \boldsymbol{s}_i^0$. To determine the effect of small parameter changes on this steady state $\boldsymbol{s}_i(\boldsymbol{p})$, we differentiate equation (8.13) with respect to $\boldsymbol{p}$ at the nominal point to yield

$$\boldsymbol{0} = \boldsymbol{N_R}\left(\frac{\partial \boldsymbol{v}}{\partial \boldsymbol{s}} \boldsymbol{L} \frac{\mathrm{d}\boldsymbol{s}_i}{\mathrm{d}\boldsymbol{p}} + \frac{\partial \boldsymbol{v}}{\partial \boldsymbol{p}} \right).$$

Solving, we arrive at

$$\frac{\mathrm{d}\boldsymbol{s}_i}{\mathrm{d}\boldsymbol{p}} = -\left(\boldsymbol{N_R} \frac{\partial \boldsymbol{v}}{\partial \boldsymbol{s}} \boldsymbol{L} \right)^{-1} \boldsymbol{N_R} \frac{\partial \boldsymbol{v}}{\partial \boldsymbol{p}}, \tag{8.14}$$

which is a vector comprising local sensitivity coefficients. To determine the effect of the redefined input on the state, we compute

$$\begin{aligned} \frac{\mathrm{d}\boldsymbol{s}_i}{\mathrm{d}\tilde{\boldsymbol{u}}} &= \frac{\mathrm{d}\boldsymbol{s}_i}{\mathrm{d}\boldsymbol{p}} \frac{\mathrm{d}\boldsymbol{p}}{\mathrm{d}\tilde{\boldsymbol{u}}} \\ &= -\left(\boldsymbol{N_R} \frac{\partial \boldsymbol{v}}{\partial \boldsymbol{s}} \boldsymbol{L} \right)^{-1} \boldsymbol{N_R} \frac{\partial \boldsymbol{v}}{\partial \boldsymbol{p}} \left(\frac{\partial \boldsymbol{v}}{\partial \boldsymbol{p}} \right)^{-1} \\ &= -\left(\boldsymbol{N_R} \frac{\partial \boldsymbol{v}}{\partial \boldsymbol{s}} \boldsymbol{L} \right)^{-1} \boldsymbol{N_R}. \end{aligned} \tag{8.15}$$

Considering the effect of the individual components of the decomposed form of $\tilde{\boldsymbol{u}}$ (that is, $\tilde{\boldsymbol{u}} = \boldsymbol{K}\boldsymbol{a}_v - \frac{\partial \boldsymbol{v}}{\partial \boldsymbol{u}} \boldsymbol{L}\boldsymbol{a}_s$), we arrive at

$$\begin{aligned}\frac{\partial \boldsymbol{s}_i}{\partial \boldsymbol{a}_s} &= \frac{\mathrm{d}\boldsymbol{s}_i}{\mathrm{d}\tilde{\boldsymbol{u}}}\frac{\partial \tilde{\boldsymbol{u}}}{\partial \boldsymbol{a}_s} \\ &= -\left(\boldsymbol{N_R}\frac{\partial \boldsymbol{v}}{\partial \boldsymbol{s}}\boldsymbol{L}\right)^{-1}\boldsymbol{N_R}\left(-\frac{\partial \boldsymbol{v}}{\partial \boldsymbol{s}}\boldsymbol{L}\right) \\ &= \boldsymbol{I}_r,\end{aligned}$$

indicating a direct effect on the independent species concentrations, while

$$\begin{aligned}\frac{\partial \boldsymbol{s}_i}{\partial \boldsymbol{a}_v} &= \frac{\mathrm{d}\boldsymbol{s}_i}{\mathrm{d}\tilde{\boldsymbol{u}}}\frac{\partial \tilde{\boldsymbol{u}}}{\partial \boldsymbol{a}_v} \\ &= -\left(\boldsymbol{N_R}\frac{\partial \boldsymbol{v}}{\partial \boldsymbol{s}}\boldsymbol{L}\right)^{-1}\boldsymbol{N_R}\boldsymbol{K} \\ &= \boldsymbol{0}_{r\times(m-r)},\end{aligned}$$

so that changes in $\boldsymbol{a}_v$ have no effect on the state, as expected from the construction in section 8.5. Extending to the complete species vector gives

$$\frac{\partial \boldsymbol{s}}{\partial \boldsymbol{a}_s} = \boldsymbol{L} \quad \text{and} \quad \frac{\partial \boldsymbol{s}}{\partial \boldsymbol{a}_v} = \boldsymbol{0}_{n\times(m-r)}.$$

Considering the steady-state response in the flux $\boldsymbol{J}$, we find

$$\begin{aligned}\frac{\mathrm{d}\boldsymbol{J}}{\mathrm{d}\tilde{\boldsymbol{u}}} &= \frac{\partial \boldsymbol{v}}{\partial \boldsymbol{s}}\frac{\mathrm{d}\boldsymbol{s}}{\mathrm{d}\tilde{\boldsymbol{u}}} + \frac{\partial \boldsymbol{v}}{\partial \boldsymbol{p}}\frac{\mathrm{d}\boldsymbol{p}}{\mathrm{d}\tilde{\boldsymbol{u}}} \\ &= \frac{\partial \boldsymbol{v}}{\partial \boldsymbol{s}}\boldsymbol{L}\frac{\mathrm{d}\boldsymbol{s}_i}{\mathrm{d}\tilde{\boldsymbol{u}}} + \frac{\partial \boldsymbol{v}}{\partial \boldsymbol{p}}\left(\frac{\partial \boldsymbol{v}}{\partial \boldsymbol{p}}\right)^{-1} \\ &= -\frac{\partial \boldsymbol{v}}{\partial \boldsymbol{s}}\boldsymbol{L}\left(\boldsymbol{N_R}\frac{\partial \boldsymbol{v}}{\partial \boldsymbol{s}}\boldsymbol{L}\right)^{-1}\boldsymbol{N_R} + \boldsymbol{I}_m. \end{aligned} \tag{8.16}$$

Then

$$\begin{aligned}\frac{\partial \boldsymbol{J}}{\partial \boldsymbol{a}_s} &= \frac{\mathrm{d}\boldsymbol{J}}{\mathrm{d}\tilde{\boldsymbol{u}}}\frac{\partial \tilde{\boldsymbol{u}}}{\partial \boldsymbol{a}_s} \\ &= \left(-\frac{\partial \boldsymbol{v}}{\partial \boldsymbol{s}}\boldsymbol{L}\left(\boldsymbol{N_R}\frac{\partial \boldsymbol{v}}{\partial \boldsymbol{s}}\boldsymbol{L}\right)^{-1}\boldsymbol{N_R} + \boldsymbol{I}_m\right)\left(-\frac{\partial \boldsymbol{v}}{\partial \boldsymbol{s}}\boldsymbol{L}\right) \\ &= \boldsymbol{0}_{m\times r},\end{aligned}$$

so that, locally, changes in $\boldsymbol{a}_s$ have no effect on the steady-state reaction rates $\boldsymbol{J}$, while

$$\begin{aligned}\frac{\partial \boldsymbol{J}}{\partial \boldsymbol{a}_v} &= \frac{\mathrm{d}\boldsymbol{J}}{\mathrm{d}\tilde{\boldsymbol{u}}}\frac{\partial \tilde{\boldsymbol{u}}}{\partial \boldsymbol{a}_v}\\ &= \left(-\frac{\partial \boldsymbol{v}}{\partial \boldsymbol{s}}\boldsymbol{L}\left(N_R\frac{\partial \boldsymbol{v}}{\partial \boldsymbol{s}}\boldsymbol{L}\right)^{-1}N_R + \boldsymbol{I}_m\right)\boldsymbol{K}\\ &= \boldsymbol{K},\end{aligned}$$

which is a local description of the direct dependence of $\boldsymbol{J}$ on $\boldsymbol{a}_v$ shown earlier in equation (8.11). In particular, the first $m - r$ rows of these matrix equations give

$$\frac{\partial \boldsymbol{J}_i}{\partial \boldsymbol{a}_s} = \boldsymbol{0}_{(m-r)\times r} \quad \text{and} \quad \frac{\partial \boldsymbol{J}_i}{\partial \boldsymbol{a}_v} = \boldsymbol{I}_{m-r}.$$

These results are summarized in the following section.

8.5.3 Separation Principle for Stoichiometric Systems

Given a stable steady state of system (8.2), if the system input is written as

$$\boldsymbol{p} = \boldsymbol{p}^0 + \left(\frac{\partial \boldsymbol{v}}{\partial \boldsymbol{p}}\right)^{-1}\left(\boldsymbol{K}\boldsymbol{a}_v - \frac{\partial \boldsymbol{v}}{\partial \boldsymbol{p}}\boldsymbol{L}\boldsymbol{a}_s\right),$$

then the local effect of the input on the steady-state independent concentration and flux is

$$\frac{\mathrm{d}\boldsymbol{s}_i}{\mathrm{d}\boldsymbol{a}_v} = \boldsymbol{0}_{r\times(m-r)}, \qquad \frac{\mathrm{d}\boldsymbol{J}_i}{\mathrm{d}\boldsymbol{a}_v} = \boldsymbol{I}_{m-r},$$

$$\frac{\mathrm{d}\boldsymbol{s}_i}{\mathrm{d}\boldsymbol{a}_s} = \boldsymbol{I}_r, \qquad \frac{\mathrm{d}\boldsymbol{J}_i}{\mathrm{d}\boldsymbol{a}_s} = \boldsymbol{0}_{(m-r)\times r}.$$

The response of the complete set of systems variables is

$$\frac{\mathrm{d}\boldsymbol{s}}{\mathrm{d}\boldsymbol{a}_v} = \boldsymbol{0}_{n\times(m-r)}, \qquad \frac{\mathrm{d}\boldsymbol{J}}{\mathrm{d}\boldsymbol{a}_v} = \boldsymbol{K}, \tag{8.17a}$$

$$\frac{\mathrm{d}\boldsymbol{s}}{\mathrm{d}\boldsymbol{a}_s} = \boldsymbol{L}, \qquad \frac{\mathrm{d}\boldsymbol{J}}{\mathrm{d}\boldsymbol{a}_s} = \boldsymbol{0}_{m\times r}. \tag{8.17b}$$

Note that, if the parameters appear linearly and specifically in the reaction rates (e.g., as enzyme activities), then $\left(\frac{\partial \boldsymbol{v}}{\partial \boldsymbol{p}}\right)^{-1}$ is simply a diagonal matrix of scaling factors.

Such separation principles are powerful aids to design in control engineering since they allow the engineer to treat two aspects of a single system independently of one another. As described at the outset of this chapter, the implications for metabolic "redesign" have been addressed in the metabolic control analysis literature (Kholodenko et al., 1998).

This separation principle recapitulates the summation and connectivity theorems of metabolic control analysis (MCA). Those results were originally derived from a rather different viewpoint, which will be treated following an illustrative example.

8.6 An Illustrative Example

Consider the simplified model of the glycolytic pathway shown in figure 8.1, modified from an example in Heinrich and Schuster (1996). The system consists of six chemical species involved in eight reactions. With the list of species identified as

$$(s_1, s_2, s_3, s_4, s_5, s_6) = (\text{G6P}, \text{F6P}, \text{TP}, \text{F2,6BP}, \text{ATP}, \text{ADP}),$$

and the reactions numbered as in figure 8.1, the stoichiometry matrix is

$$N = \begin{bmatrix} -1 & -1 & 0 & 1 & 0 & 0 & 0 & 0 \\ 1 & 0 & 1 & 0 & -1 & 0 & -1 & 0 \\ 0 & 0 & 0 & 0 & 2 & -1 & 0 & 0 \\ 0 & 0 & -1 & 0 & 0 & 0 & 1 & 0 \\ 0 & 0 & 0 & -1 & -1 & 2 & -1 & -1 \\ 0 & 0 & 0 & 1 & 1 & -2 & 1 & 1 \end{bmatrix}.$$

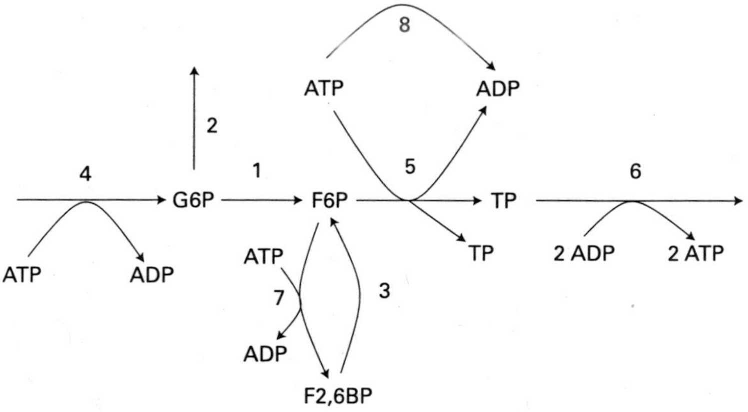

Figure 8.1
Simplified glycolytic reaction scheme. Abbreviations: G6P, glucose 6-phosphate; F6P, fructose 6-phosphate; TP, triose phosphate; F2,6BP fructose 2,6-bisphosphate; ATP, adenosine triphosphate; ADP, adenosine diphosphate. The source (glucose) and sinks (glucose 1-phosphate and pyruvate) are not included in the model.

This 6×8 matrix has rank 5, indicating that there is one dependent species and three independent reactions.

8.6.1 Row Reduction

The reaction scheme reveals a conserved moiety. ATP and ADP are interchanged, but are neither produced nor consumed, and so their total concentration is constant throughout the motion of the system. Either of these can be chosen as the dependent species. In this case, the species have been numbered so that ADP (s_6) corresponds to the last row of $\boldsymbol{N}$. As a result, the choice of ADP as the dependent species allows us to truncate this row to reach the 5×8 (full row rank) reduced stoichiometry matrix $\boldsymbol{N_R}$. We set $\boldsymbol{s}_i = (s_1, s_2, s_3, s_4, s_5)^T$ and $\boldsymbol{s}_d = (s_6)$. The row link matrix takes the form

$$\boldsymbol{L} = \begin{bmatrix} \boldsymbol{I}_5 \\ \boldsymbol{L}_0 \end{bmatrix}, \qquad \text{where} \qquad \boldsymbol{L}_0 = [0 \quad 0 \quad 0 \quad 0 \quad -1].$$

The structural constraint $s_5 + s_6 = \tilde{T}$ is then formalized as $\boldsymbol{s}_d = \boldsymbol{L}_0 \boldsymbol{s}_i + \tilde{T}$.

8.6.2 Column Reduction

To exploit redundancy in the columns, we begin by identifying independent and dependent reactions. Again, the numbering has been chosen with this goal in mind—the first three columns of $\boldsymbol{N}$ are dependent on the remaining five. Consequently, we can set

$$\boldsymbol{v}_i = (v_1, v_2, v_3) \quad \text{and} \quad \boldsymbol{v}_d = (v_4, v_5, v_6, v_7, v_8),$$

and proceed with the reduction as outlined in section 8.4.

We begin by finding a basis of the nullspace of $\boldsymbol{N}$ with the appropriate form:

$$\boldsymbol{K} = \begin{bmatrix} 1 & 0 & 0 \\ 0 & 1 & 0 \\ 0 & 0 & 1 \\ 1 & 1 & 0 \\ 1 & 0 & 0 \\ 2 & 0 & 0 \\ 0 & 0 & 1 \\ 2 & -1 & -1 \end{bmatrix}.$$

The columns of $\boldsymbol{K}$ correspond to pathways through the network in which flux could be altered without affecting the species concentrations. Specifically, an appropriately coordinated change in v_4, v_1, v_5, v_6, and v_8 could increase flux through the central pathway without any effect on the state dynamics. Likewise for the triples v_4, v_2, v_8

and v_3, v_7, v_8. (This last would change the flux, not through the network, but rather through the cycle composed of v_3 and v_7.)

This illustrates the ability to manipulate the reaction rates without affecting the state dynamics. From the form of $\boldsymbol{K}$, each of these pathways involves exactly one independent reaction rate. Consequently, if one can manipulate the reaction rates directly (through $\tilde{\boldsymbol{u}} = \boldsymbol{v}$), then, with any decomposition of $\tilde{\boldsymbol{u}}$ in the from (8.9), the choice of $\boldsymbol{a}_v$ influences the independent reaction rates directly since $\boldsymbol{v}_i = \boldsymbol{a}_v$. With $\boldsymbol{a}_s = \boldsymbol{0}$, we have

$$\boldsymbol{v}(t) = \boldsymbol{K}\boldsymbol{a}_v(t) = \begin{bmatrix} \boldsymbol{a}_v(t) \\ -\boldsymbol{P}_0\boldsymbol{a}_v(t) \end{bmatrix}.$$

Having addressed reaction rate control by the choice of $\boldsymbol{K}$, we now consider two possibilities for treating manipulation of the species concentrations.

8.6.3 State Dynamics

The first possibility is to complete the decomposition (8.9) by choosing

$$\boldsymbol{M} = \overline{\boldsymbol{M}} = \begin{bmatrix} \boldsymbol{0}_{(m-r)\times r} \\ (\boldsymbol{N}_{\boldsymbol{R}}^{\boldsymbol{C}})^{-1} \end{bmatrix},$$

where $\boldsymbol{N}_{\boldsymbol{R}}^{\boldsymbol{C}}$ is the top-right 5×5 submatrix of $\boldsymbol{N}$. In this case,

$$(\boldsymbol{N}_{\boldsymbol{R}}^{\boldsymbol{C}})^{-1} = \begin{bmatrix} 1 & 0 & 0 & 0 & 0 \\ 0 & -1 & 0 & -1 & 0 \\ 0 & -2 & -1 & -2 & 0 \\ 0 & 0 & 0 & 1 & 0 \\ -1 & -3 & -2 & -4 & -1 \end{bmatrix}.$$

The entries of this matrix provide a procedure for controlling the independent species concentrations individually, using only the dependent reactions as inputs. In the product $\overline{\boldsymbol{M}}\boldsymbol{a}_s$, each coefficient of $\boldsymbol{a}_s$ is multiplied by a column of $(\boldsymbol{N}_{\boldsymbol{R}}^{\boldsymbol{C}})^{-1}$. The columns of this matrix thus specify which dependent reaction rates should be perturbed to effect a change in each species concentration. For instance, the first column indicates that the concentration of s_1 can be manipulated by increasing rate v_4 and simultaneously decreasing rate v_8 by the same amount. The amount by which v_4 is increased corresponds to the rate of increase of s_1 (which would be the first entry of $\boldsymbol{a}_s$ following the notation in section 8.5.2). The corresponding decrease in v_8 is required to balance the increased consumption of ATP. The other columns of $(\boldsymbol{N}_{\boldsymbol{R}}^{\boldsymbol{C}})^{-1}$ indicate corresponding procedures for manipulating the other four independent species.

8.6.4 Local Steady-State Behavior

Rather than manipulate the reaction rates directly, the second possibility is to define locally the input $\tilde{\boldsymbol{u}}$ by equation (8.12), in which case the decomposition (8.9) prescribes the choice of

$$\boldsymbol{M} = -\frac{\partial \boldsymbol{v}}{\partial \boldsymbol{s}}\boldsymbol{L}.$$

The specific value of $\frac{\partial \boldsymbol{v}}{\partial \boldsymbol{s}}$ depends on the reaction kinetics, but the form of this matrix can be attained from knowledge of which metabolites influence which reactions. To illustrate, we will consider the simplest case in which reactions depend only on their substrates. In this case,

$$-\frac{\partial \boldsymbol{v}}{\partial \boldsymbol{s}}\boldsymbol{L} = -\begin{bmatrix} \frac{\partial v_1}{\partial s_1} & 0 & 0 & 0 & 0 \\ \frac{\partial v_2}{\partial s_1} & 0 & 0 & 0 & 0 \\ 0 & 0 & 0 & \frac{\partial v_3}{\partial s_4} & 0 \\ 0 & 0 & 0 & 0 & \frac{\partial v_4}{\partial s_5} \\ 0 & \frac{\partial v_5}{\partial s_2} & 0 & 0 & \frac{\partial v_5}{\partial s_5} \\ 0 & 0 & \frac{\partial v_6}{\partial s_3} & 0 & -\frac{\partial v_6}{\partial s_6} \\ 0 & \frac{\partial v_7}{\partial s_2} & 0 & 0 & \frac{\partial v_7}{\partial s_5} \\ 0 & 0 & 0 & 0 & \frac{\partial v_8}{\partial s_5} \end{bmatrix}.$$

As before, each column of this matrix indicates a set of perturbations which will influence the steady-state concentration of exactly one independent species, in this case leaving all reaction fluxes unchanged at steady state. The coefficients in each column indicate the relative strengths of the simultaneous perturbations that are required to elicit a change in the corresponding species concentration. For instance, to effect an increase of Δ in species s_1, the first column of this matrix indicates that we can make a decrease of $\Delta(\partial v_1/\partial s_1)$ in reaction v_1 and a simultaneous decrease of $\Delta(\partial v_2/\partial s_1)$ in reaction v_2. So long as these perturbations are small enough (i.e., the local linear approximation remains valid), then these changes will elicit the predicted system response.

This local analysis takes an elegant form when posed in terms of *relative* perturbations and responses, for which, for example, a $\Delta(\partial v_1/\partial s_1)\%$ decrease in v_1 and a $\Delta(\partial v_2/\partial s_1)\%$ decrease in reaction v_2 lead to a $\Delta\%$ increase in the steady-state concentration of s_1. These relative sensitivities are the primary objects of study in metabolic control analysis, to which we now turn.

8.7 Metabolic Control Analysis

The field of metabolic control analysis (MCA) was born in the mid-1970s out of the work of Kacser and Burns (1973) and Heinrich and Rapoport (1974a,b). These two groups independently arrived at an analytical framework for addressing questions of control and regulation of metabolic networks. Specifically, their papers outline a parametric sensitivity analysis around a steady state and address linear reaction chains in detail. In addition to deriving sensitivities, the authors also present relationships between the sensitivity coefficients. These relations, known as the summation and connectivity theorems, have been used to provide valuable insights into the behavior of metabolic networks. Mathematically, they amount to descriptions of sensitivity invariants (as described in Rosenwasser and Yusupov, 2000), and are consequences of the stoichiometric nature of the system.

Because these theorems were originally derived by intuitive arguments rather than rigorous mathematical analysis, they were not immediately generalized to more complicated networks. Such generalizations first appeared in Fell and Sauro (1985). Subsequently, a number of papers provided the theorems with a rigorous foundation (Cascante et al., 1989a,b; Giersch, 1988a,b; Reder, 1988). In particular, Reder (1988), provides a general mathematical framework in which to address sensitivity analysis and the resulting sensitivity invariants. The historical development of metabolic control analysis (both theoretical and experimental) has been treated in Fell (1997; and more concisely, in Fell, 1992).

The primary motivation for the development of MCA was the need to describe how biochemical pathways respond to perturbation. In particular, the results provided by metabolic control analysis were instrumental in defeating the notion of "rate-limiting step," in which control over the rate of a single reaction was perceived to allow authority over an entire reaction chain. In its place was installed an understanding that system behavior is dependent on all of the components of the network. Since its inception, MCA has been used successfully in the study of a great many metabolic systems. In addition to elucidating these biochemical mechanisms, this sensitivity analysis allows prediction of the effects of intervention. As such, it is a powerful design tool that has been adopted by the metabolic engineering community (Cornish-Bowden and Cárdenas, 1999; Kholodenko and Westerhoff, 2004; Stephanopoulos et al., 1998) and has been used in rational drug design (Cascante et al., 2002; Cornish-Bowden and Cárdenas, 1999).

The main theorems of metabolic control analysis will be addressed in section 8.7.1. The required preliminaries, which follow below, amount to a straightforward parametric sensitivity analysis. Although the material itself is standard, the MCA community makes use of a specialized terminology and notation, which we will introduce, following the formalism developed by Reder (1988; see also Heinrich and Schuster, 1996, and Hofmeyr, 2001).

As in section 8.5.2, we assume given a nominal parameter input $\boldsymbol{p}^0$ and a corresponding (asymptotically) stable steady state $\boldsymbol{s}^0$ for system (8.2). Repeating the derivation in section 8.5.2, we arrive at the sensitivities in species concentration (equation (8.14)), referred to as *unscaled independent concentration-response coefficients*:

$$\bar{\boldsymbol{R}}_{\boldsymbol{p}}^{s_i} = \frac{\mathrm{d}}{\mathrm{d}\boldsymbol{p}} \boldsymbol{s}_i(\boldsymbol{p}) = -\left(\boldsymbol{N_R}\frac{\partial \boldsymbol{v}}{\partial \boldsymbol{s}}\boldsymbol{L}\right)^{-1} \boldsymbol{N_R}\frac{\partial \boldsymbol{v}}{\partial \boldsymbol{p}}, \tag{8.18}$$

which can be extended to the complete concentration vector $\boldsymbol{s}$ by defining $\bar{\boldsymbol{R}}_{\boldsymbol{p}}^{s} = \boldsymbol{L}\bar{\boldsymbol{R}}_{\boldsymbol{p}}^{s_i}$. The use of the specialized term *response* for the coefficients in (8.18) was introduced to distinguish these system sensitivities (total derivatives) from component sensitivities (partial derivatives) that will be introduced below as *elasticities*.

In addition to this sensitivity in the state variables, the sensitivities of the steady-state fluxes, referred to as the *unscaled flux-response coefficients*, are also of interest:

$$\bar{\boldsymbol{R}}_{\boldsymbol{p}}^{J} = \frac{\mathrm{d}}{\mathrm{d}\boldsymbol{p}} \boldsymbol{v}(\boldsymbol{s}(\boldsymbol{p}), \boldsymbol{p}) = -\frac{\partial \boldsymbol{v}}{\partial \boldsymbol{s}}\boldsymbol{L}\left(\boldsymbol{N_R}\frac{\partial \boldsymbol{v}}{\partial \boldsymbol{s}}\boldsymbol{L}\right)^{-1} \boldsymbol{N_R}\frac{\partial \boldsymbol{v}}{\partial \boldsymbol{p}} + \frac{\partial \boldsymbol{v}}{\partial \boldsymbol{p}}.$$

The first $m - r$ rows of this matrix constitute the *independent* unscaled flux-response coefficients $\bar{\boldsymbol{R}}_{\boldsymbol{p}}^{J_i} = \frac{\mathrm{d}}{\mathrm{d}\boldsymbol{p}} \boldsymbol{v}_i(\boldsymbol{s}(\boldsymbol{p}), \boldsymbol{p})$.

These response coefficients represent absolute sensitivities. In application, it is the relative sensitivities, reached through scaling by the values of the related variables, that provide more useful measures of system behavior. These *scaled* concentration- and rate-response coefficients are given by

$$\boldsymbol{R}_{\boldsymbol{p}}^{s} = (\boldsymbol{D}^{s})^{-1}\bar{\boldsymbol{R}}_{\boldsymbol{p}}^{s}\boldsymbol{D}^{p} = \frac{\mathrm{d}\ln \boldsymbol{s}}{\mathrm{d}\ln \boldsymbol{p}},$$

$$\boldsymbol{R}_{\boldsymbol{p}}^{J} = (\boldsymbol{D}^{J})^{-1}\bar{\boldsymbol{R}}_{\boldsymbol{p}}^{J}\boldsymbol{D}^{p} = \frac{\mathrm{d}\ln \boldsymbol{J}}{\mathrm{d}\ln \boldsymbol{p}},$$

where $\boldsymbol{D}^{z}$ is the diagonal matrix composed from the vector z. Although, in the biochemical literature (and especially in addressing experimental data), the scaled versions are preferred, from a mathematical point of view, the unscaled sensitivities can be seen as more fundamental. For that reason, we will deal primarily with unscaled coefficients in what follows, allowing the interested reader to translate the results to relative sensitivities through the appropriate scaling factors.

The response coefficients describe the asymptotic response of the linearized system to (step) changes in the parameter vector $\boldsymbol{p}$. As such, they can be used to predict the steady-state effect of small changes in the parameter values. On the other hand, in using sensitivity analysis to address the inherent behavior of a network, it is often more useful to ignore the details of the actuation and identify the reaction rates directly with the parameters, as was done in section 8.5 through the input redefinition $\tilde{\boldsymbol{u}}$. In addressing absolute (unscaled) sensitivities, this amounts to supposing that the reaction rates depend on the parameter inputs specifically and directly: $\frac{\partial \boldsymbol{v}}{\partial \boldsymbol{p}} = \boldsymbol{I}_m$ (i.e., $\boldsymbol{p} = \tilde{\boldsymbol{u}}$, in the notation of section 8.5). As mentioned earlier, if the parameter inputs appear linearly and specifically in the reaction rates (e.g., as enzyme activities), then this condition holds automatically for the scaled sensitivities. Under this assumption, the response coefficients defined above are referred to as the unscaled *control coefficients* of the system. The control coefficients are the primary objects of interest in metabolic control analysis because they provide a means to quantify the dependence of system behavior on the individual reactions in the network. The unscaled control coefficients are defined by

$$\bar{\boldsymbol{C}}^s = -\boldsymbol{L}\left(\boldsymbol{N}_R \frac{\partial \boldsymbol{v}}{\partial \boldsymbol{s}} \boldsymbol{L}\right)^{-1} \boldsymbol{N}_R, \tag{8.19a}$$

$$\bar{\boldsymbol{C}}^J = -\frac{\partial \boldsymbol{v}}{\partial \boldsymbol{s}} \boldsymbol{L}\left(\boldsymbol{N}_R \frac{\partial \boldsymbol{v}}{\partial \boldsymbol{s}} \boldsymbol{L}\right)^{-1} \boldsymbol{N}_R + \boldsymbol{I}_m, \tag{8.19b}$$

and were derived in section 8.5.2 as the sensitivities to the redefined input $\tilde{\boldsymbol{u}}$, equation (8.15) and equation (8.16). The scaled control coefficients

$$\boldsymbol{C}^s = (\boldsymbol{D}^s)^{-1}\bar{\boldsymbol{C}}^s\boldsymbol{D}^J \quad \text{and} \quad \boldsymbol{C}^J = (\boldsymbol{D}^J)^{-1}\bar{\boldsymbol{C}}^J\boldsymbol{D}^J$$

are used to address relative sensitivities.

The description of system behavior in terms of response and control coefficients has proven immensely useful in the analysis of biochemical systems. These sensitivities can be measured directly from observations of the intact system or can be derived from measurements of the component sensitivities, that is, the partial derivatives of $\boldsymbol{v}$. When it is possible to reproduce individual reactions in vitro, these component sensitivities can often be measured with a high degree of accuracy. The system sensitivities can then be derived from the definitions given above.

To make the distinction between component and system sensitivities explicit, the partial derivatives of $\boldsymbol{v}$ are referred to as the *elasticities* of the system. Specifically, we define the scaled and unscaled *substrate elasticity* ε_s and *parameter elasticity* ε_p by

$$\bar{\varepsilon}_s = \frac{\partial \boldsymbol{v}}{\partial \boldsymbol{s}}, \qquad \varepsilon_s = (\boldsymbol{D}^J)^{-1}\frac{\partial \boldsymbol{v}}{\partial \boldsymbol{s}}\boldsymbol{D}^s,$$

$$\bar{\varepsilon}_p = \frac{\partial v}{\partial p}, \qquad \varepsilon_p = (D^J)^{-1}\frac{\partial v}{\partial p}D^p.$$

Using this notation, we can organize the relation between response coefficients and parameter elasticities into the *partitioned-response equations*

$$\bar{R}_p^s = \bar{C}^s \bar{\varepsilon}_p \quad \text{and} \quad \bar{R}_p^J = \bar{C}^J \bar{\varepsilon}_p, \tag{8.20}$$

which hold in the scaled variables as well.

Although an important tool for the study of biochemical networks, this sensitivity analysis fails to distinguish metabolic control analysis as a theoretical field of study. It is in the treatment of sensitivity invariants, to which we now turn, that MCA provides an extension of standard sensitivity analysis.

8.7.1 Sensitivity Invariants: The Theorems of Metabolic Control Analysis

In applications of sensitivity analysis, it is sometimes found that the structure of the system imposes restrictions on the sensitivity coefficients. When these restrictions take the form of algebraic relations among the sensitivities that do not depend on the state or parameter values they are referred to as *sensitivity invariants* (Rosenwasser and Yusupov, 2000).

The stoichiometric nature of a biochemical network imposes sensitivity invariants on the system. Descriptions of these invariants originally appeared in the work of Kacser and Burns (1973) and Heinrich and Rapoport (1974a,b) and have since been generalized and extended. These relations are described by the summation theorem and the connectivity theorem, which will be addressed below. In each case, the classical statement will be given before the general result is stated.

8.7.2 The Summation Theorem

The original development of the summation theorem (Heinrich and Rapoport, 1974a; Kacser and Burns, 1973) addresses an unbranched chain of reactions, as in figure 8.2. At steady state, mass balance dictates that the reaction rates are all equal. If the reaction rates are simultaneously increased by a factor α (e.g., by increasing each enzyme's activity by the same relative amount), then the species concentrations will not change because the difference between the rates of production and consumption for each species is unaltered. Moreover, the reaction rates themselves will have

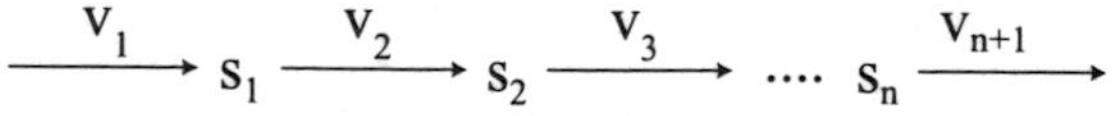

Figure 8.2
Unbranched reaction chain.

increased precisely by α because there is no systemic response to this perturbation. Describing this situation in the language of sensitivities, we address a perturbation in a scalar parameter p satisfying

$$\varepsilon_p = (\boldsymbol{D}^{\boldsymbol{v}})^{-1} \frac{\partial \boldsymbol{v}}{\partial p} p = \begin{bmatrix} \alpha \\ \alpha \\ \vdots \\ \alpha \end{bmatrix}.$$

That is, changes in the parameter p correspond to simultaneous coordinated changes in all of the enzyme activity levels. Consider the scaled version of the partitioned-response property (8.20). Addressing metabolite s_j, we have

$$0 = \boldsymbol{R}_p^{s_j} = \boldsymbol{C}^{s_j} \varepsilon_p = \boldsymbol{C}_1^{s_j} \alpha + \boldsymbol{C}_2^{s_j} \alpha + \cdots + \boldsymbol{C}_{n+1}^{s_j} \alpha = \alpha \sum_{l=1}^{n+1} \boldsymbol{C}_l^{s_j},$$

where $\boldsymbol{C}_l^{s_j}$ is the lth element of the vector of control coefficients associated with species s_j. Because the relative change in the flux through each reaction is α, we find, considering reaction v_k,

$$\alpha = \boldsymbol{R}_p^{J_k} = \boldsymbol{C}^{J_k} \varepsilon_p = \boldsymbol{C}_1^{J_k} \alpha + \boldsymbol{C}_2^{J_k} \alpha + \cdots + \boldsymbol{C}_{n+1}^{J_k} \alpha = \alpha \sum_{l=1}^{n+1} \boldsymbol{C}_l^{J_k}.$$

Dividing by α, we arrive at the *concentration* and *flux summation theorems*: for each $j = 1, \ldots, n$ and $k = 1, \ldots, n+1$:

$$\sum_{l=1}^{n+1} \boldsymbol{C}_l^{s_j} = 0 \quad \text{and} \quad \sum_{l=1}^{n+1} \boldsymbol{C}_l^{J_k} = 1. \tag{8.21}$$

The flux summation theorem was instrumental in helping to clarify the identification of "control points" along a pathway. Because all control coefficients are nonnegative in many cases of interest, there is only rarely a true "rate-limiting step" with control coefficient of one. More commonly, the coefficients range between zero and one, with sensitivity distributed throughout the pathway.

General statements of the summation theorems are given in Reder (1988), where it is observed that, if the matrix $\boldsymbol{K}$ is chosen as in section 8.3 so that the columns of $\boldsymbol{K}$ lie in the nullspace of $\boldsymbol{N}$, then

$$\bar{\boldsymbol{C}}^s \boldsymbol{K} = \boldsymbol{0}_{n \times (m-r)} \quad \text{and} \quad \bar{\boldsymbol{C}}^J \boldsymbol{K} = \boldsymbol{K}, \tag{8.22}$$

which follow directly from the definitions of the control coefficients (8.19). These observations are explicit generalizations of the classical statement of the summation theorems, as follows. Given the unbranched chain in figure 8.2, the stoichiometry matrix has a one-dimensional kernel spanned by the vector of ones. Expanding the matrix multiplication in (8.22) gives, for each species s_j, $j = 1, \dots, n$, and each reaction flux J_k, $k = 1, \dots, n+1$,

$$\sum_{l=1}^{n+1} \bar{C}_l^{s_j} = 0 \quad \text{and} \quad \sum_{l=1}^{n+1} \bar{C}_l^{J_k} = 1.$$

Scaling gives the statements in equations (8.21). In general, equations (8.22) capture all such constraints imposed by the stoichiometry, regardless of the network structure.

8.7.3 The Connectivity Theorem

Kacser and Burns (1973) describe the relationship between flux control coefficients and substrate elasticities. The result is illustrated with the system shown in figure 8.3.

Consider a perturbation that has the effect of simultaneously increasing v_2 and decreasing v_1. This will lead to a decrease in the concentration of s_1, which will, in turn, lead to a decrease in v_2 and an increase in v_1. If the perturbation were chosen appropriately, the steady-state effect could be that v_1 and v_2 would return to their preperturbation levels, the concentration of s_1 would remain depressed, and there would be no effect on the steady state of the rest of the network. Locally, this effect is achieved by perturbation of a parameter p satisfying $\partial \mathbf{v}/\partial p = (s_1/p)(\partial \mathbf{v}/\partial s_1)$, that is: $\varepsilon_p = \varepsilon_{s_1}$.

Because such a perturbation has no steady-state effect on the flux, we have

$$0 = \boldsymbol{R}_p^{J_1} = \boldsymbol{C}_1^{J_1}\varepsilon_p^1 + \boldsymbol{C}_2^{J_1}\varepsilon_p^2 = \boldsymbol{C}_1^{J_1}\varepsilon_{s_1}^1 + \boldsymbol{C}_2^{J_1}\varepsilon_{s_1}^2,$$

where ε^j is the j-th component of ε. An analogous statement holds for $\boldsymbol{R}_p^{J_2}$. These are statements of the *flux connectivity theorem*. Written in the form

$$\frac{\boldsymbol{C}_1^{J_1}}{\boldsymbol{C}_2^{J_1}} = \frac{\boldsymbol{C}_1^{J_2}}{\boldsymbol{C}_2^{J_2}} = -\frac{\varepsilon_{s_1}^2}{\varepsilon_{s_1}^1},$$

this constraint indicates that the ratios of these control coefficients depend only on the ratio of elasticities of the species s_1 that "connects" the reactions of interest.

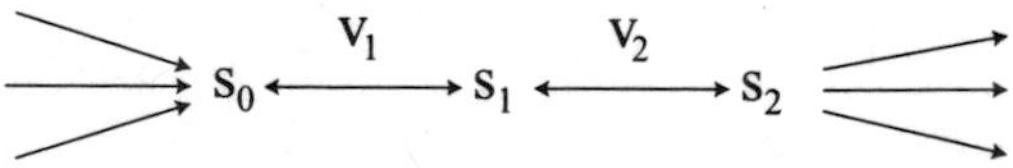

Figure 8.3
Two reactions linked by a single species.

Westerhoff and Chen (1984) provided the analogous statement for concentration control coefficients, which says that for the same choice of parameter p:

$$-1 = \boldsymbol{R}_p^{s_1} = \boldsymbol{C}_1^{s_1}\varepsilon_p^1 + \boldsymbol{C}_2^{s_1}\varepsilon_p^2 = \boldsymbol{C}_1^{s_1}\varepsilon_{s_1}^1 + \boldsymbol{C}_2^{s_1}\varepsilon_{s_1}^2,$$

while

$$0 = \boldsymbol{R}_p^{s_j} = \boldsymbol{C}_1^{s_j}\varepsilon_p^1 + \boldsymbol{C}_2^{s_j}\varepsilon_p^2 = \boldsymbol{C}_1^{s_j}\varepsilon_{s_1}^1 + \boldsymbol{C}_2^{s_j}\varepsilon_{s_1}^2,$$

for $j \neq 1$.

Reder (1988) generalized these results into the algebraic statements

$$\bar{\boldsymbol{C}}^s\bar{\varepsilon}_s\boldsymbol{L} = -\boldsymbol{L} \quad \text{and} \quad \bar{\boldsymbol{C}}^J\bar{\varepsilon}_s\boldsymbol{L} = \boldsymbol{0}, \tag{8.23}$$

which follow directly from equations (8.19). As with the summation theorem, these matrix equations can be scaled and written out as sums to recover the classical statements.

8.7.4 Relation to Separation Principle

The separation principle derived in section 8.5.2 recapitulates the summation and connectivity theorems from a design perspective. Reder's statements of the metabolic control analysis theorems (8.22) and (8.23) describe the result of postmultiplying the control coefficients with specific matrices. Referring to the partitioned-response equation (8.20), these can be understood as response coefficients for perturbations of specific types. Interpreted in this manner, these theorem statements indicate the response to perturbations $\boldsymbol{p}$ for which $\varepsilon_p = \boldsymbol{K}$ or $\varepsilon_p = \varepsilon_s\boldsymbol{L}$.

In the notation of section 8.4.2, those two cases are achieved precisely when the parameter input is perturbed through $\boldsymbol{a}_v$ or $\boldsymbol{a}_s$ respectively. Consequently, the separation principle (8.17) can be seen as a recapitulation of the theorem statements (8.22) and (8.23). When stated together as in (8.17), the theorem statements are referred to as the control matrix equation (Hofmeyr and Cornish-Bowden, 1996).

8.8 Conclusion

This chapter has highlighted the role of stoichiometry in the control of metabolic systems. The separation principle derived above is a control-theoretic statement of the avenues for redesign in these networks. Such results can be valuable both in the manipulation of system behavior and in the investigation of inherent network regulation. The derivation in this chapter also serves to bridge the gap between the fields of metabolic control analysis and engineering control theory, with the intent of encouraging further cross-fertilization between these complementary fields.

9 Robustness and Sensitivity Analyses in Cellular Networks

Jason E. Shoemaker, Peter S. Chang, Eric C. Kwei, Stephanie R. Taylor, and Francis J. Doyle III

Mathematical models provide a convenient framework to collect and store experimental observations and ultimately allow experimenters to probe biological systems for nonintuitive and unexpected behaviors. To understand biochemical networks predictively, however, such models must overcome the innate complexity of cellular signaling. Although it is unlikely researchers will ever produce a model capable of predicting all feasible phenotypic outputs to all possible environments, conditions, and genomic perturbations, models do exist that can reproduce phenotypic expression in response to several thousand genetic perturbations. Typically, models are built and their parameters are fit to time trace experiments. To complement the exercise of model building, model analyses are performed to gauge the accuracy of the model and to help determine points for efficient control. Secondary characteristics such as sensitivity analysis can identify model inconsistencies, optimal points to perturb (and control) network response, and optimal experimental conditions to minimize parameter variation and optimize the information content for each measurement. By applying cost functions with sensitivity analyses, researchers can identify regulation and optimization motifs that may provide further insight into the principles guiding network evolution. In this chapter, we introduce the useful model-analytic tools of both sensitivity analysis and structured singular-value analysis and their application to cellular networks.

9.1 Sensitivity Analysis and the Fisher Information Matrix

Maximizing information extracted from a biological system is important because biological experiments are often time consuming and costly. When a preliminary model structure and a parameter set exist for a system, optimal experimental design can be used to maximize parameter information from that system; two ways to do this are by manipulating the input to the system and by choosing a proper set of system states for measurement. Sensitivity analysis, specifically the Fisher information matrix (FIM), can be used to distinguish between a variety of input profiles and measurement selections for optimal design of experiments. Sensitivity analysis quantitatively investigates the change in response of a system with changes in parameter values. For a system of ordinary differential equations, perturbation of a parameter p_j may cause a change in state x_i; the magnitude of this change is captured by a sensitivity coefficient, S_{ij}:

$$S_{ij}(t) = \frac{\partial x_i(t)}{\partial p_j}. \tag{9.1}$$

Taking these arrays of sensitivity coefficients and estimations of measurement error and assuming that measurements have Gaussian distributions, the Fisher information matrix generates a set of metrics of the parameter information that can be extracted from these measurements. For discrete time steps, the FIM is calculated as follows (Zak et al., 2003):

$$FIM = \sum_{k=1}^{N_t} \boldsymbol{S}^T(t_k)\boldsymbol{V}^{-1}(t_k)\boldsymbol{S}(t_k), \tag{9.2}$$

where N_t is the total number of time steps and $\boldsymbol{V}$, which describes estimates of measurement covariance, is a diagonal matrix with elements

$$V_i = \begin{bmatrix} \sigma_i^2(t_1) & 0 & 0 \\ 0 & \ddots & 0 \\ 0 & 0 & \sigma_i^2(t_k) \end{bmatrix}, \tag{9.3}$$

$$\sigma_{x_i}^2(t_k) = RE_i x_i(t_k) + AE_i.$$

RE_i is a relative error in state i, while AE_i (absolute error) is a small but nonzero number to prevent the matrices $\boldsymbol{V}$ from being singular.

One of the important uses of the Fisher information matrix is to optimize parameter estimation accuracy from a set of proposed experimental protocols, by using the FIM to estimate lower bounds on the variance of parameter estimation, $\sigma_{p_j}^2$ (Zak et al., 2003):

$$\sigma_{p_j}^2 \geq FIM_{jj}^{-1}. \tag{9.4}$$

One can then use the size of standard deviations for each parameter of interest to select between experimental protocols. In the application below, if the 95% confidence interval around a nominal parameter value, $[p_{\text{nominal}} - 1.96\sigma_p, p_{\text{nominal}} + 1.96\sigma_p]$ does not contain zero, a parameter is taken to be identifiable. For equivalent numbers of identifiable parameters, one possible optimization is to choose the experimental protocol that minimizes the average normalized 95% confidence interval $((1/N_p)\sum_j 1.96\sigma_{p_j}/p_j)$ over all identifiable parameters, a condition known as A-optimality. Other identifiability and optimality conditions may be used as well. Using formulations of identifiability and optimality, one can then design experiments to maximize the accuracy of parameter estimation.

9.1.1 Application to Insulin Signaling

The optimal input and measurement selection outlined above was applied to a detailed published model of the insulin-signaling pathway (Sedaghat et al., 2002). Two variations of the model were proposed—one without feedback mechanisms and one with feedback mechanisms. Differential equations, largely mass action in nature, were used to describe the concentrations or relative abundances of 21 state variables, as illustrated in figure 9.1.

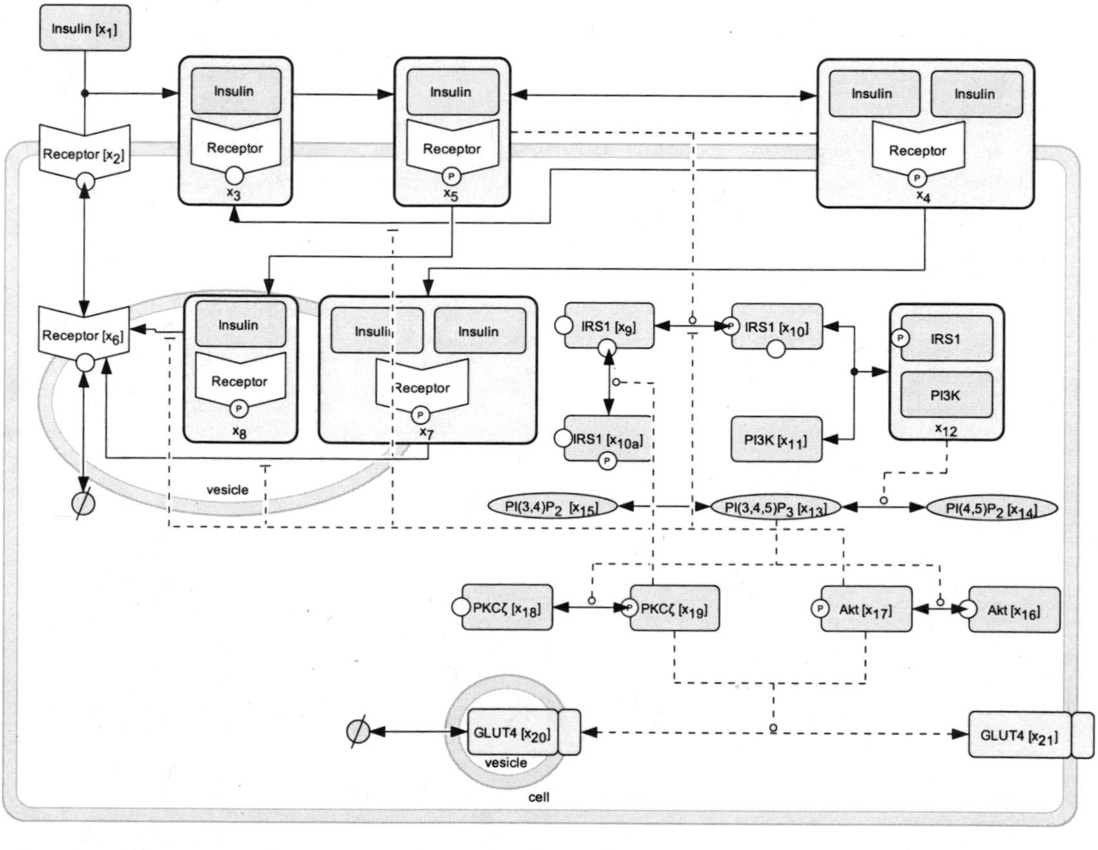

Figure 9.1
Model with feedback, adapted from Sedaghat et al., 2002.

The Sedaghat model comprises three submodels. The first one describes both insulin-receptor binding and recycling. The second submodel describes the post-receptor signaling cascade and also contains both positive and negative feedback loops. The third submodel describes the effects of the postreceptor signaling on GLUT4 translocation between vesicles and the cell surface.

The Sedaghat model, with slight modifications for ease of sensitivity analysis, was compiled in XPP and solved with numerical integration over a 60-minute "experiment" (Kwei et al., 2008). Following Sedaghat, the nominal "experimental" insulin input concentration went from 0 M to 10^{-7} M and returned to 0 M in a 15-minute pulse. Thirty parameters for the variant model with feedback and 28 parameters for the one without were perturbed by 1% of their nominal values to calculate sensitivities for all states for each time step. This sensitivity analysis was conducted using BioSens (Taylor et al., 2008c). For models in XPP, BioSens uses a centered difference approximation to calculate sensitivity coefficients numerically:

$$S_{ij}(t_k) \approx \frac{x_i(t_k)|_{p_j+\Delta p_j} - x_i(t_k)|_{p_j-\Delta p_j}}{2\Delta p_j}. \tag{9.5}$$

In general, when all 21 states were "measured," the relative error for each state (RE) was set to 1%, while absolute error (AE) was set to 10^{-7} to ensure that V could be inverted. When necessary to remove state i from the set of measured states (see below), RE_i was set to 10^7% to simulate a highly noisy measurement.

Optimal Input Selection

Varying insulin input to the system is one method to maximize parameter estimation accuracy. Simple insulin input profiles were analyzed for maximum parameter identification in the model with feedback (table 9.1). The peak value for each insulin in-

Table 9.1
Parameter identification from different input selections

Input description	Parameters	$\frac{1}{N_p}\sum_j 1.96\sigma_{p_j}/p_j$
1-minute pulse	21	9.84%
5-minute pulse	21	9.98%
0.5-minute pulse	21	11.2%
15-minute pulse	20	6.58%
Ramp up	20	7.61%
Ramp down	20	11.3%
Two 1-minute pulses	19	7.37%
Step	19	15.1%

Note: Input selections are ranked by number of identifiable parameters followed by average width of normalized 95% confidence interval for identifiable parameters put was chosen to be 10^{-7} M, as in the Sedaghat model, to compare similar insulin dosages. The inputs were ranked first by number of identifiable parameters, then by A-optimality, with a Fisher information matrix including all 21 states, measured nearly continuously.

The number of identifiable parameters ranges from 19 to 21 for this variety of insulin inputs, with the best result being the 1-minute pulse (table 9.1). Generally, the same parameters are found to be identifiable for each input profile considered. We therefore conclude that input dynamics should have a small but quantifiable effect on the identifiability of model parameters for this system.

Although the model with feedback is larger than the one without, the model with feedback is more readily identifiable. For example, for the 15-minute insulin pulse input, 65% of the parameters can be identified in the model with feedback, compared to 62% for the model without feedback. This observation is also true for the insulin step input, with 61% of parameters identifiable for the model with feedback compared to 52% for the model without. The model with feedback is more identifiable because measurements of early states in the signaling pathway contain information about parameter values for reactions involved in feedback that occur further down the pathway.

Optimal Measurement Selection

For ease of calculation, because insulin input dynamics did not seem to have a significant effect on parameter identifiability, measurement selection was carried out only on the 1-minute pulse, which had 21 identifiable parameters. Measurements of as many states as possible were removed from the Fisher information matrix while maintaining the same number of identified parameters. The FIMs including all permutations of the remaining states were then calculated, with the results ranked first by number of identified parameters and then by A-optimality. The optimal measurement selection for each number of allowed measurements is given in table 9.2.

A measurement of five states (x_{15}, x_{17}, x_{19}, x_{20}, and x_{21}) gives 21 identifiable parameters; nearly all of the parameter information content available is included in

Table 9.2
Parameter identification from optimized measurement selections

State measurement	Parameters	$\frac{1}{N_p}\sum_j 1.96\sigma_{p_j}/p_j$
$x_2, x_3, \ldots, x_{21}$	21	9.84%
$x_{15}, x_{17}, x_{19}, x_{20}, x_{21}$	21	11.7%
$x_{15}, x_{17}, x_{19}, x_{20}$	21	16.5%
x_{15}, x_{17}, x_{20}	20	23.0%
x_{15}, x_{17}	14	25.1%
x_{17}	9	46.3%

these five states, all of which are near the end of the signaling pathway (see Kwei et al., 2008, for more information). It makes sense that a sparse measurement selection of a signaling cascade can yield high parameter information content because measurements of later states in the signaling pathway can contain parameter information from reactions involving previous states. Indeed, measuring just one state (x_{17}) allows one to identify 9 parameters. Note that there is no guarantee that the above measurement selections will actually yield the most parameter information for an arbitrary insulin input profile.

9.2 Phase-Sensitivity Analysis for Biological Oscillators

Although classical sensitivity analysis has many advantages, it falls short when phase behavior of biological oscillators becomes a more appropriate performance metric (Bagheri et al., 2007). Sensitivity measures of the phase of a system due to state and parametric perturbations have been developed for limit cycle models of biological oscillators. Taylor et al. (2008a) developed the parametric impulse phase-response curve (pIPRC) to reliably predict responses to stimuli manifested as perturbations in the parameters.

9.2.1 Parametric-Impulse Phase-Response Curve

A multistate oscillator can be reduced to a single ordinary differential equation limit cycle oscillator model based on phase (Kuramoto, 1984; Winfree, 2001), called the phase-evolution equation. This approach has been used to track single and multiple oscillators. A limit cycle oscillator model consisting of a set of ODEs with an attracting orbit γ can be represented by

$$\dot{\boldsymbol{x}}(t) = \boldsymbol{f}(\boldsymbol{x}(t), \boldsymbol{p}), \tag{9.6}$$

where $\boldsymbol{x}$ is the vector of states and $\boldsymbol{p}$ represents the vector of parameters. The solution about the limit cycle, $\boldsymbol{x}^{\gamma}(t)$, has period τ, that is, $\boldsymbol{x}^{\gamma}(t) = \boldsymbol{x}^{\gamma}(t+\tau)$. The limit cycle is a one-dimensional structure; movement along the limit cycle can be reduced to a single variable, ϕ, that tracks the internal time of the oscillator and progresses at the same rate as time. However, a stimulus changes the rate of progression with respect to time, leading to a separation between internal and external time. The phased-based model follows the phase of the system, taking into account perturbations in one of two forms—either to the state's dynamics directly, or to a parameter. The effects of state perturbations are predicted by the state impulse phase-response curve (sIPRC), given for the kth state by

$$sIPRC_k(t) = \frac{\partial \phi}{\partial x_k^{\gamma}(t)}, \tag{9.7}$$

where an infinitesimal perturbation in state k at time t leads to an infinitesimal change in phase. The sIPRC is computed by solving the adjoint linear variational equation corresponding to the system represented by equation (9.6) (Brown et al., 2004; Kramer et al., 1984). The phase equation incorporating the sIPRC is

$$\frac{d\phi}{dt} = 1 + sIPRC(\phi) \times G(\phi, t), \tag{9.8}$$

where G is the stimulus. Equations (9.7) and (9.8) can be used to arrive at the phase equation accounting for perturbations in parameters:

$$pIPRC_j(t) = \frac{d}{dt}\frac{d\phi}{dp_j}(t), \tag{9.9}$$

which leads to the following phase equation:

$$\frac{d\phi}{dt} = 1 + pIPRC_j(t)\Delta p_j(t). \tag{9.10}$$

The pIPRC can be expressed in terms of the sIPRC by

$$pIPRC_j(t) = \sum_{k=1}^{N} sIPRC_k(t)\frac{\partial f_k}{\partial p_j}(t). \tag{9.11}$$

Equation (9.11) has been used to predict the response of circadian oscillators to arbitrary stimuli that ultimately lead to changes in parameter values (Taylor et al., 2008a).

9.2.2 Application to Circadian Clocks

Organisms have evolved to adapt to the light/dark cycle caused by the revolution of the Earth. In order to survive, they have developed sustained internal oscillators with periods of approximately 24 hours. Environmental cues, such as light, entrain circadian oscillators that, in turn, influence organism behavior. The heart of these circadian clocks is believed to lie in gene regulatory networks involving transcription/translation feedback loops.

Analysis of circadian clock models reveals that circadian systems are relatively insensitive to parametric perturbations, and that the phase appears to be a key attribute that is modulated to entrain to the local environment. Models such as that of Becker-Weimann et al. account for the influence of light on the circadian clock (Becker-Weimann et al., 2004; Geier et al., 2005). Light is incorporated into this model by modulating the induction rate of *Per* mRNA.

The Becker-Weimann model uses molecular information about the circadian clock in mammals. The molecular clock consists of genes, mRNA, and proteins involved in a transcription/translation feedback loop (Reppert and Weaver, 2002). *Per1*, *Per2*, *Cry1*, and *Cry2* genes are lumped into one state, *Per*/*Cry*. Per/Cry protein inhibits the induction of its mRNA, forming a negative feedback loop, and also promotes the induction of *Bmal1* mRNA, whose protein promotes induction of *Per*/*Cry*, which forms the positive feedback loop. In the Becker-Weimann model, the light input promotes the induction of *Per*/*Cry* through the inclusion of a parameter $L(t)$, whose effect is gated by a clock component, which allows the light to have an effect at certain times of the day.

A standard tool in circadian study is the phase-response curve (PRC), which measures the phase shift in response to a specific signal. Typically, creating a numerical PRC requires stimulating the system at different circadian times, simulating the tra-

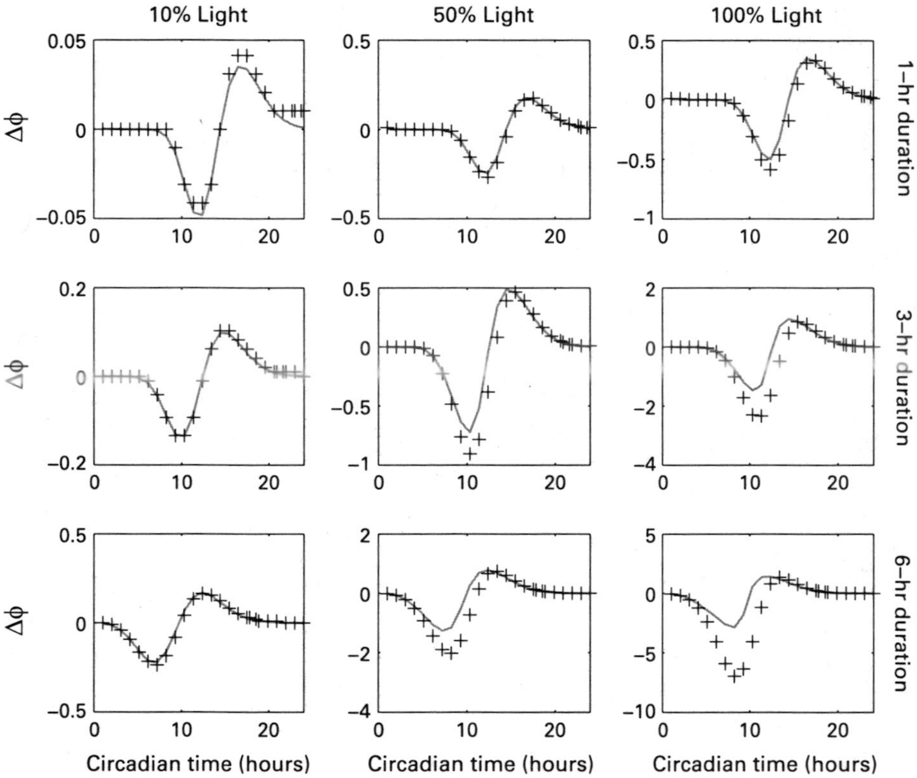

Figure 9.2
Phase-response curves (PRCs) to light. Phase shifts were calculated for varying intensity and duration of light. The light signal is sinusoidal, lasting one hour for the first row, three hours for the second row, and six hours for the third row. The signal maximum is at 10%, 50%, and 100% of full light, respectively.

jectory of all the states in the model, and calculating the resulting phase shift after the oscillator has returned to its limit cycle orbit. The phase evolution equation simplifies the task by requiring the solution of only one ordinary differential equation. Taylor et al. (2008a) reduced the above model to a phase evolution equation (as in equation (9.10)) to predict the phase response to a half-sinusoidal light signal. Parametric impulse phase-response curve analysis provides additional insight into the timekeeping properties of intracellular processes. By computing the pIPRC for each parameter in a model, we learn which processes are most capable of shifting the clock at what times during the cycle. Comparing relative pIPRCs reveals which parameters dominate the clock's timing and which play minimal roles. Taylor et al. (2008b) used pIPRC analysis to predict which feedback loops were critical and which were unnecessary for proper behavior of the mammalian clock. Figure 9.2 compares the results from the traditional method and the phase evolution equation. Note that the phase-response curves predicted by the phase equation (solid gray line) are in good quantitative agreement with the numerical PRCs (pluses). Although the magnitude of the maximum delay is overestimated as the intensity and duration of light are increased, the timing of the maximum delay and advance agree well with the numerical PRC.

9.3 Target Identification Using Structured Singular-Value Analysis

A primary goal of systems biology is to better understand biochemical networks and to manipulate them for therapeutic applications, yet realizing this goal is complicated by the innate difficulties of system identification in a highly variable environment. With generation-to-generation variability, possibly severe differences between macro- and microenvironments, stochastic noise due to low copy number, parameter variability, and so on, one might suppose that the innate uncertainty in biology would prevent any meaningful model development. Yet, despite these challenges, by properly exploiting the robust features that have evolved in biological systems to protect and maintain critical network features, models with great predictive power can be derived.

Although the definition of robustness varies in the systems biology literature (Kitano, 2007), where the term is often generalized, in control engineering, it does not: *robust performance* (RP) is defined as the ability to maintain performance despite network and environmental uncertainties (Stelling et al., 2004b). Application of robust performance requires the proper development of performance specifications as well as a suitable description of system uncertainties. For example, healthy phenotype maintenance is a robust process because, even though the human genome suffers approximately 120 irreparable mutations each generation, these mutations generally do not affect the health of the individual. If robustness to deletions is the performance metric, more the 80% of the yeast genome is robust (Tong et al., 2001). At the intracellular signaling level, an example of robust performance is precise adaptation in *Escherichia coli* chemotaxis, where adaptation precision was shown to be robust to parameter uncertainty even though adaptation time was fragile (Barkai and Leibler, 1997). A tool widely used for evaluating system performance in the face of uncertainty is the structured singular value (SSV), which has become crucial to understanding and designing proper flight control algorithms, and which has recently been applied to understanding parametric uncertainty in biochemical networks (Ma and Iglesias, 2002; Schmidt and Jacobsen, 2004). When applied to an apoptotic signaling model, SSV analysis identified known fragilities in the FasL apoptosis architecture and determined that the apoptotic output is best manipulated by targets upstream of apoptosome formation (Shoemaker and Doyle, 2008).

Before introducing structured singular value analysis for robust performance below, let us review the Nyquist stability criterion and extend it to conditions guaranteeing *robust stability* (RS), which may be viewed as the absolute minimum performance metric for any given dynamical system. Indeed, it can be shown that RP for any definable performance specification is identical to robust stability with an added perturbation block. Once we have established conditions for RP, we will apply SSV analysis to a generalized kinase cascade model.

9.3.1 Nyquist Stability Criteria

The origins of structured singular-value analysis are rooted in the Nyquist stability criterion, which determines whether an open-loop system is stable when closed under negative or positive feedback. Both the Nyquist stability criterion and SSV analysis are applied in the frequency domain (readers unfamiliar with frequency domain analysis who wish to delve deeper into the topic are referred to Skogestad and Postlethwaite, 2005, or to Seborg et al., 1989, although unfamiliarity with this topic should not limit the general understanding of robust performance analysis). Suffice it to say, as discussed in chapter 1, the frequency response is the system's oscillatory response to a sine wave input of fixed amplitude. The frequency response of a dynamical system offers analytical advantages over step- or impulse-based analyses because dynamic behavior is considered over all frequencies.

To create a Nyquist plot, the real part of the frequency response is plotted against the imaginary part over frequencies from $-\infty$ to $+\infty$ (the amplitude of the input remains fixed). If the open-loop system, G_{OL}, has P unstable poles (unstable eigenvalues), then the closed-loop system is stable under negative feedback if the Nyquist plot encircles -1 precisely P times (Skogestad and Postlethwaite, 2005). Figure 9.3 illustrates the use of the Nyquist stability criterion. The open-loop system (details not shown) has one unstable eigenvalue and must encircle -1 once to ensure stability under negative feedback. A gain (k) of 0.8 is insufficient to stabilize the system, but, by increasing the gain to 1.5, stability can be ensured.

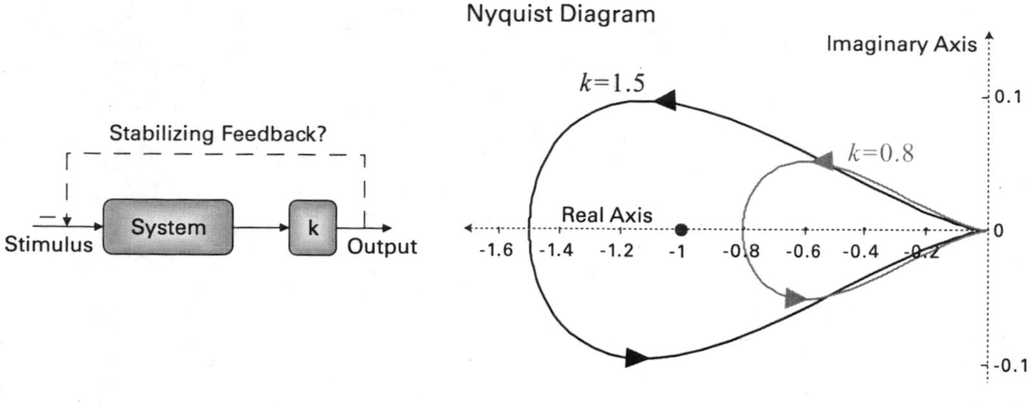

Figure 9.3
Nyquist stability. For an open-loop system with one unstable eigenvalue, the Nyquist plot must encircle −1 exactly one time. A feedback gain of 0.8 is not sufficient to stabilize the system, but a feedback gain of 1.5 ensures closed-loop stability.

9.3.2 Robust Stability

Considering figure 9.3, let us assume the open-loop transfer function is stable and allow some uncertainty in the gain of the negative feedback such that $k \in [k_{\text{low}}, k_{\text{high}}]$. If one starts with a stable, nominal system G_{OL} ($P = 0$), robust stability (RS) is guaranteed so long as the Nyquist plots for all possible values of the perturbation between k_{low} and k_{high} never encircle the critical point -1. This is the concept of robust stability. A system is robustly stable if stability is maintained despite variations due to system uncertainty. Now one needs only a method to determine the size of the perturbation that destabilizes the feedback system.

The first step to applying structured singular-value analysis is the proper construction of the uncertainty system. The *structure* of the uncertainty comes from applying uncertainty perturbations to specific interactions within the network. Figure 9.4 illustrates how a nominal system with uncertainty about two of the internal transfer functions is shaped into the **PΔ** block in which the uncertainties are lumped into the diagonal **Δ** block. The uncertainties are first designed to be of size 1 ($\|\delta_i\|_{\text{inf}} \leq 1$), and the frequency-dependent weighting blocks (W_1 and W_2) are used to manipulate the effective magnitude of the uncertainty. Producing the necessary weighting blocks to distribute the uncertainties properly about the system is not always a trivial procedure, and toolboxes exist to assist with the construction and wiring (Balas et al., 2001).

Assuming the nominal system without uncertainty is stable, then the only way for instabilities to enter the system is via the uncertainty feedback through the $\mathbf{P}_{11}$ subblock. Thus stability is maintained in the full, uncertain system so long as the

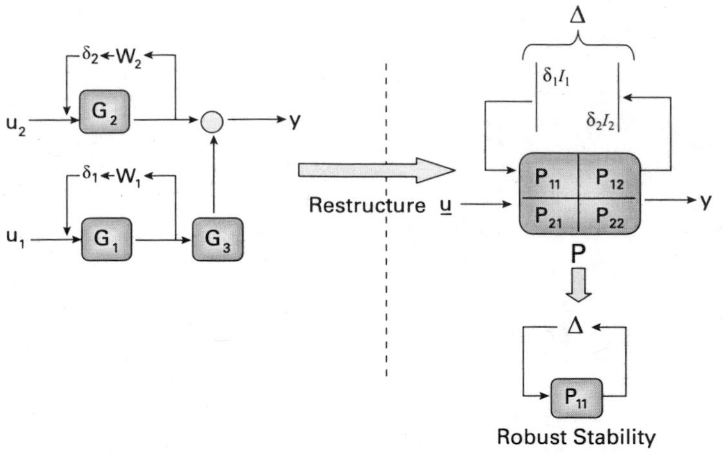

Figure 9.4
Nominal feedback system with uncertainty, δ_i, assigned to separate components. The system is restructured into the $\mathbf{P\Delta}$ structure for robust stability and robust performance analysis.

closed-loop $\mathbf{P_{11}\Delta}$ system is stable. Applying the Nyquist stability criterion to the $\mathbf{P_{11}\Delta}$ block structure, one finds the $\mathbf{\Delta}$ that shifts at least one stable eigenvalue to the imaginary axis, and reports μ, defined formally as

$$\mu(\mathbf{P}_{11})^{-1} = \min_{\mathbf{\Delta}}\{\bar{\sigma}(\mathbf{\Delta}) \mid \det(\mathbf{I} - \mathbf{P}_{11}\mathbf{\Delta}) = \mathbf{0}, \text{ for structured } \mathbf{\Delta}\}, \tag{9.12}$$

where $\bar{\sigma}(\mathbf{\Delta})$ is the maximum singular value of $\mathbf{\Delta}$. Remembering that the system has been designed such that $\|\delta_i\|_{\text{inf}} \le 1$, we see that a value of $\mu < 1$ means the destabilizing perturbation is outside of the predefined uncertainty ranges (the destabilizing perturbation necessarily satisfies $\|\mathbf{\Delta}\|_{\text{inf}} > 1$). The calculation of μ is an NP-hard problem (interested readers are referred to the original citations, Braatz et al., 1994; Doyle et al., 1992, for details). For 2×2 input-output systems, exact solutions exist, but for larger systems, lower and upper bounds on μ are calculated to bound the true value of μ.

9.3.3 Robust Performance

Extending μ-analysis to robust performance is a simple matter because robust performance, under the proper construction, is equivalent to robust stability for linear systems (for explicit proofs of the equivalence of RP and RS, see Skogestad and Postlethwaite, 2005). To apply structured singular-value analysis for robust performance, we first normalize the input-output channels to be of size 1, and apply some performance weight, $\mathbf{W_p}$. The input-output channels are then closed through a full-block

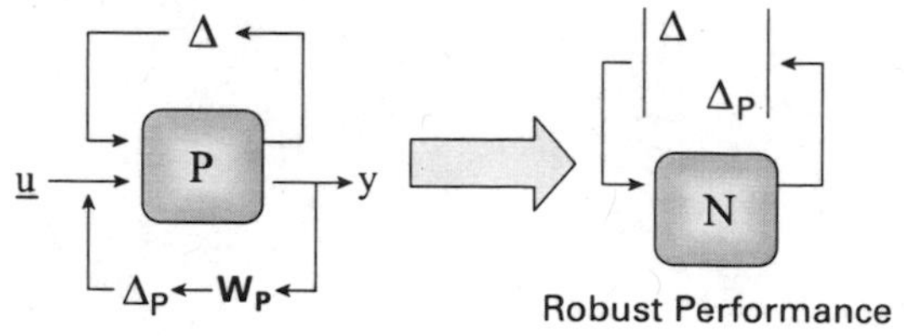

Figure 9.5
By closing the input-output channels of the uncertain system and redesigning the system into the **NΔ** block structure, robust performance of the uncertain system can be assessed by evaluating the robust stability of the **NΔ** block structure.

uncertainty matrix, $\mathbf{\Delta_p}$, whose size has again been designed to satisfy $\|\Delta_p\|_{\inf} < 1$. This system can be lifted (through linear fractional transformation) to the **NΔ** block structure where the $\mathbf{\Delta}$ is now a block-diagonal matrix composed of the original $\mathbf{\Delta}$, which accounts for the system uncertainties, and $\mathbf{\Delta_p}$, which bounds the system performance (figure 9.5). The same criterion for robust stability is then applied to the **NΔ** block structure:

$$\mu(\mathbf{N})^{-1} = \min_{\mathbf{\Delta}}\{\bar{\sigma}(\mathbf{\Delta}) \mid \det(\mathbf{I} - \mathbf{N\Delta}) = 0, \text{ for structured } \mathbf{\Delta}\}.$$

Whereas, in engineering, performance weight design is generally determined by safety standards, product quality specification, and so on, there is no general or universal method for choosing performance specifications for a biological system. Ideally, variation within the data can provide meaningful bounds on the input-output behavior. Allowing variation in protein counts can account for stochasticity inherent in biochemical networks, and stochastic models may be used to understand better the allowable extremes in performance by correlating the variability with reaction volume. For biological systems, it is often more significant to analyze the potential, observable performance in order to generate or invalidate hypotheses or, for a fixed performance, to determine the maximum variability in parameter space which may be allowed while maintaining the desired performance. These applications require the use of skewed-μ, in which the uncertainty and performance weights are iteratively adjusted until performance is precisely met ($\mu = 1$).

As with any analytical tool, structured singular-value analysis is limited in its application. It is generally applicable to linear systems; its extensions to nonlinear systems only apply to a limited class of problems. Furthermore, SSV analysis is a conservative tool, both in its calculation and its application. Because the distance between the upper and lower bounds means additional uncertainties must be considered during the performance analysis, the calculation of μ is conservative. Because performance criteria in the time domain do not easily translate into the frequency domain, time

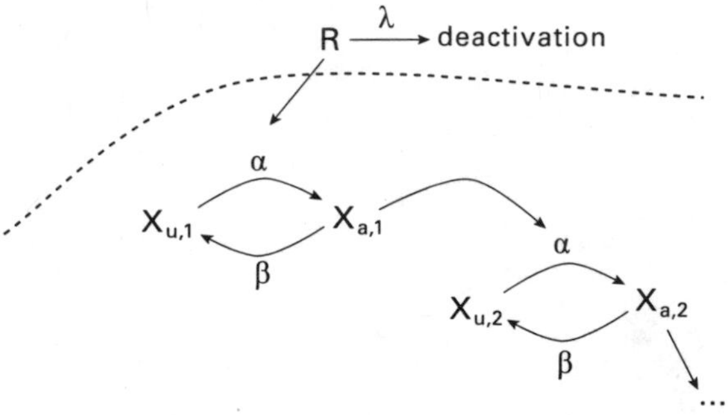

Figure 9.6
Receptor (R) activation at the cell surface (*dashed line*) activates a cascade of phosphorylation and dephosphorylation steps, ultimately resulting in a cellular response. Adapted from Heinrich et al., 2002.

domain behaviors must be approximated. Yet it remains a powerful tool for analyzing systems that may be well approximated by a linear description, and for analyzing regions of behavior for which linear descriptions work well.

9.3.4 Application to a Simplified Kinase Cascade

During signal transduction, a receptor signal at the cell surface must be detected, verified, and amplified to ensure proper downstream transcription factor activation. As shown in figure 9.6, the signal processing consists of a sequence of kinase and phosphatase mediated reaction steps of phosphorylation and dephosphorylation, respectively (Heinrich et al., 2002). In this application (Doyle and Stelling, 2005), the key performance attributes are (1) speed at which the signal arrives to destination, (2) duration of signal, and (3) signal strength. Ultimately, translating these three attributes into formal control specifications is possible, but only as an approximation because structured singular-value analysis is in the frequency domain.

If one assumes a low degree of phosphorylation, individual steps in the kinase cascade obey linear kinetics:

$$\frac{\mathrm{d}X_i}{\mathrm{d}t} = \alpha_i X_{i-1} - \beta_i X_i,$$

where α_i is the phosphorylation rate constant, β_i is the dephosphorylation rate constant, and X_i is the phosphorylated version of kinase i. If one further assumes that the cascade is fourth-order, that the rates for phosphorylation and dephosphorylation are constant for each step and that, at the cell surface, receptor deactivation is suitably approximated as a simple exponential decay, one can simplify the overall kinase signaling response to the following transfer function:

$$Y(s) = \left(\frac{s}{\lambda^{-1}s + 1}\right)\left(\frac{\alpha^4}{(s+\beta)^4}\right)R(s), \tag{9.13}$$

where $R(s)$ is the signal input at the cell surface, and λ is the exponential decay time constant.

Under these assumptions, one can derive explicit expressions for the performance metrics in question (Heinrich et al., 2002). The signal duration is found to be

$$\theta = \sqrt{\frac{1}{\lambda^2} + \sum_{i=1}^{n} \frac{1}{\beta_i^2}},$$

the signal amplitude is given by

$$S = \frac{S_0 \sum_{k=1}^{n} \frac{\alpha_k}{\beta_k}}{\sqrt{1 + \lambda^2 \sum_{i=1}^{n} \frac{1}{\beta_i^2}}},$$

and the signaling time through the entire network is

$$\tau = \frac{1}{\lambda} + \sum_{i=1}^{n} \frac{1}{\beta_i}.$$

We consider parametric uncertainty by assigning relative uncertainty to α and β such that

$$\alpha = \tilde{\alpha} \times (1 + r_\alpha \delta_\alpha),$$
$$\beta = \tilde{\beta} \times (1 + r_\beta \delta_\beta). \tag{9.14}$$

Recalling that $\|\Delta\|_{\text{inf}} \leq 1$, we apply perturbations r_α and r_β to manipulate the effective magnitudes of the uncertainties. The remaining step is the design of our performance weight/filter. Although, ideally, data are available to define the appropriate bounding volumes of the system response, in the case of biological systems, they are often too scarce to develop truly meaningful filters. In this scenario, one can design a simple performance weight that allows a small difference margin between the uncertain response of the system (the actual cellular response) and the system nominal response. Here the performance filter is defined as a tracking error:

$$W_p = \frac{(s + 0.2)^2}{0.3(s + 0.001)}.$$

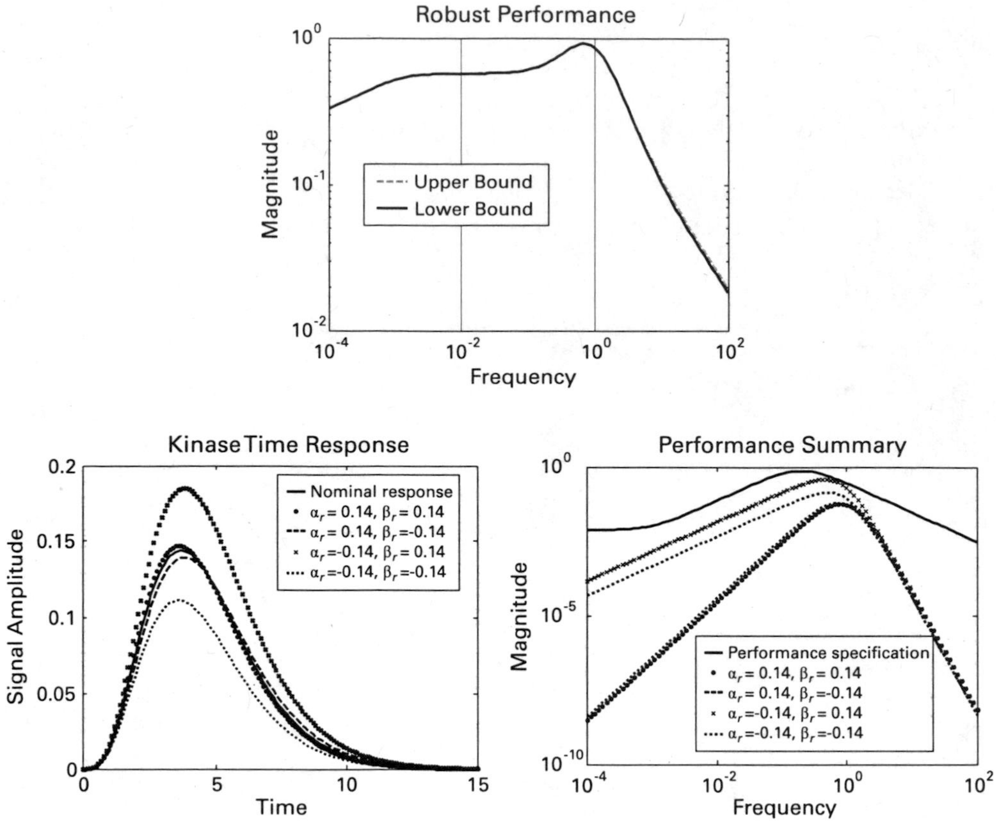

Figure 9.7
Results from μ-analysis of a simplified kinase cascade. The upper panel plots the value of μ versus input frequency. The lower two panels plot the system's performance in the time and frequency domain, respectively.

At frequencies for which the cascade signal is most active, the performance demands are tight, demanding an absolute error difference of 1.3 between the cellular response and nominal response, although performance demands are loosened for low- and high-frequency signals.

Structured singular-value analysis is applied for the parameter set ($\alpha = 1.0$, $\beta = 1.1$, and $\lambda = 1$) and the relative error for the rates of phosphorylation and dephosphorylation, r_α and r_β, is 14.0%. For these conditions, we see that the magnitude of μ is less than 1 for all frequencies, and thus we are guaranteed robust performance when there is a 14.0% uncertainty in the rates of phosphorylation/dephosphorylation (figure 9.7). We then further analyze performance by plotting the permutations of the extreme ends of the uncertainty to observe how the perturbations affect signaling behavior. As can be seen in the kinase time response, the cascade signal is most sensitive when the dephosphorylation and phosphorylation steps are perturbed in opposite directions. In the frequency domain, these perturbations push the system performance closest to the allowed specifications. If skewed-μ is applied, we find that the system can support a maximal 16.2% parameter uncertainty and maintain performance.

9.4 Hypothesis Generation, Validation, and Invalidation

Biological intuition complements the tools previously mentioned. Once a model is built and the associated parameter values calibrated, the model can be used to do *in silico* experiments. With a model, one can explore the effect of a mechanism on a given system's behavior by comparing the results in the absence of the mechanism. Indeed, one can perform a wider range of "experiments" than one could on the real system, facilitating generation of hypotheses, predictions, and explanations of experimental results. *In silico* experiments involve changing parameter values, model structures, and initial conditions mimicking gene knockouts, over- and underexpression of genes, creating variability in populations, and generating disease states. Because the resulting hypotheses or explanations can be highly model dependent, one should carefully consider the assumptions and simplifications used to create the model when interpreting results. The interplay of model results with experiment is critical in assessing the validity of the model and the assumptions and simplifications used to create it.

9.4.1 Application to Population of Synchronized Neurons

The master clock in mammals is believed to reside in the area of the hypothalamus called the suprachiasmatic nucleus. Synchronization of the cellular rhythms is critical in establishing a coherent phase to output signals to other peripheral oscillators in the mammal. Applying the genetic regulatory feedback network model created by Leloup and Goldbeter (2003), To et al. (2007) generated a mathematical model of a population of circadian oscillators that synchronized their rhythms through intercellular coupling. Their model included more detail, such as protein phoshorylation and transport between the nucleus and the cytosol, than the Becker-Weimann model.

Coupling between neurons in the suprachiasmatic nucleus (SCN) was modeled through release and binding of vasoactive intestinal polypeptide (VIP) among cells in the SCN (Maywood et al., 2006). This model incorporates signaling pathways initiated by the binding between VIP and its putative receptor, VPAC2. The VIP binding event is modeled in rapid equilibrium. The VIP/VPAC2 complex initiates an increase in intracellular calcium, which activates the transcription factor CREB.

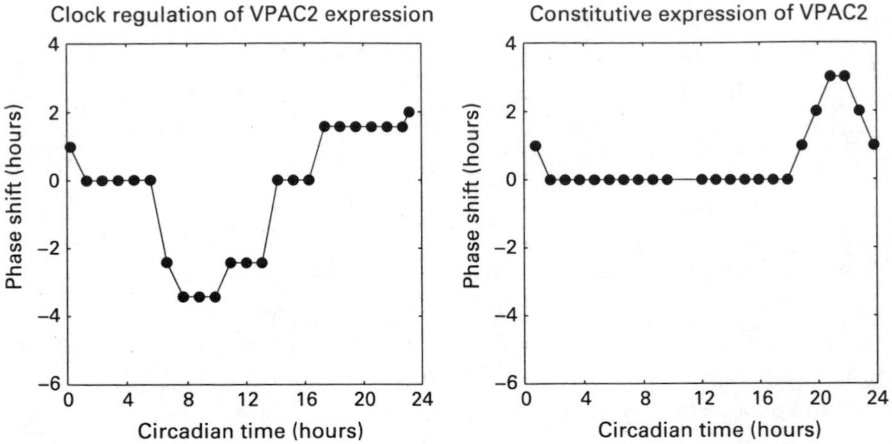

Figure 9.8
Population phase-response curves for vasoactive intestinal polypeptide (VIP). Phase shifts of synchronized population were calculated for 10 μM pulse of VIP for one hour. VPAC2 expression is regulated by the clock (*left*) or unregulated and expressed constitutively (*right*).

Binding of CREB causes induction of *Per* mRNA. Release of VIP from a cell was assumed to have a profile similar to its concentration of *Per* mRNA. A population of cells was modeled on a two-dimensional square grid where the contribution of VIP released from a cell to target cells is inversely related to the distance from those target cells.

Rhythmic expression of VPAC2 mRNA in slices of the suprachiasmatic nucleus was observed by Cagampang et al. (1998). Although modulation of VPAC2 receptor is not needed for synchronization of circadian rhythms in the model, this variation in receptor expression may play a role in the response of the population to exogenous VIP. The model was altered to include rhythmic expression of VPAC2 to the cell surface, dependent on a clock component. Numerical phase-response curves were generated for 1-hour pulses of 10 μM VIP. In figure 9.8, the resulting phase-response curve shows regions of phase advance and delay, and a region where little response to VIP stimulation is observed. Omitting clock control of VPAC2 expression resulted in phase-response curves with a large region where no phase shift is observed while an advance is observed at approximately 18–24 hours in circadian time. These results yield testable hypotheses that could be verified by experiment.

One can also explore the response of a single cell to applications of VIP. Currently, no data is available that characterizes the single cell behavior of suprachiasmatic nucleus neurons under VIP application. Although a single clock component was modeled to influence expression of VPAC2, the identity of the clock component is

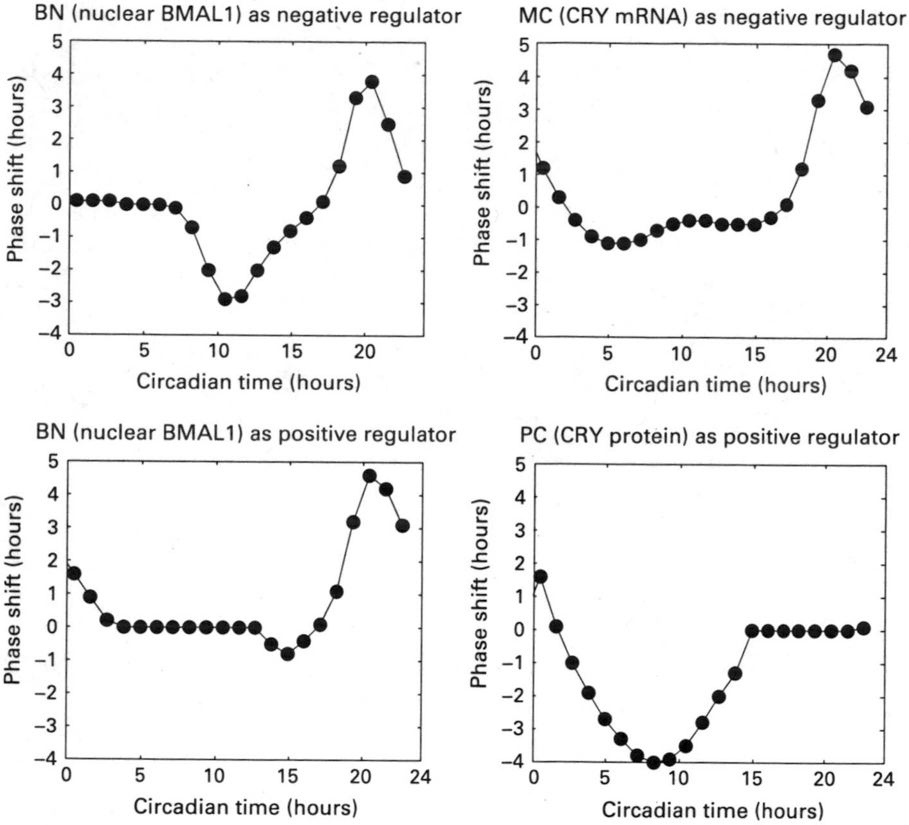

Figure 9.9
Single-cell VIP phase-response curves. Phase shifts to VIP were calculated for models that had different clock components positively or negatively regulating the expression of VPAC2 receptor. Cells were pulsed with 10 μM VIP for one hour.

unknown. Here the identity and the type of regulation on VPAC2 expression was varied. Figure 9.9 shows the resulting single-cell phase-response curves to 10 μM VIP for one hour.

The VIP phase-response curves are qualitatively different; presumably this difference arises from the varied phase differences between the time traces of total Per protein and VPAC2 expression. Regions of phase advance and delay can be eliminated and the location of nearly no response can be shifted in circadian time. On the other hand, the order of the regions does not appear to change. These single-cell models can now be expanded to population models to explore the impact of the character of the coupling on synchronization and response properties.

9.5 Conclusion

Control-theoretic tools, such as sensitivity analysis and phase sensitivity, provide powerful means for network elucidation and manipulation in biological systems. As in experimentation, however, every tool has its limitations. Proper network identification requires a collection of tools for optimal experimental design, proper data filtering (e.g., for microarray analyses), target identification and hypothesis generation. Although sensitivity analysis identifies the network components most sensitive to small perturbations, these results are not guaranteed for large perturbations (gene knockdowns, knockouts) and do not consider network components being perturbed simultaneously. Structured singular-value analysis can overcome both of these limitations but is constrained by both the complexity and conservativeness of the calculation. Thus one may use sensitivity analysis to constrain the number of scenarios considered during SSV analysis. In conclusion, using tools from control theory to guide both mathematical modeling and experimental design can facilitate the iterative paradigm of systems biology and shed light on the complex network behavior underlying biological organisms.

10 Structural Robustness of Biochemical Networks: Quantifying Robustness and Identifying Fragilities

Camilla Trané and Elling W. Jacobsen

In contrast to traditional parametric robustness analysis, this chapter considers explicit perturbations of the network structure. Applying dynamic perturbations to the direct interactions between the network nodes, we compute the smallest relative perturbation that will qualitatively change the network behavior, using results from robust control theory. Structural robustness analysis plays an important role both in determining the potential importance of unmodeled phenomena, such as intermediate reaction steps and transport mechanisms, and in identifying specific fragilities of given biological functions. We show how it can also be used to reduce nonlinear models and to elucidate the mechanisms underlying a given function. To illustrate our method, we consider models of the MAPK signaling cascade, metabolic oscillations in white blood cells, and the mammalian circadian clock.

10.1 Background

Robustness, the ability to maintain functionality in the presence of internal and external perturbations, is a fundamental property of biological systems (Kitano, 2004; Stelling et al., 2004b). Uncovering the mechanistic and structural properties of biological robustness is a key issue in systems biology (Kitano, 2007). Detecting specific network fragilities will help us determine not only the sources of disease states but also strategies for fighting those which have developed robustness.

Robustness has important implications for modeling the biochemical networks that underlie biological functions. First, it implies that even models with relatively high levels of parametric and structural uncertainty can describe the target function well. Biochemical network models typically postulate a set of biochemical components and reactions from which a corresponding set of differential equations is formulated. Because the knowledge of the underlying processes is incomplete, the set of equations will not completely match the biological system and this structural uncertainty will in most cases be significant. Moreover, the formulated model will usually involve a number of unknown parameters that need to be fitted to limited amounts of experimental data. Second, that a real biological function is known to be robust implies that robustness analysis can be used to validate a model of that function. In particular, if a model is found to be unrobust to biologically probable perturbations, then it must be refined to determine whether the lack of robustness is caused by a flaw in the model or by an important fragility of the biological function itself.

Robustness analysis is well established as a validation tool in modeling biochemical networks. Moreover, it has been used to elucidate the principles underlying robustness of specific biological functions (Locke et al., 2008; Stelling et al., 2004a; Wagner, 2005). Thus far, however, the focus has been almost exclusively on parametric robustness; that is, on the impact of uncertainty in model parameters such as kinetic rate constants (Chen et al., 2005; Cross, 2003; Morohashi et al., 2002; Wilhelm et al., 2004). In contrast, the impact of structural uncertainty is rarely addressed, most likely because methods for this purpose are not readily available, or perhaps because models robust to parametric perturbations are also believed to be robust to structural perturbations. As we demonstrate in this chapter, however, parametric robustness does not by any means imply structural robustness.

We start by providing a clear definition of robustness, a general formulation of the structural robustness problem for nonlinear ODE models, and the motivation for using linearized perturbed models for robustness analysis. We then transform the perturbed model into an input-output feedback model, consistent with the most common system representation employed in classical and robust control theory. Using results from control theory on robust stability of feedback systems, we determine the smallest structural perturbation that will change the qualitative behavior of the model. With computations performed in the frequency domain, we show how to translate the computed perturbations into a set of differential equations that, when combined with the original nonlinear model, can verify the results of the linear analysis. We employ a simple example, the Goodwin oscillator, to explain the method in a biological context. We then present three case studies to demonstrate the usefulness of our proposed method.

10.2 Defining Robustness of Biological Functions

Kitano (2007) defines *biological robustness* as "a property that allows a system to maintain its functions against internal and external perturbations." To clarify this definition, we need to determine what is implied by "function," "maintaining function," and "perturbations."

10.2.1 Biological Functions in the Context of Dynamical Systems

Here we consider functions generated by biochemical reaction networks, described by a set of ordinary differential equations, that is,

$$\frac{\mathrm{d}\boldsymbol{x}(t)}{\mathrm{d}t} = \boldsymbol{f}(\boldsymbol{x}(t), \boldsymbol{p}), \quad \boldsymbol{x} \in \mathbf{R}^n,\ \boldsymbol{p} \in \mathbf{R}^m, \tag{10.1}$$

where $\boldsymbol{x}(t)$ denotes a vector of state variables, such as concentrations of biochemical components, and $\boldsymbol{p}$ denotes the parameters, such as kinetic rate constants. In this context, a biological function is defined to correspond to a particular stationary behavior on an attractor of system (10.1) Local stationary behaviors of relevance to biological functions include steady states, corresponding to equilibrium attractors for which $\mathrm{d}\boldsymbol{x}(t)/\mathrm{d}t = \mathbf{0}$, and stable periodic oscillations, corresponding to limit cycle attractors for which $\boldsymbol{x}(t+T) = \boldsymbol{x}(t)$, for $T > 0$. The maintenance of stable steady states is typical for homeostatic functions, aimed at providing specific intracellular conditions, and corresponds to the setpoint control problem typically considered in technical control systems. Functions such as heat-shock response and bacterial chemotaxis are other examples of steady-state setpoint control problems. Contrary to technical control systems, biological systems frequently employ sustained oscillations, corresponding to limit cycles, to provide functionality. Typical examples include circadian timekeeping, oscillatory calcium signaling, and embryonic cell cycle control.

Steady states and limit cycles are examples of local stationary behaviors. A global behavior involving multiple attractors of significant relevance to biological systems is the existence of multiple stable steady states for a given value of the parameter vector $\boldsymbol{p}$. In particular, bistability is employed to provide irreversible switching between distinct steady states. An example is apoptosis, programmed cell death, in which the final death decision corresponds to a bistable switch (see, for example, Eissing et al., 2005).

10.2.2 Robust Stability and Robust Performance

Having defined what is implied by a function in the context of dynamical systems, we next need to define robustness in the sense of function maintenance. First, it is important to distinguish between robustness and sensitivity, which are often used interchangeably in the literature. *Sensitivity* quantifies the effects of a given perturbation on the function properties, such as steady-state concentrations or oscillation periods, within a biological system, whereas robustness quantifies the largest perturbations, within a given class, the system can tolerate. In the latter case, one should distinguish between robust stability and robust performance, corresponding to the effect on qualitative and quantitative behavior, respectively.

Robust stability, the focus of this chapter, refers to persistence of a qualitative behavior, corresponding to the biological function in question, in the face of perturbations. An analysis of robustness in terms of robust stability quantifies the largest perturbations the model (10.1) can tolerate, while the qualitative stationary behavior, for example, a stable steady state, stable limit cycle, or bistability, persists.

Robust performance refers to the attainment of a desired quantitative behavior of a biological function in the face of perturbations. Any biological function has quantifiable function properties, such as response time and steady-state concentration changes in the case of steady-state homeostasis, or period, amplitude, and phase shift in the case of sustained oscillations. Given a performance specification, such as maintaining an oscillation period within a given range, robust performance quantifies the largest perturbations the model (10.1) can tolerate while satisfying the performance specifications.

Defining Perturbations

Living cells face numerous external perturbations, such as changes in the physicochemical environment, as well as internal perturbations, notably in the form of gene mutations. Ideally, all relevant perturbations can be specified together with a probability of occurrence; by applying these, the robustness of the modeled function can then be evaluated. In reality, however, relatively little is known about most of the specific perturbations the function should withstand. It is therefore usually more relevant to define a general class of perturbations and then determine the subset within this class that the function will withstand. The size of the subset then provides a quantitative measure of the robustness. As stated above, robustness analysis of biochemical network models is frequently used in the context of model validation. In this case, it is the robustness of the model predictions to model uncertainties that is of interest. Again, because the precise uncertainties are not known, it is most relevant to define a class of uncertainties and then determine the subset within this class that satisfies the robustness criteria.

To define properly which classes of perturbations to consider, one must address the modeling process and its relation to the underlying biochemical system. Models of intracellular functions are usually based on postulating the biochemical components, cellular compartments, biochemical reactions, and transport mechanisms involved in generating the function. The reaction kinetics are usually described using standardized models, such Michaelis-Menten or Hill kinetics. Based on mass balances, a set of differential equations is then formulated. In principle, the presence of spatial gradients implies that partial differential equations should be used, while the relatively small copy number of the involved components implies that stochastic models should be used. In practice, however, these effects are usually neglected, and deterministic ordinary differential equations employed. The result is then a set of ordinary differential equations with a number of unknown parameters that need to be fitted to experimental data or prior knowledge of the system behavior. The outcome of this modeling process is thus a model structure that is uncertain due to incomplete knowledge of the involved components and reaction kinetics as well as to assumptions concerning spatial distributions and stochastic effects. Moreover, the fitted model parameters will be uncertain since the available data are limited and corrupted by measurement uncertainty. Thus model uncertainty is, in general, a combination of parametric and structural uncertainty, implying that robustness analysis for model validation should consider both parametric and structural perturbations.

Because the external and internal perturbations affecting a function can be seen as corresponding to changes in both the parameters and the structure of the model used to describe the function, parametric and structural uncertainty should also be considered when using validated models to analyze the robustness of a biological function. For instance, a temperature change will affect most kinetic parameters. Similarly, a gene mutation may fundamentally change the reaction kinetics or modify the transport properties of a protein, thereby effectively changing the corresponding model structure. In the next section, we discuss appropriate frameworks for analyzing the robustness to parametric and structural uncertainty.

10.3 Parametric and Structural Perturbations

Consider again the nominal model

$$\frac{\mathrm{d}\boldsymbol{x}(t)}{\mathrm{d}t} = \boldsymbol{f}(\boldsymbol{x}(t), \boldsymbol{p}), \quad \boldsymbol{x} \in \mathbf{R}^n,\ \boldsymbol{p} \in \mathbf{R}^m.$$

Parametric robustness analysis involves perturbing the parameter vector $\boldsymbol{p}$. That is, the perturbed model can be written

$$\frac{\mathrm{d}\boldsymbol{x}(t)}{\mathrm{d}t} = \boldsymbol{f}(\boldsymbol{x}(t), \boldsymbol{p}^* + \Delta\boldsymbol{p}), \tag{10.2}$$

where $\boldsymbol{p}^*$ is the nominal parameter vector and $\Delta\boldsymbol{p} = \boldsymbol{p} - \boldsymbol{p}^*$ is the parameter perturbation. Considering persistence of function, that is, robust stability, the aim of the robustness analysis is to determine the range of $\Delta\boldsymbol{p}$ such that the qualitative behavior of equation (10.2) remains unchanged. The robustness is usually quantified by the smallest norm $\|\Delta\boldsymbol{p}\|$ for which the perturbation qualitatively changes the behavior of the model.

As most network models are nonlinear and too complex to allow for any rigorous analytical studies, numerical approaches based on continuation methods combined with bifurcation analysis are usually employed in parametric robustness analysis (Seydel, 1994). These numerical methods cannot, in general, deal with multiple parameter perturbations, implying that the robustness analysis is employed for each individual parameter p_i in the parameter vector $\boldsymbol{p}$, that is, varying only one parameter at a time (Cross, 2003; Forger and Peskin, 2003; Leloup and Goldbeter, 2004; Ma and Iglesias, 2002; Wong et al., 2007).

Ma and Iglesias (2002) and Kim et al. (2006) proposed an approach to robustness analysis for biological oscillators based on multiple simultaneous parameter perturbations using methods from robust control theory and hybrid optimization techniques (see chapter 11). As they demonstrate on a model of cAMP oscillations, models that are robust to single parameter variations can be unrobust to multiple parameter perturbations.

To account for structural perturbations of the model (10.1), we propose a perturbed model of the form

$$\frac{\mathrm{d}\boldsymbol{x}(t)}{\mathrm{d}t} = \boldsymbol{f}(\boldsymbol{x}(t), \boldsymbol{p}) + \boldsymbol{f}_{\Delta}(\boldsymbol{x}(t), z(t), \boldsymbol{p}), \quad \boldsymbol{x} \in \mathbf{R}^n, \tag{10.3a}$$

$$\frac{\mathrm{d}z(t)}{\mathrm{d}t} = \boldsymbol{g}_{\Delta}(\boldsymbol{x}(t), z(t)), \quad z \in \mathbf{R}^m. \tag{10.3b}$$

This type of perturbation is termed a *dynamic structural perturbation* because the perturbation introduces new dynamic states $z(t)$ in the perturbed model. The extra states account for dynamics, such as delays and intermediate reactions steps, not captured by the original model. If no additional dynamic states are introduced by the perturbation, that is $m = 0$, then $\boldsymbol{f}_{\Delta}(\boldsymbol{x}(t), \boldsymbol{p})$ represents a static structural perturbation. Considering persistence of function, the aim of the robustness analysis is now to determine the smallest distance between the perturbed model (10.3) and the nominal model (10.1) such that their qualitative behaviors differ; in other words, if the nominal model produces stable sustained oscillations, for example, to determine the smallest perturbation that removes the oscillations. Solving this problem for a general class of nonlinear functions $\boldsymbol{f}_{\Delta}$ and $\boldsymbol{g}_{\Delta}$ is in general not feasible. As we show below, however, the problem can be reduced to a linear problem for which there exist powerful solution methods in classic and robust control theory. Also, within the linear framework, it is easy to impose restrictions on the structure of the perturbations $\boldsymbol{f}_{\Delta}$ and $\boldsymbol{g}_{\Delta}$ so as to not introduce biologically irrelevant network interactions in the perturbed model. Before presenting the robustness analysis, let us briefly consider the relation between persistence of function, stability, and bifurcations of dynamical systems.

10.4 Robustness Analysis Based on Inducing Bifurcations

In order to show how structural robustness of biochemical network models can be formulated as a linear robust stability problem, we need to briefly review bifurcations of nonlinear dynamic models (for a more detailed theoretical exposition of bifurcation theory, see Guckenheimer and Holmes, 1986; for a more practical introduction, see Seydel, 1994). Considering robustness in the sense of persistence of function, we are concerned with finding perturbations of the nominal model $d\boldsymbol{x}/dt = \boldsymbol{f}(\boldsymbol{x}(t), \boldsymbol{p})$ such that the qualitative behavior differs from the nominal behavior. As discussed above, the nominal behavior will, in general, correspond to a specific attractor in state space, such as an equilibrium point or a limit cycle, or a set of attractors, such as multiple stable equilibria. A qualitative change of behavior thus corresponds to a change in the number and type of attractors. The critical point in state-parameter space at which a nonlinear dynamic system changes its qualitative behavior is called a bifurcation point. Bifurcations can be either local or global, depending on the extent of behavior from which they can be determined. We are only concerned with local bifurcations here, in particular with those involving equilibrium points and limit cycles.

As discussed in chapter 1, a local bifurcation point corresponds to a point at which a local stationary behavior, or attractor, changes stability. For equilibrium points, local stability is determined from the eigenvalues of the Jacobian, $\boldsymbol{A}$, obtained from linearization of the model about the equilibrium point $\boldsymbol{x}^*$:

$$\boldsymbol{A} = \left.\frac{\partial \boldsymbol{f}}{\partial \boldsymbol{x}}\right|_{\boldsymbol{x}^*, \boldsymbol{p}}, \qquad \boldsymbol{f}(\boldsymbol{x}^*, \boldsymbol{p}) = \boldsymbol{0}. \tag{10.4}$$

A bifurcation occurs when some eigenvalues of $\boldsymbol{A}$ cross the imaginary axis in the complex plane as the parameter $\boldsymbol{p}$ is changed. If a real eigenvalue crosses through the origin, then it corresponds to a static bifurcation at which there is a change in the number of equilibrium attractors. The generic form of a static bifurcation is the so-called saddle-node bifurcation, at which an unstable and a stable equilibrium meet and collapse. Saddle-node bifurcations lead to bistability in biochemical networks, such as those underlying apoptosis (Eissing et al., 2005). A complex pair of eigenvalues crosses the imaginary axis at a Hopf bifurcation point where a limit cycle emerges from an equilibrium point, resulting in coexistence of an equilibrium point and a limit cycle close to the bifurcation point. Hopf bifurcations underlie the oscillations seen in many models of intracellular oscillators, such as circadian clock models.

For sustained oscillations, corresponding to limit cycle attractors, there exist two principal types of local bifurcations. The first is the multicycle bifurcation, at which multiple limit cycle attractors emerge. Analysis of such bifurcations requires linearization of $\boldsymbol{f}(\boldsymbol{x}, \boldsymbol{p})$ along a limit cycle, leading either to a discrete map or to a time-varying Jacobian $\boldsymbol{A}(t)$. The second type of bifurcation is the Hopf bifurcation at which a limit cycle collapses into an equilibrium point. This bifurcation can be analyzed based on the time-invariant Jacobian $\boldsymbol{A}$, equation (10.4), obtained at the equilibrium point.

From the discussion above, we conclude that robustness of a biochemical network model (10.1) can be analyzed based on determining the smallest perturbation, within a given class, that translates a steady-state solution of equation (10.1) into a bifurcation point.

Inducing a bifurcation at the nominal steady state of equation (10.1) corresponds to perturbing the Jacobian A so that it has one or two eigenvalues at the imaginary axis, with the remaining eigenvalues in the complex left half plane so that the bifurcation point corresponds to a shift in stability, corresponding to a saddle-node or Hopf bifurcation, respectively. For the structurally perturbed model (10.3), it is thus sufficient to consider the linearization

$$\frac{dx(t)}{dt} = Ax(t) + B_\Delta x(t) + F_\Delta z(t), \qquad x \in \mathbf{R}^n, \tag{10.5a}$$

$$\frac{dz(t)}{dt} = G_\Delta z(t) + H_\Delta x(t), \qquad z \in \mathbf{R}^m, \tag{10.5b}$$

where A, B_Δ, F_Δ, G_Δ and H_Δ are constant matrices of appropriate dimensions. The robustness problem is now reduced to determining B_Δ, F_Δ, G_Δ, H_Δ and m such that the perturbed system with Jacobian

$$A_\Delta = \begin{bmatrix} A + B_\Delta & F_\Delta \\ H_\Delta & G_\Delta \end{bmatrix} \tag{10.6}$$

has eigenvalues on the imaginary axis, with the remaining eigenvalues in the complex left half plane, while the distance between equation (10.5) and the nominal system $dx(t)/dt = Ax(t)$ is minimized. As shown below, this problem can be solved by casting it as a linear feedback control problem and employing frequency response-based methods. However, to make the problem as formulated in equation (10.5) meaningful in the context of biochemical networks, it is necessary to impose restrictions on the perturbation matrices. In particular, a typical biochemical network is sparse, in the sense that there exist relatively few direct interactions between the biochemical components, corresponding to a sparse Jacobian matrix A. If no restrictions are imposed on the matrices B_Δ, F_Δ, G_Δ, and H_Δ, then the perturbed model will, in general, contain direct interactions between all components corresponding to a full A_Δ. To impose reasonable restrictions on the allowable class of perturbations, we consider first putting the perturbed model (10.5) in input-output form so as to see the effect of the perturbations directly on the network interactions.

10.5 Putting the Perturbed Network Model in Feedback Form

Using the Laplace transform, the linear perturbed model (10.5) can be written in input-output form

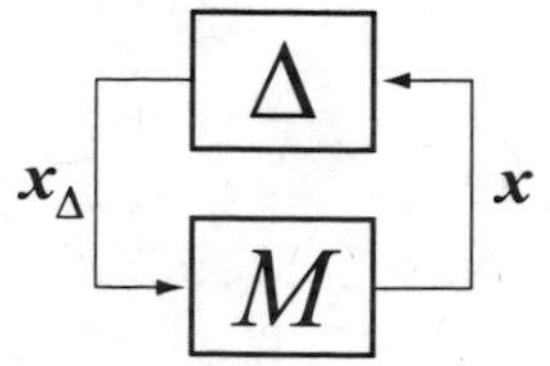

Figure 10.1
Perturbed model in feedback form with loop transfer function $\boldsymbol{M}(s)\boldsymbol{\Delta}(s)$, where $\boldsymbol{M}(s)$ is the nominal model and $\boldsymbol{\Delta}(s)$ is the perturbation.

$$\boldsymbol{x}(s) = \underbrace{(s\boldsymbol{I} - \boldsymbol{A})^{-1}\boldsymbol{A}}_{\boldsymbol{M}(s)}\, \boldsymbol{x}_\Delta(s), \tag{10.7a}$$

$$\boldsymbol{x}_\Delta(s) = \underbrace{\boldsymbol{A}^{-1}(\boldsymbol{B}_\Delta + \boldsymbol{F}_\Delta(s\boldsymbol{I} - \boldsymbol{G}_\Delta)^{-1}\boldsymbol{H}_\Delta)}_{\boldsymbol{\Delta}(s)}\, \boldsymbol{x}(s), \tag{10.7b}$$

where $\boldsymbol{x}_\Delta(t) = \boldsymbol{A}^{-1}(\boldsymbol{B}_\Delta \boldsymbol{x}(t) + \boldsymbol{F}_\Delta \boldsymbol{z}(t))$. We have assumed that the nominal Jacobian $\boldsymbol{A}$ has no eigenvalues at the origin. The perturbed model hence corresponds to a closed feedback loop with loop transfer function $\boldsymbol{M}(s)\boldsymbol{\Delta}(s)$ in which $\boldsymbol{M}(s)$ is the nominal model and $\boldsymbol{\Delta}(s)$ is the perturbation (see figure 10.1). Note that the eigenvalues of the Jacobian $\boldsymbol{A}$ in equation (10.4) now correspond to the poles of $\boldsymbol{M}(s)$, while the eigenvalues of the perturbed Jacobian $\boldsymbol{A}_\Delta$ (equation [10.6]) are given by the poles of the closed-loop transfer function $(\boldsymbol{I} - \boldsymbol{M}(s)\boldsymbol{\Delta}(s))^{-1}$.

The robustness problem now corresponds to determining the smallest size perturbation $\|\boldsymbol{\Delta}\|$ such that the closed-loop in figure 10.1 becomes marginally stable with a real eigenvalue at zero, corresponding to a saddle-node bifurcation in the corresponding nonlinear model, or a conjugate pair of imaginary eigenvalues, corresponding to a Hopf bifurcation point. Although results from classical and robust control theory can be used to solve this problem, we can significantly simplify the task if we first write the feedback loop in a form in which all loop transfer functions are stable. For this purpose, we impose that the perturbation $\boldsymbol{\Delta}(s)$ should be stable, and rewrite the transfer function for the nominal model $\boldsymbol{M}(s)$ as a feedback system composed of stable subsystems.

To rewrite the nominal model $\boldsymbol{M}(s)$ as a feedback system with a stable open-loop transfer function, we rewrite the nominal linearized model

$$\frac{\mathrm{d}\boldsymbol{x}(t)}{\mathrm{d}t} = \boldsymbol{A}(\boldsymbol{x}(t) + \boldsymbol{x}_\Delta(t)),$$

corresponding to $\boldsymbol{M}(s)$, in the form

$$\frac{\mathrm{d}\boldsymbol{x}(t)}{\mathrm{d}t} = \tilde{\boldsymbol{A}}\boldsymbol{x}(t) + (\boldsymbol{A} - \tilde{\boldsymbol{A}})(\boldsymbol{x}(t) + \boldsymbol{x}_\Delta(t)), \tag{10.8}$$

where $\tilde{\boldsymbol{A}}$ is a diagonal matrix with negative elements. All eigenvalues thus lie strictly in the left half plane. In most biochemical models, the self-dynamics of the biochemical components (states) will be stable due to self-degradation and lack of autocatalytic effects, and hence the diagonal elements of $\boldsymbol{A}$ will all be negative. A natural choice for $\tilde{\boldsymbol{A}}$ is then to let its diagonal be equal to the diagonal of $\boldsymbol{A}$. Note that equation (10.8), with $\tilde{\boldsymbol{A}}$ equal to the diagonal of $\boldsymbol{A}$, corresponds to letting the perturbation $\boldsymbol{x}_\Delta$ affect the interactions between the components only, while the self-dynamics are left unperturbed. The Laplace transform of equation (10.8) yields

$$\boldsymbol{x}(s) = \underbrace{(s\boldsymbol{I} - \tilde{\boldsymbol{A}})^{-1}(\boldsymbol{A} - \tilde{\boldsymbol{A}})}_{\boldsymbol{L}(s)}(\boldsymbol{x}(s) + \boldsymbol{x}_\Delta(s)), \tag{10.9}$$

where $\boldsymbol{L}(s)$ is stable. As shown in figure 10.2, the system (10.9) corresponds to a feedback loop with $\boldsymbol{L}(s)$ as the loop transfer function. The corresponding nominal model is then

$$\boldsymbol{M}(s) = (\boldsymbol{I} - \boldsymbol{L}(s))^{-1}\boldsymbol{L}(s).$$

Note that the formulation in equation (10.9) and figure 10.2 is a representation of the perturbed model from which one can easily make interpretations in terms of relative perturbations of the direct network interactions. In the unperturbed model, the direct interactions between the network components are given by $\boldsymbol{L}(s)$. For instance, element $L_{ij}(s)$ gives the effect of a change in component x_j on component x_i when all other components are held constant. Note that we are using the terms "state" and "component" interchangeably, assuming that the states correspond to concentrations of biochemical components. Thus $L_{ij}(s)$ is nonzero only if there exists a direct connection from x_j to x_i. For the perturbed model, the direct interactions are given by $\boldsymbol{L}_\Delta = \boldsymbol{L}(s)(\boldsymbol{I} + \boldsymbol{\Delta}(s))$. Thus, $\boldsymbol{\Delta}(s)$ can be interpreted as a relative perturbation of the

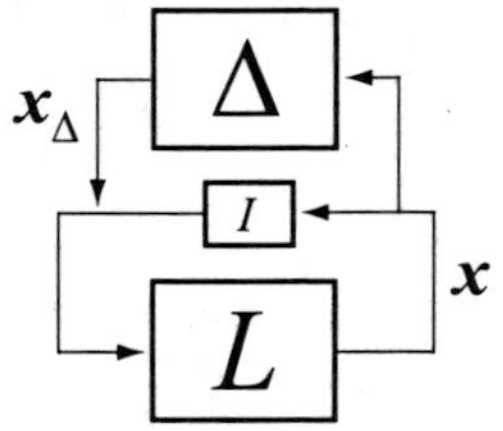

Figure 10.2
Perturbed model as a feedback system with a stable open-loop transfer function $\boldsymbol{L}(s)$. $\boldsymbol{\Delta}(s)$ is a stable perturbation.

direct interactions. Moreover, if $\boldsymbol{\Delta}(s)$ is restricted to be a diagonal matrix, only existing interactions are perturbed and network connections that do not exist in the nominal network model are not introduced in the perturbed model. Having put the perturbed network model in the form shown in figure 10.2, the smallest $\|\boldsymbol{\Delta}\|$ that will translate the nominal steady state of equation (10.1) into a saddle-node or Hopf bifurcation point can be determined using the generalized Nyquist criterion introduced below.

10.6 The Generalized Nyquist Stability Criterion for Biochemical Networks

Consider the linearized network model in feedback form in figure 10.2, which is a multivariable feedback loop with loop transfer function

$$\boldsymbol{L}_{\boldsymbol{\Delta}}(s) = \boldsymbol{L}(s)(\boldsymbol{I} + \boldsymbol{\Delta}(s)).$$

The generalized Nyquist stability criterion is based on the frequency response description $\boldsymbol{L}_{\boldsymbol{\Delta}}(j\omega)$ of the loop transfer function (Skogestad and Postlethwaite, 2005).

Theorem 10.1: Generalized Nyquist Stability Let P_{ol} denote the number of unstable poles in the loop transfer function $\boldsymbol{L}_{\boldsymbol{\Delta}}(s)$. Then the closed-loop system with positive unity feedback around $\boldsymbol{L}_{\boldsymbol{\Delta}}$ is stable if and only if the image of $\det(\boldsymbol{I} - \boldsymbol{L}_{\boldsymbol{\Delta}}(j\omega))$ in the complex plane, for $\omega \in [-\infty, \infty]$,

1. makes P_{ol} counterclockwise encirclements of the origin, and
2. does not pass through the origin. ■

Because the open-loop system $\boldsymbol{L}_{\boldsymbol{\Delta}}(s)$ here is stable by construction, that is, $P_{ol} = 0$, the condition for stability is that the image of $\det(\boldsymbol{I} - \boldsymbol{L}(j\omega))$ should not encircle 0. To refine the criterion, note that the determinant of any square matrix $\boldsymbol{L}$ can be written in terms of the eigenvalues $\lambda_i(\boldsymbol{L})$ as:

$$\det(\boldsymbol{L}) = \prod_i \lambda_i(\boldsymbol{L}).$$

Then, since

$$\lambda_i(\boldsymbol{I} - \boldsymbol{L}) = 1 - \lambda_i(\boldsymbol{L}),$$

the stability condition becomes that no eigenvalue locus $\lambda_i(\boldsymbol{L}_{\boldsymbol{\Delta}}(j\omega))$ for $\omega \in [-\infty, \infty]$ should encircle the point $+1$ on the positive real line in the complex plane. Note that the classical Nyquist theorem usually has -1, rather than $+1$, as the critical point since a negative feedback structure is assumed for technical control systems. Because any real system is strictly proper, implying that $\lim_{\omega \to \infty} \boldsymbol{L}_{\boldsymbol{\Delta}}(j\omega) = 0$, an encirclement of the +1 point implies that the image crosses the real axis to the right of +1 for $\omega \in [0, \infty]$.

We are mainly interested in determining a structural perturbation $\mathbf{\Delta}(s)$ such that the perturbed system (10.7) becomes marginally stable, corresponding to a bifurcation point in the nonlinear system (10.3). Thus we seek a perturbation $\mathbf{\Delta}$ such that one eigenvalue locus satisfies

$$\lambda_i(\boldsymbol{L}_{\Delta}(j\omega^*)) = 1, \tag{10.10}$$

for some i and some frequency ω^*. If condition (10.10) is satisfied at frequency $\omega = 0$, corresponding to steady state, then the perturbed system has a Jacobian $\boldsymbol{A}_{\Delta}$ with a real eigenvalue at zero, that is, $\lambda_k(\boldsymbol{A}_{\Delta}) = 0$ for some k. This corresponds to a saddle-node bifurcation in the corresponding perturbed nonlinear model (10.3). If instead, condition (10.10) is fulfilled at a nonzero frequency $\omega = \omega^* > 0$, then the perturbed system has a Jacobian $\boldsymbol{A}_{\Delta}$ with a pair of purely imaginary eigenvalues at $\lambda_{k,l}(\boldsymbol{A}_{\Delta}) = \pm j\omega^*$. This corresponds to a Hopf bifurcation in the corresponding perturbed nonlinear model (10.3).

If the nominal steady state is stable, corresponding to a stable $\boldsymbol{M}(s)$, then equation (10.10) is a necessary and sufficient condition for inducing a bifurcation in the system. If the nominal steady state is unstable, which is the case when considering robustness of limit cycles, then the condition is necessary but not sufficient. This is because some locus of the nominal loop $\lambda_j(\boldsymbol{L}(j\omega))$ is encircling the +1 point, and a perturbation $\mathbf{\Delta}$ may move some other locus $\lambda_i(\boldsymbol{L}(j\omega))$, $i \neq j$ to the +1 point, implying that there still is encirclement. As shown below, however, this can easily be checked and, if it is the case, then the computed perturbation will provide a lower bound on the size of the perturbation required to make the system marginally stable.

By introducing $\boldsymbol{M} = (\boldsymbol{I} - \boldsymbol{L})^{-1}\boldsymbol{L}$, we may write the loop in figure 10.2 in the form of figure 10.1, and the condition for marginal stability becomes

$$\lambda_i(\boldsymbol{M}(j\omega)\mathbf{\Delta}(j\omega)) = 1, \tag{10.11}$$

for some ω and some i. Based on equation (10.11) and classical results from robust control theory, we can now compute the smallest perturbation $\mathbf{\Delta}(j\omega)$ at each frequency ω required to make the system marginally stable, corresponding to inducing a bifurcation point.

If $\mathbf{\Delta}(j\omega)$ is allowed to be a full complex matrix, then the smallest $\mathbf{\Delta}$ at each frequency that will satisfy equation (10.11) will have size (see, for example, Skogestad and Postlethwaite, 2005):

$$\bar{\sigma}(\mathbf{\Delta}(j\omega)) = \frac{1}{\bar{\sigma}(\boldsymbol{M}(j\omega))},$$

where $\bar{\sigma}$ denotes the maximum singular value. The smallest size perturbation is then

$$\|\boldsymbol{\Delta}_{\text{min}}\| = \min_{\omega} \bar{\sigma}(\boldsymbol{\Delta}(j\omega)) = \min_{\omega} \frac{1}{\bar{\sigma}(\boldsymbol{M}(j\omega))}.$$

As discussed above, a full $\boldsymbol{\Delta}$ corresponds to introducing interactions between all network components in the perturbed model and is usually not biologically relevant.

If $\boldsymbol{\Delta}$ is restricted to be a diagonal matrix $\boldsymbol{\Delta}_I$, it corresponds to relative individual perturbations of the activities of each component in the network

$$x_i^{\Delta} = (1 + \Delta_{I,ii})x_i,$$

where x_i and x_i^{Δ} denote the unperturbed and perturbed activity of component i, respectively, and $\Delta_{I,ii}$ is the ith diagonal element of $\boldsymbol{\Delta}_I$. For this case, the smallest $\|\boldsymbol{\Delta}\|$ that moves an eigenlocus to $+1$ can be computed using the structured singular value μ (Skogestad and Postlethwaite, 2005):

$$\frac{1}{\mu_{\boldsymbol{\Delta}_I}(\boldsymbol{M})} = \min_{\boldsymbol{\Delta}_I}\{\bar{\sigma}(\boldsymbol{\Delta}_I) \mid \det(\boldsymbol{I} - \boldsymbol{M}\boldsymbol{\Delta}_I) = 0\}, \tag{10.12}$$

where $\boldsymbol{\Delta}_I$ is a diagonal complex matrix. The corresponding smallest perturbation is then

$$\|\boldsymbol{\Delta}_{\text{min}}\| = \min_{\omega} \frac{1}{\mu_{\boldsymbol{\Delta}_I}(\boldsymbol{M})}. \tag{10.13}$$

Computations of the structured singular value μ only provide lower and upper bounds on the size of $\boldsymbol{\Delta}$. For the case of a diagonal complex $\boldsymbol{\Delta}$, however, the bounds are in general tight (Skogestad and Postlethwaite, 2005). Note that a complex perturbation $\boldsymbol{\Delta}(j\omega)$ in the frequency domain corresponds to a dynamic perturbation in the time domain. This corresponds to the state-space matrices $\boldsymbol{G}_{\Delta}$ and $\boldsymbol{H}_{\Delta}$ being nonzero in the perturbed model (10.5). One advantage of using frequency domain computations is that the dynamics of a system then are uniquely determined by the amplification and phase lag at each frequency, and hence the order m of the perturbation does not need to be specified or computed. Note that the computations of $\boldsymbol{\Delta}$ are performed frequency by frequency and that the smallest stabilizing $\boldsymbol{\Delta}$ corresponds to the minimal $\boldsymbol{\Delta}$ over all frequencies. As shown below, the corresponding frequency response can easily be fitted to a general transfer function, which can then be realized as a set of differential equations corresponding to $\boldsymbol{B}_{\Delta}$, $\boldsymbol{F}_{\Delta}$, $\boldsymbol{G}_{\Delta}$, and $\boldsymbol{H}_{\Delta}$ in equation (10.5).

10.6.1 Perturbing Specific Network Interactions

Although the case with a diagonal perturbation $\boldsymbol{\Delta}_I$ is relevant when considering the overall robustness of a biochemical network, to obtain more detailed information on

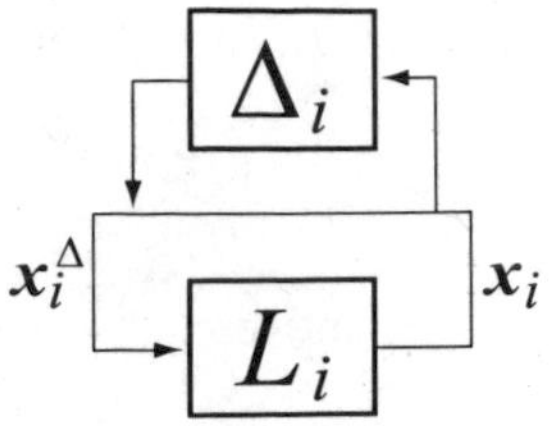

Figure 10.3
Perturbing the activity of a single component i.

which components and specific interactions are least robust, we need also to consider more strategic perturbations of specific components and interactions. We first perturb the activity of a single component i, as illustrated in figure 10.3. Here we obtain the open-loop transfer function $L_i(s)$ by lifting out the effect of state x_i on all other states x_j, $j \neq i$ and then reintroducing the effect by closing the loop. We can then perturb the effect of x_i on the other components with a relative dynamic perturbation, Δ_i, as shown in figure 10.3.

To obtain $L_i(s)$ from the nominal linearized model $\frac{dx(t)}{dt} = \boldsymbol{A}\boldsymbol{x}(t)$, introduce the matrix $\boldsymbol{P}^{1i}$, which is obtained from the $n \times n$ identity matrix by letting element $P_{ii}^{1i} = 0$. Similarly, let $\boldsymbol{P}^{2i}$ be a $1 \times n$ vector with $P_i^{2i} = 1$ and with all other elements zero to obtain the loop transfer function

$$L_i(s) = \boldsymbol{P}^{2i}(\boldsymbol{I} - \boldsymbol{L}(s)\boldsymbol{P}^{1i})^{-1}\boldsymbol{L}(s)(\boldsymbol{P}^{2i})^T, \tag{10.14}$$

where $\boldsymbol{L}(s)$ is given by equation (10.9). In this case, the nominal loop transfer function $L_i(s)$ is a scalar. Provided $L_i(s)$ is stable, the condition for marginal stability of the perturbed system with loop transfer function $L_{\Delta,i} = L_i(1 + \Delta_i)$ is that the frequency response $L_{\Delta,i}(j\omega)$ must pass through the point $+1$ on the positive real axis in the complex plane. The Δ_i that satisfies this is given at each frequency by

$$\Delta_i(j\omega) = \frac{1}{L_i(j\omega)} - 1. \tag{10.15}$$

The smallest stabilizing perturbation is obtained by minimizing $|\Delta_i(j\omega)|$ over all frequencies

$$\|\Delta_{i,\min}\| = \min_{\omega}\left|\frac{1}{L_i(j\omega)} - 1\right|.$$

It is also of interest to consider perturbing the direct interactions between two specific components x_i and x_j. The corresponding perturbed feedback loop is shown in figure 10.4. Here $L_{ij}(s)$ is obtained from the full network by removing the effect of

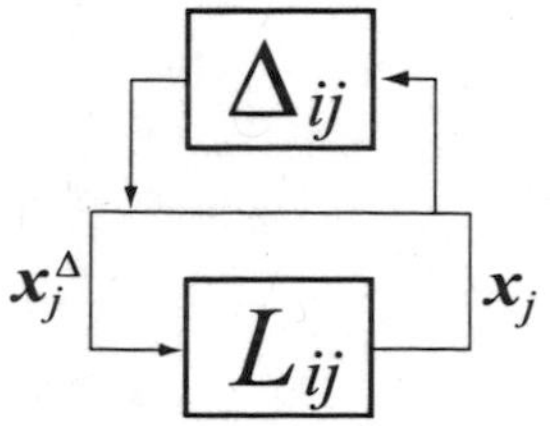

Figure 10.4
Perturbing the specific interaction between two components i and j.

changes in x_j on x_i, and then the loop is closed by introducing a perturbed x_j. Introduce the $n \times n$ matrix $\boldsymbol{C}^{ij}$ with $C_{ij}^{ij} = 1$ and with all other elements zero to obtain the scalar loop transfer function

$$L_{ij}(s) = \boldsymbol{P}^{2j}(s\boldsymbol{I} - \boldsymbol{A} + \boldsymbol{C}^{ij}\boldsymbol{A}^T\boldsymbol{C}^{ij})^{-1}\boldsymbol{C}^{ij}\boldsymbol{A}^T\boldsymbol{C}^{ij}(\boldsymbol{P}^{2j})^T. \tag{10.16}$$

Provided $L_{ij}(s)$ is stable, the perturbation

$$\Delta_{ij}(j\omega) = \frac{1}{L_{ij}(j\omega)} - 1 \tag{10.17}$$

will make the network steady state marginally stable. The smallest stabilizing perturbation is obtained by minimizing $|\Delta_{ij}(j\omega)|$ over all frequencies:

$$\|\Delta_{ij,\min}\| = \min_{\omega}\left|\frac{1}{L_{ij}(j\omega)} - 1\right|. \tag{10.18}$$

The use of these more targeted structural perturbations is important both in detecting and locating specific network fragilities and in elucidating the most important interactions underlying the modeled function, as will be demonstrated in the three case studies below (see sections 10.8–10.10). To illustrate the practical use and interpretations of the proposed robustness analysis method, we first consider a simple gene regulatory feedback loop model.

10.7 The Goodwin Oscillator

In this classic model describing autonomous periodic oscillations in a simple gene regulatory network (Goodwin, 1965), a gene transcribes mRNA, with concentration x_1, which is translated into protein in the cytoplasm, x_2, and subsequently transported back into the nucleus to act as a transcription factor, x_3, regulating gene expression (see figure 10.5). The Goodwin oscillator can be described by the following set of equations (Gonze et al., 2005):

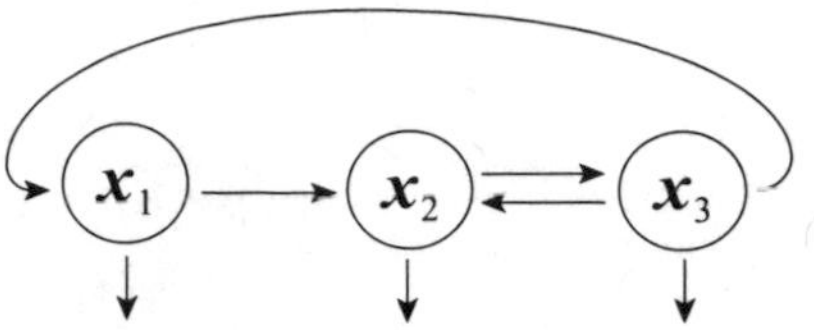

Figure 10.5
Goodwin oscillator describing autonomous periodic oscillations in a gene regulatory network. The state x_1 corresponds to mRNA, x_2 to protein in cytoplasm, and x_3 to protein in the nucleus, which acts as a transcription factor for the gene, thereby closing the feedback loop.

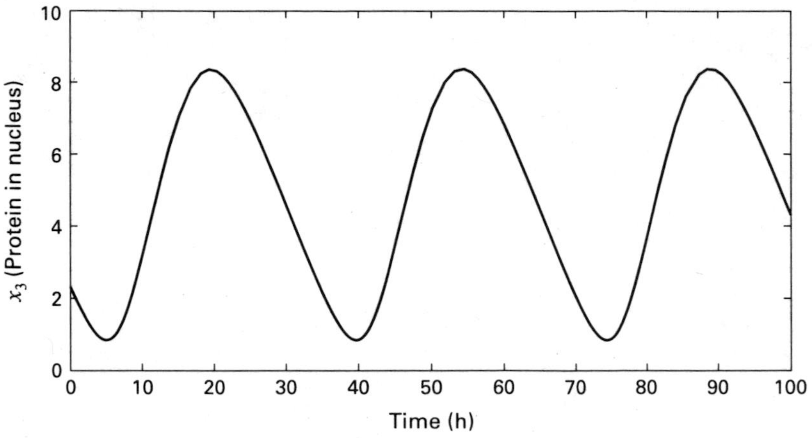

Figure 10.6
Periodic oscillations in the concentration of transcription factor x_3 of the Goodwin oscillator.

$$\frac{dx_1(t)}{dt} = v_1 \frac{K_1^n}{K_1^n + x_3^n(t)} - v_2 \frac{x_1(t)}{K_2 + x_1(t)}, \tag{10.19a}$$

$$\frac{dx_2(t)}{dt} = k_3 x_1(t) - v_4 \frac{x_2(t)}{K_4 + x_2(t)} + \epsilon x_3(t), \tag{10.19b}$$

$$\frac{dx_3(t)}{dt} = k_5 x_2(t) - v_6 \frac{x_3(t)}{K_6 + x_3(t)}. \tag{10.19c}$$

For illustrative purposes, an ϵx_3 term has here been added to the dynamics of x_2. For the parameter values $v_1 = 0.7$, $K_1 = 1$, $n = 4$, $v_2 = 0.2$, $K_2 = 1$, $k_3 = 0.7$, $v_4 = 1.2$, $K_4 = 1$, $k_5 = 0.7$, $\epsilon = 0.01$, $v_6 = 0.7$, and $K_6 = 1$, the Goodwin oscillator generates self-sustained periodic oscillations, as shown in figure 10.6.

A bifurcation diagram for the Goodwin oscillator is shown in figure 10.7. For $v_1 = 0.22$, there is a Hopf bifurcation at which the steady state exchanges stability,

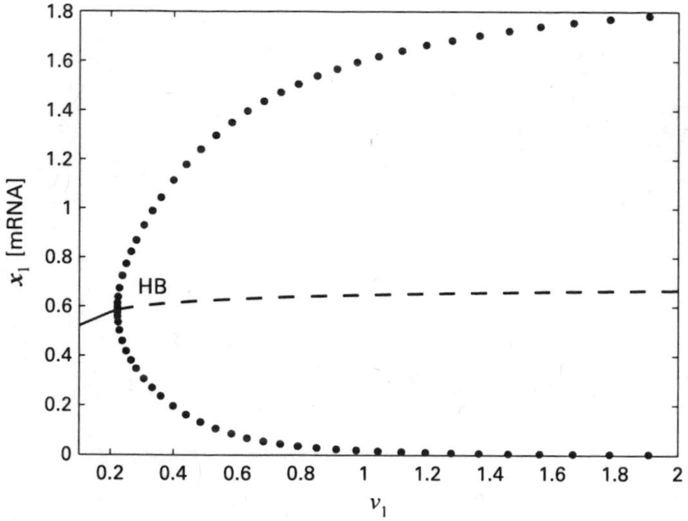

Figure 10.7
Bifurcation diagram for the Goodwin oscillator, showing stable (*solid line*) and unstable (*dashed line*) steady states, as well as amplitudes of stable limit cycles (*filled circles*). The nominal value of the bifurcation parameter is $v_1 = 0.7$. Hopf bifurcations (HB) connect branches of steady states and limit cycles.

and a branch of limit cycles emerges. The Hopf bifurcation is supercritical in this case and the limit cycle is hence stable. For the nominal value of the bifurcation parameter, $v_1 = 0.7$, there is consequently a stable limit cycle coexisting with an unstable steady state. The robustness analysis addresses perturbations that could translate the unstable steady state into a Hopf bifurcation point onto which the limit cycle collapses, thus resulting in removal of the autonomous periodic solution. Note that in terms of parametric robustness, the parameter v_1 can be changed by 70% without losing the sustained oscillations. A parametric robustness analysis, perturbing individual parameters, reveals that the least robust parameter is v_6, which can be changed by only 35% before the sustained oscillations disappear.

The open-loop network and the recovered closed-loop network of the Goodwin oscillator are illustrated in figure 10.8. The arrows connecting the boxes represent direct interactions between the components, and the robustness analysis involves perturbing these interactions. Note that the self-dynamics of the components, such as their degradation, will not be perturbed. Perturbing an interaction, such as the direct effect of component x_3 on x_1, corresponds to perturbing the dynamic effect of changes in component x_3 on changes in component x_1. The resulting change in x_1 then propagates to affect all other components in the closed-loop network and some of these changes are fed back to give a secondary change in x_3. We turn next to the minimal dynamic perturbation that changes the steady-state stability by inducing a Hopf

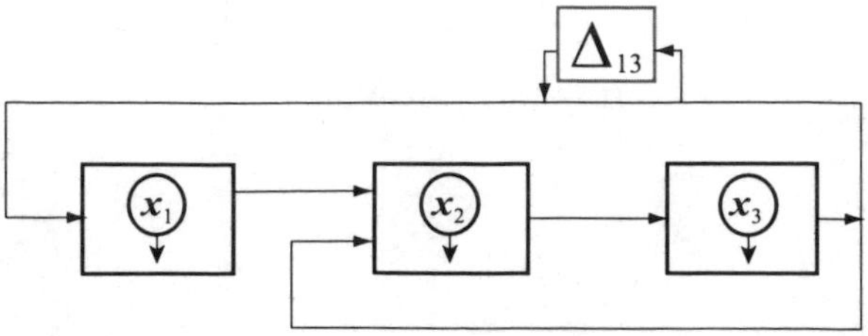

Figure 10.8
Addition of a relative dynamic perturbation to the direct effect of the transcription factor x_3 on the gene activity x_1 in the Goodwin oscillator.

bifurcation at the nominal conditions, and hence changes the qualitative behavior of the nominal model.

Consider first the case in which the direct interactions between two components are perturbed. Figure 10.8 illustrates the specific perturbation of the pairwise interaction between transcription factor x_3 and gene activity x_1 for the Goodwin oscillator. The perturbation can represent any change in the dynamic effect of the transcription factor on the mRNA concentration, such as an amplification or a delay of the effect. The aim of the analysis is to determine the smallest such perturbation that removes the oscillatory behavior by inducing a Hopf bifurcation at the underlying steady state.

In terms of the loop transfer function $L_{13}(s)$, a marginally stabilizing perturbation at a given frequency will move the Nyquist curve to the point $+1$ on the positive real axis at that frequency. The size of the required perturbation as a function of the frequency, computed from equation (10.17), is shown in figure 10.9, which shows the inverse of the size of Δ to highlight the minimum-size perturbation. The minimum-size perturbation is $\Delta_{\text{min},13} = -0.112 + 0.174j$ with size $|\Delta_{\text{min},13}| = 0.207$ at the frequency $\omega = 0.345$ rad/h, corresponding to a sinusoidal perturbation with period $T = 2\pi/\omega = 18$ hours.

Figure 10.10 shows the nominal Nyquist locus for the loop transfer function L_{13}, as well as three instances of marginally stabilizing perturbations that move the locus to $+1$ at three specific frequencies. Figure 10.10b shows a static (real) perturbation that moves the point at which the nominal locus crosses the real axis to the $+1$ point. The size of this relative perturbation is $\|\Delta_{13}\| = 0.43$. Figure 10.10c shows a pure phase shift of L_{13}, corresponding to Δ introducing a pure time delay in the loop. The size of this perturbation is $\|\Delta_{13}\| = 2$. Finally, figure 10.10d shows the effect of the minimum-size perturbation with $\|\Delta_{13}\| = 0.207$ at frequency $\omega = 0.345$, computed using equation (10.17).

The smallest stabilizing perturbation of this specific interaction corresponds to a 20% relative change in the effect of the transcription factor on the gene activity. The

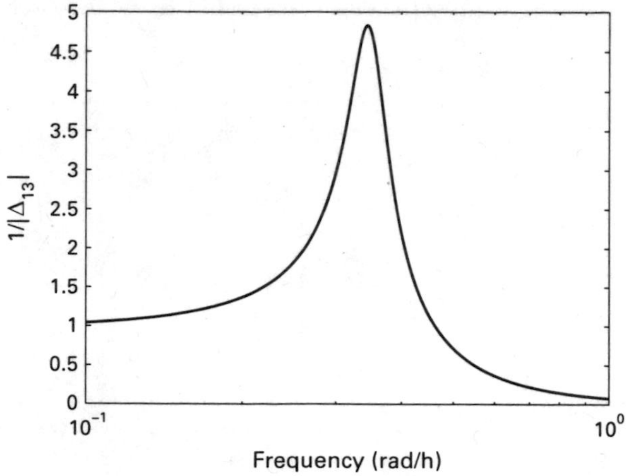

Figure 10.9
Size of stabilizing perturbation as a function of frequency when perturbing the specific interaction of x_3 on x_1 for the Goodwin oscillator.

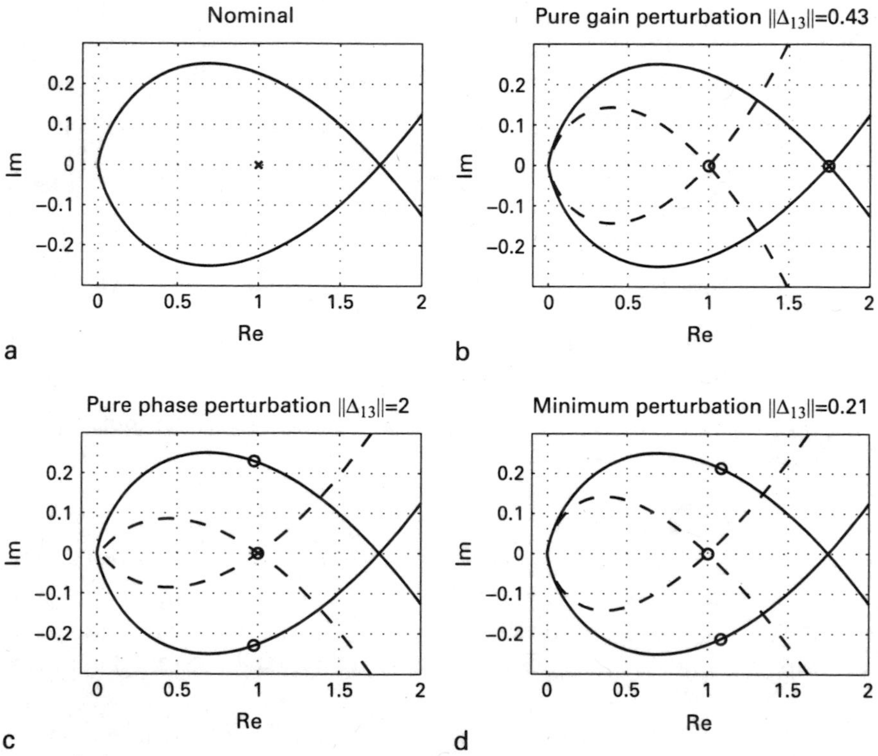

Figure 10.10
Nyquist curve $L_{13}(j\omega)$ for the Goodwin oscillator, showing unperturbed (*solid lines*) and perturbed (*dashed lines*) network with different stabilizing perturbations Δ_{13}.

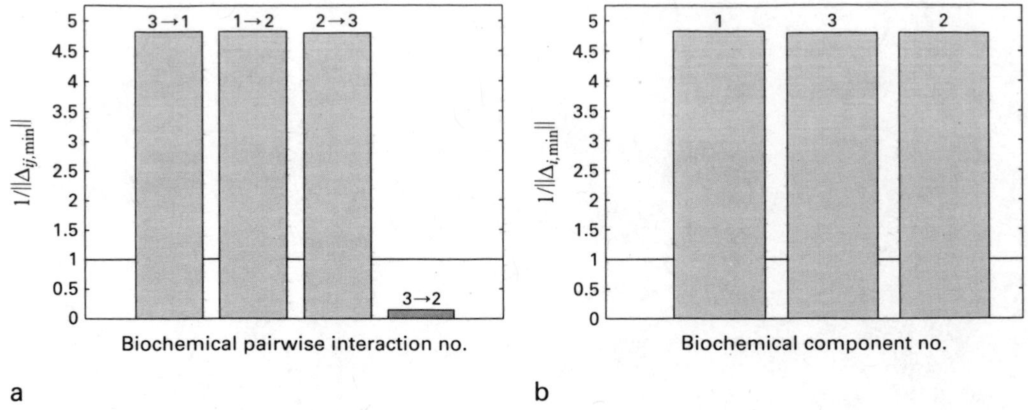

Figure 10.11
Size of pairwise perturbations (a) and individual component perturbations (b) required to induce a Hopf bifurcation in the Goodwin oscillator.

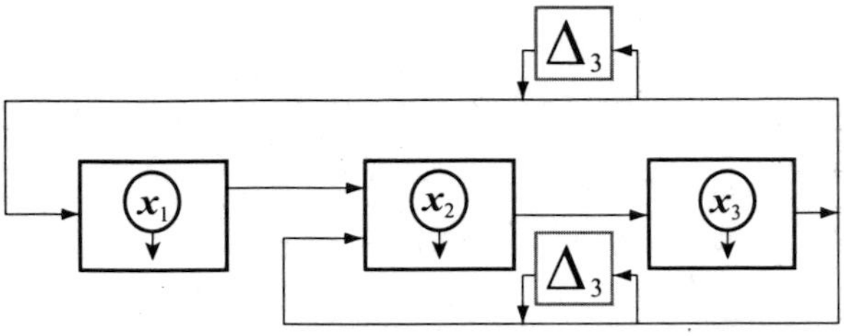

Figure 10.12
Addition of a perturbation to the activity of a single component, x_3, for the Goodwin oscillator.

minimum perturbation affects both the amplification of changes in the transcription factor as well as the phase lag of the effect.

Figure 10.11a shows the minimum-size relative perturbation for all pairwise interactions in the Goodwin oscillator. As can be seen, the interactions that form the main feedback loop all require the same-size perturbations, which is as expected for a single scalar feedback loop. The local feedback from x_3 to x_2 requires a significantly larger relative perturbation, indicating that this specific interaction has a relatively small impact on the nominal steady-state instability and hence the limit cycle oscillation.

In figure 10.12, a single component x_3 is perturbed by a relative perturbation Δ_3. The corresponding dynamic network perturbation computed according to equation (10.15) is shown in figure 10.13. As can be seen, the required perturbation is almost identical to the one for the specific interaction from x_3 to x_1 above. This is also as

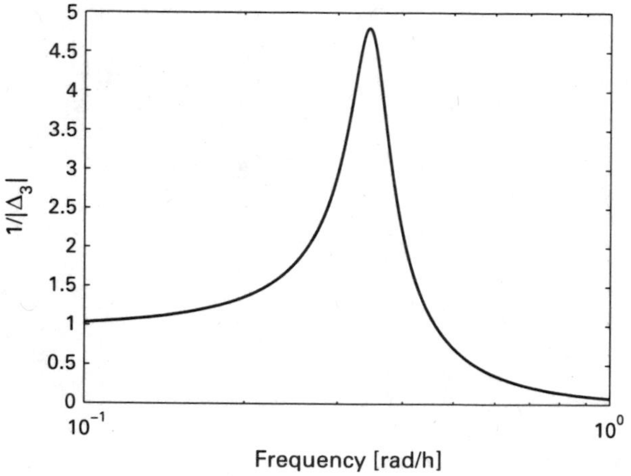

Figure 10.13
Size of stabilizing perturbation when perturbing the activity of the transcription factor x_3 in the Goodwin oscillator.

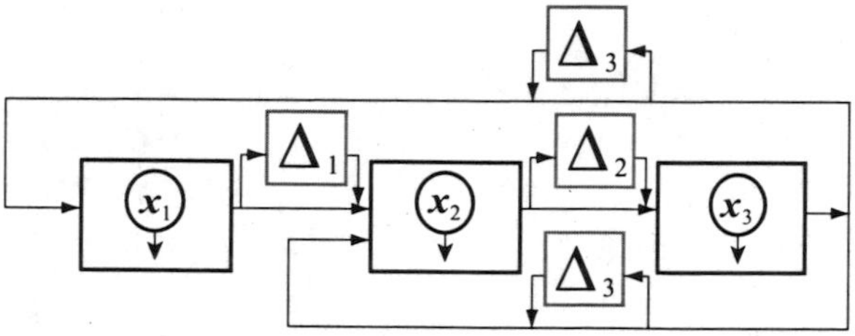

Figure 10.14
Addition of perturbations to the activity of all components simultaneously for the Goodwin oscillator.

expected since the perturbation now only perturbs the effect of x_3 on x_2 in addition to the effect of x_3 on x_1, and since the former interaction was found to be relatively robust to perturbations. As shown in figure 10.11b, the required perturbations for all components when perturbed individually are seen to be 20%. Again, this is expected since they all form a single feedback loop.

Perturbations affecting the activity of all components simultaneously are illustrated in figure 10.14. The required size of the dynamic network perturbation as a function of frequency, computed according to equation (10.12), is shown in figure 10.15. As can be seen, the smallest stabilizing perturbation has $\|\Delta\| = 0.072$, corresponding to a 7.2% simultaneous perturbation in the activity (concentration) of all network components.

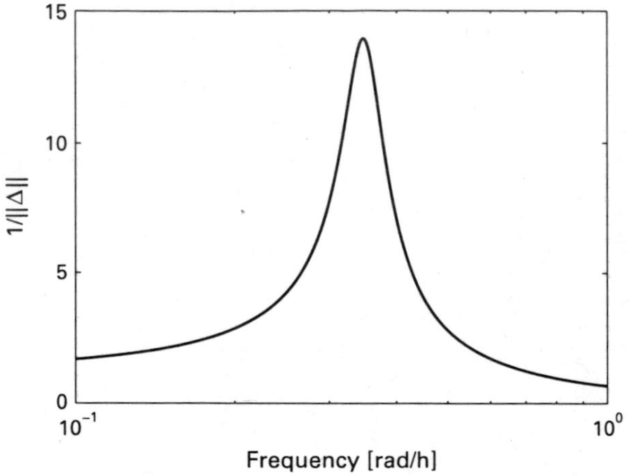

Figure 10.15
Smallest stabilizing perturbations for simultaneous perturbations of all component activities in the Goodwin oscillator.

10.7.1 Perturbing the Nonlinear Model

The dynamic network perturbations $\Delta(j\omega)$ computed above apply to linear dynamical systems with the amplification and phase lag given at a single frequency only. To implement the perturbations in the nonlinear model, for example, to verify their impact on the network function, they must first be fitted to a system description consisting of a set of linear differential equations.

Any scalar perturbation $\Delta_{\min}(j\omega^*)$ at a single frequency ω^* can be fitted to a stable second-order transfer function. Since only purely dynamic effects affect the existence of a limit cycle, the steady-state gain of the transfer function can always be chosen to be zero such that the steady-state properties of the network are unaffected by the perturbations. Imposing also that the transfer function should be strictly proper, we obtain the third-order transfer function

$$\Delta_T(s) = K\frac{-Ts+1}{Ts+1}\frac{\tau_1 s}{\tau_1 s+1}\frac{1}{\tau_2 s+1}, \tag{10.20}$$

where K adjusts amplification, $(-Ts+1)/(Ts+1)$ adjusts phase, $\tau_1 s/(\tau_1 s+1)$ provides zero steady-state gain and $1/(\tau_2 s+1)$ ensures zero gain at high frequencies. The parameters K, T, τ_1, and τ_2 can be chosen to obtain the correct frequency response at ω^*:

$$\Delta_T(j\omega^*) = \Delta_{\min}(j\omega^*),$$

and ensure that the norm of Δ_T equals the norm of $\Delta_{\min}(j\omega^*)$:

$$\max_{\omega} |\Delta_T(j\omega)| = |\Delta_{\min}(j\omega^*)|.$$

The fitted transfer function $\Delta_T(s)$ can be realized as a set of three first-order linear differential equations, which can then be combined with the nonlinear differential equations of the original model (10.1). Note that the input to the linear perturbation should be deviations of the network state variables x from their steady-state values, and that the output of the perturbation system are perturbations added to the state variables x. The smallest pairwise perturbation, $\Delta_{\min,13}$, to transform the unstable steady state into a Hopf bifurcation point for the Goodwin oscillator was implemented in the nonlinear model (10.19) according to the above procedure. The response $\Delta_{13}(j\omega^*) = -0.112 + 0.174j$ at $\omega^* = 0.345$ was fitted to the transfer function (10.20) with $K = -0.24$, $T = 2.12$, $\tau_1 = 5.80$ and $\tau_2 = 0.58$. The transfer function Δ_T can be realized as a set of linear differential equations:

$$\frac{dz_1(t)}{dt} = \frac{1}{\tau_1}(-z_1(t) + x_3(t)),$$

$$\frac{dz_2(t)}{dt} = \frac{1}{T}(-z_2(t) + 2K(x_3(t) - z_1(t))),$$

$$\frac{dz_3(t)}{dt} = \frac{1}{\tau_2}(-z_3(t) + z_2(t) - K(x_3(t) - z_1(t))),$$

$$x_3^{\Delta}(t) = x_3(t) - z_1(t),$$

and the perturbation introduced in the differential equation for x_1 as

$$\frac{dx_1(t)}{dt} = v_1 \frac{K_1^n}{K_1^n + (x_3(t) + x_3^{\Delta}(t))^n} - v_2 \frac{x_1(t)}{K_2 + x_1(t)}. \tag{10.21}$$

The resulting bifurcation diagram is shown in figure 10.16. As can be seen, the formerly unstable steady state at $v_1 = 0.7$ of the nominal Goodwin oscillator (figure 10.7) has been translated into a Hopf bifurcation point, and the limit cycle has collapsed into the Hopf bifurcation. The qualitative behavior of the model has changed and the autonomous periodic solution is removed by the perturbation $\Delta_{\min,13}$.

A simulation of the perturbed nonlinear Goodwin oscillator is shown in figure 10.17. Because, for the first 50 hours, the feedback loop through the perturbation is not closed, the perturbation does not exhibit its full effect on the nonlinear model. As can be seen, over this time interval, the perturbation reduces the effect of transcription

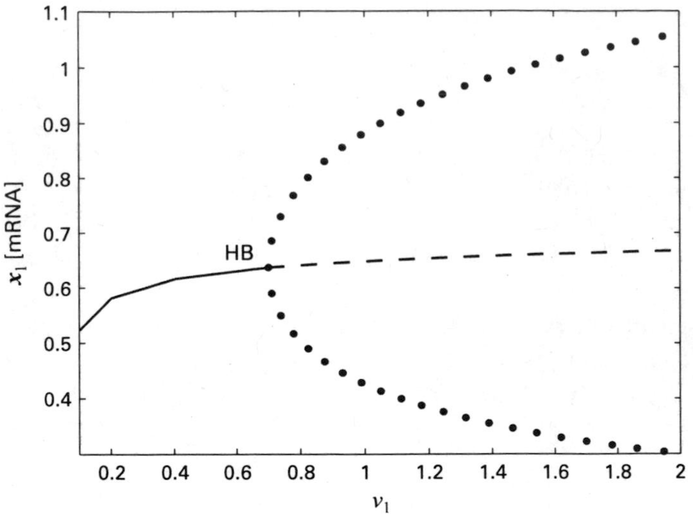

Figure 10.16
Bifurcation diagram for the perturbed Goodwin oscillator. The perturbation has induced a Hopf bifurcation (HB) at the nominal parameter value $v_1 = 0.7$. Compare with the nominal bifurcation diagram before the perturbation in figure 10.7.

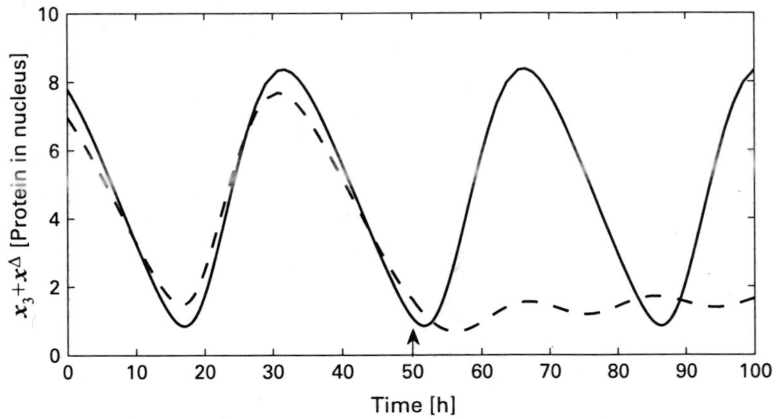

Figure 10.17
Simulation of the perturbed Goodwin oscillator, showing nominal (*solid line*) and perturbed (*dashed line*) state. After 50 hours, the feedback loop is closed, and the perturbation starts affecting the nonlinear model, resulting in dampened oscillations.

factor x_3 on gene activity x_1 by approximately 10% while advancing the effect about one hour. The phase advance can be seen as a reduction in the delay of the transcription reaction. When closing the loop with the perturbation after 50 hours, the periodic oscillations dampen out and the system goes to a stable steady state.

The Goodwin oscillator was presented to shed light on the biological interpretation of applied perturbations and their impact on the network function. We now turn to more complex networks to demonstrate the usefulness of our proposed method in analyzing robustness, detecting fragilities, elucidating key mechanisms, and reducing models.

10.8 Case Study 1: The MAPK Signaling Cascade

One of the most studied signaling systems in the literature (Qiao et al., 2007), the mitogen-activated protein kinase (MAPK) signaling cascade is conserved in most eukaryotic cells. An early mathematical model of the cascade (Huang and Ferrell, 1996) postulates 22 biochemical components and 10 biochemical reactions, and consists of 22 ordinary differential equations with 37 parameters. The presence of 7 moiety conservation relationships implies that the model can be reduced to 15 ODEs combined with 7 algebraic equations (Qiao et al., 2007). The signaling network is illustrated in figure 10.18.

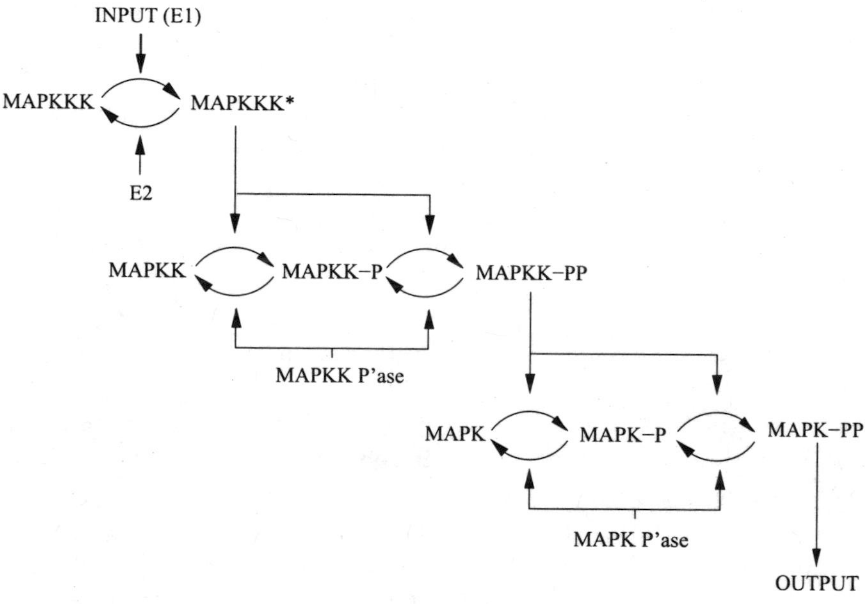

Figure 10.18
Mitogen-activated protein kinase (MAPK) signaling cascade.

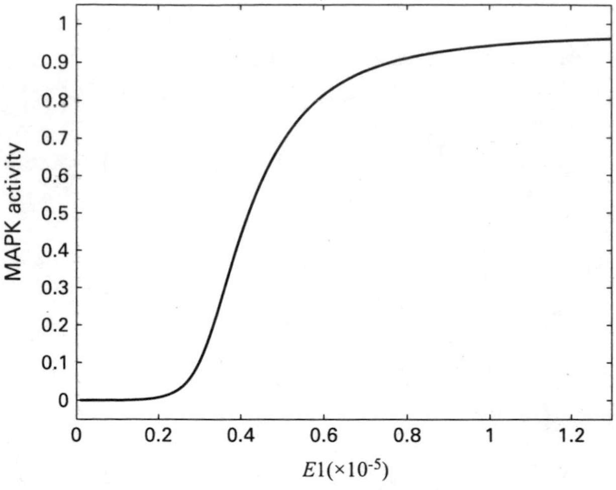

Figure 10.19
Input-output response curve of the nominal Huang and Ferrell model (1996).

Huang and Ferrell (1996) show that the MAPK cascade can produce an input-output response curve that is ultrasensitive, having a large slope in the transition from low- to high-output response. The input-output response curve for the nominal model is shown in figure 10.19. In later papers, Xiong and Ferrell (2003) demonstrate that adding a positive feedback around the ultrasensitive response curve, from output to input, can result in a bistable response, and Kholodenko (2000) shows that ultransensitivity combined with negative feedback can cause sustained oscillations in the MAPK cascade. More recently, Qiao et al. (2007) show, through an extensive multi-dimensional parameter search evaluating 20,000 different parameter sets, that the original Huang and Ferrell model without external feedback is also capable of displaying bistability and sustained oscillations for specific choices of parameter combinations. Here we analyze the nominal Huang and Ferrell model to determine its structural robustness: how large the structural perturbations must be to translate the ultrasensitive response curve into a bistable and oscillating response curve, respectively. We choose $E1_{\text{tot}} = 4 \times 10^{-6}$ as the nominal input value, corresponding to an output activity of MAPK-PP $= 0.44$.

Let us first consider the overall robustness of the network when all 15 states are perturbed simultaneously, corresponding to a diagonal perturbation matrix $\mathbf{\Delta}_I$. The corresponding minimum-size perturbation, as computed using the structured singular value μ in equation (10.12), is shown as a function of frequency in figure 10.20. Note that only lower and upper bounds for μ can be computed, but that these are so tight that they cannot be distinguished. The minimum perturbation, corresponding to the maximum $\mu_{\mathbf{\Delta}_I}$-value, is obtained at frequency $\omega = 0$, corresponding to

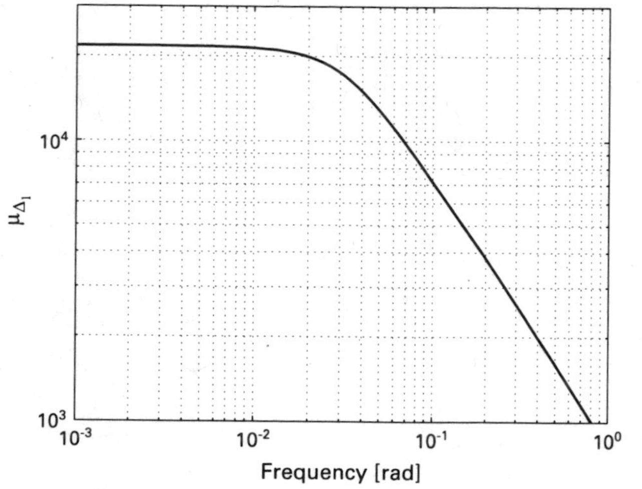

Figure 10.20
Measure of the overall structural robustness of the Huang and Ferrell model (1996), using a complex diagonal perturbation $\mathbf{\Delta}_I$ corresponding to perturbing all components simultaneously. The minimum-size perturbation that induces a bifurcation in the network at $E1_{\text{tot}} = 4 \times 10^{-6}$ is $\|\mathbf{\Delta}_I\| = 1/\mu_{\Delta_I}$, thus a large μ_{Δ_I} implies weak robustness.

steady state. Thus inducing a saddle-node bifurcation requires the smallest perturbation. At $\omega = 0$, the value of μ_{Δ_I} is 2.2×10^4, implying that a relative perturbation of only 0.005% in the activities of all components is sufficient to induce a saddle-node bifurcation. Thus, we can conclude that the model is highly fragile to structural perturbations.

To obtain more detailed information on the structural fragility of the model, we compute the required perturbations in single components and specific pairwise interactions as given by equation (10.15) and equation (10.17) respectively. We first consider saddle-node bifurcations, and hence frequency $\omega = 0$ only. The results are shown in figure 10.21. As can be seen, there are 6 components that require small perturbations in their steady-state effect on the other components of the network to induce a saddle-node bifurcation. For the pairwise interactions, there are 25 out of a total of 61 existing direct interactions that can be perturbed by less than 10% to induce a saddle-node bifurcation. For instance, the effect of component 8 (MAPKK-PP) on component 13 (MAPKK-PP $\times$ MAPK-P complex) requires a relative perturbation $\Delta_{13,8} = 1.63 \times 10^{-4}$ to induce a saddle-node bifurcation for the nominal input. Implementing this as a static perturbation in the original nonlinear model, that is, perturbing the differential equation for state x_{13} according to

$$\frac{dx_{13}}{dt} = f_{13}(x_1, x_2, \dots, x_8 + \Delta_{13,8}(x_8 - x_8^{ss}), x_9, \dots, x_{15}, \boldsymbol{p})$$

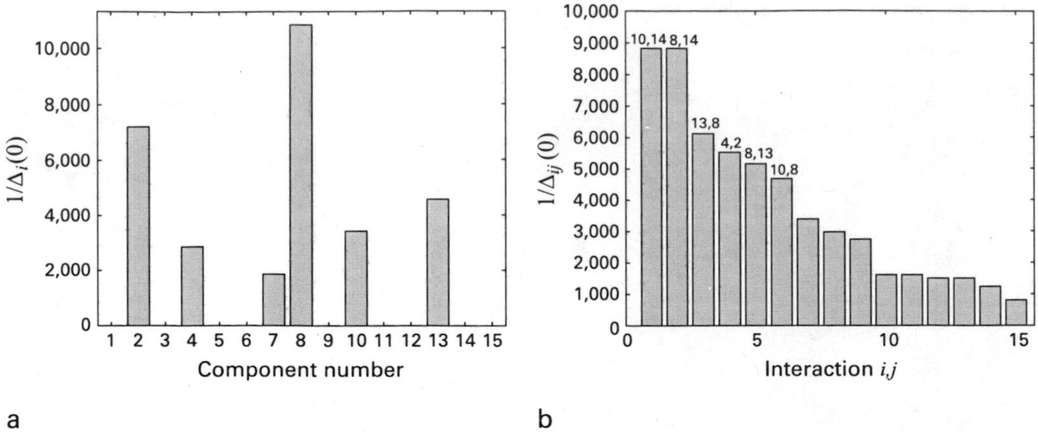

Figure 10.21
Relative static perturbations in single components (a) and pairwise interactions (b) of the MAPK network required to induce a saddle-node bifurcation at the nominal input value $E1_{\text{tot}} = 4 \times 10^{-6}$.

yields the response curve shown in figure 10.22a. As can be seen, the small perturbation induces a saddle-node bifurcation at $E1_{\text{tot}} = 4 \times 10^{-6}$ and another one close to this point, leading to a small range of bistability. If we double the perturbation, that is, let $\Delta_{13,8} = 3.26 \times 10^{-4}$, then the bifurcation diagram in figure 10.22b is obtained. Now the nominal steady state is unstable, with the Jacobian A having a real eigenvalue in the right half plane, and the response curve displays a larger region of bistable behavior.

We next consider the size of a dynamic perturbation required to induce a Hopf bifurcation, and thus a limit cycle behavior, in the MAPK model. The results, not shown here, reveal that several components of the MAPK model can be dynamically perturbed by a relatively small amount to induce a Hopf bifurcation. For component 7 (MAPKKK-P $\times$ MAPKK-P complex), we find that a relative perturbation $\Delta_7(j\omega^*) = 5.10 \times 10^{-4} + j1.67 \times 10^{-4}$ at the frequency $\omega^* = 0.0145$ is sufficient to induce a Hopf bifurcation at the nominal input $E1_{\text{tot}} = 4 \times 10^{-6}$. Fitting this frequency response to the transfer function $\Delta_T(s)$ equation (10.20) and implementing the resulting linear differential equations in the Huang and Ferrell model yields the bifurcation diagram in figure 10.23a. As can be seen, the small structural perturbation induces a Hopf bifurcation at the nominal point and makes the steady state unstable for inputs up to $E1_{\text{tot}} = 9.8 \times 10^{-6}$ at which point there is another Hopf bifurcation. As can be seen from the bifurcation diagram, large amplitude oscillations exist between the two Hopf points. Figure 10.23b shows a simulation of the oscillations for $E1_{\text{tot}} = 4.2 \times 10^{-6}$.

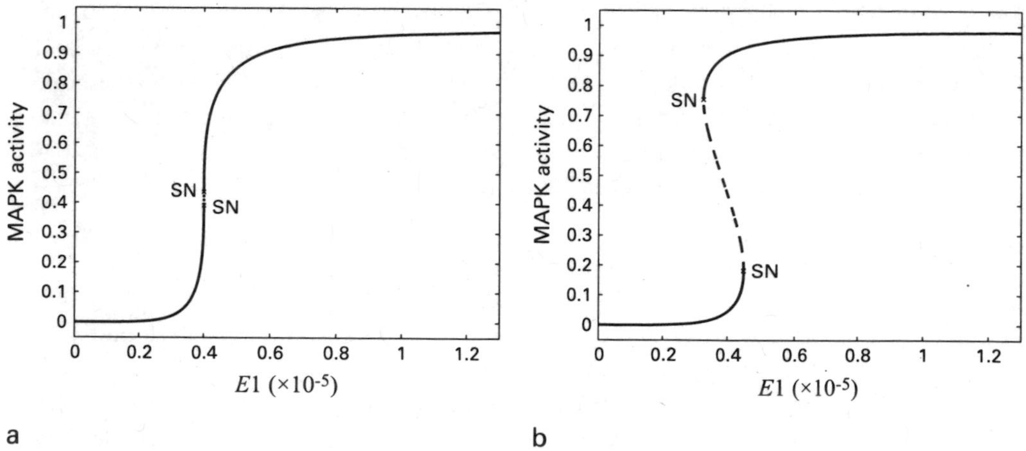

Figure 10.22
Response curve of MAPK network after perturbing the effect of component 8 (MAPKK-PP) on component 13 (MAPKK-PP × MAPK-P complex) by a static relative perturbation (a) $\Delta_{13,8} = 1.63 \times 10^{-4}$ and (b) $\Delta_{13,8} = 3.26 \times 10^{-4}$.

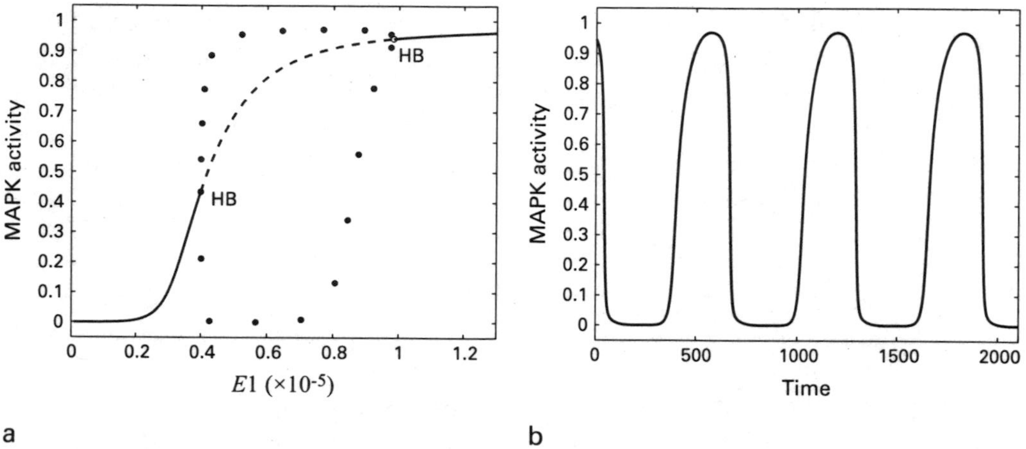

Figure 10.23
(a) Response curve of MAPK network after dynamic perturbation of component 7 (MAPKKK-P × MAPKK-P) by 0.05% at the frequency $\omega = 0.0145$, showing amplitude of oscillations between Hopf points (*filled circles*). (b) Simulation of oscillations for $E1_{tot} = 4.2 \times 10^{-6}$.

In conclusion, using the proposed robustness analysis method, we find the structure of the MAPK model in Huang and Ferrell (1996), to be highly unrobust, and that small perturbations of the network interactions can induce either saddle-node bifurcations, leading to bistability, or Hopf bifurcations, leading to sustained oscillations. Whether this is a weakness of the model, or reflects a feature of the real biological system is a question we leave for future studies.

10.9 Case Study 2: Metabolic Oscillations in Activated Neutrophils

The model of the central metabolism of an activated neutrophil (white blood cell) presented by Olsen et al. (2003) will be referred to here simply as the neutrophil model. The model's 16 states, represented by 16 ordinary differential equations with 25 parameters, correspond to concentrations of biochemical substances, divided between two compartments: the cytosol and the phagosome. That the model has two conserved moieties implies that it has only 14 independent states. Most of its reactions are well documented and the rate constants have been determined. A main feature is the model's ability to reproduce the temporal oscillations observed in vitro, as when measuring NAD(P)H through fluorescence techniques. All components of the model oscillate with the same period, approximately 25 seconds (Olsen et al., 2003). Although spatial waves that move along the cell interior have also been observed in experiments (Kindzelskii and Petty, 2002; Petty, 2000), because the neutrophil model contains only two compartments, these cannot be captured. When researchers added diffusion to the cytosol compartment in an attempt to reproduce the spatiotemporal oscillations (Cedersund, 2004), they were unable to reproduce any oscillations in the model, even for relatively large diffusion coefficients. This was surprising because the model had been shown to be highly robust to parametric perturbations (Cedersund, 2004; Jacobsen and Cedersund, 2008). That even small spatial effects completely alter its predicted behavior, however, indicates a severe structural fragility.

Although Jacobsen and Cedersund (2008) present a complete structural robustness analysis of the neutrophil model, here we show only the most important results. Figure 10.24 shows the overall robustness result in terms of μ_{Δ_I} for the case of simultaneous perturbations of all 14 states. We can see that the model has some severe structural fragility from the maximum value $\mu_{\Delta_I} = 397$ at the frequency $\omega = 0.26$ rad/s, corresponding to a 0.25% simultaneous change in the effect of all network metabolites to induce a Hopf bifurcation.

To identify the specific fragilities, we compute the robustness to specific pairwise interactions as given by equation (10.17). The results, in terms of the minimum perturbation over all frequencies are shown in figure 10.24b: the most severe fragilities

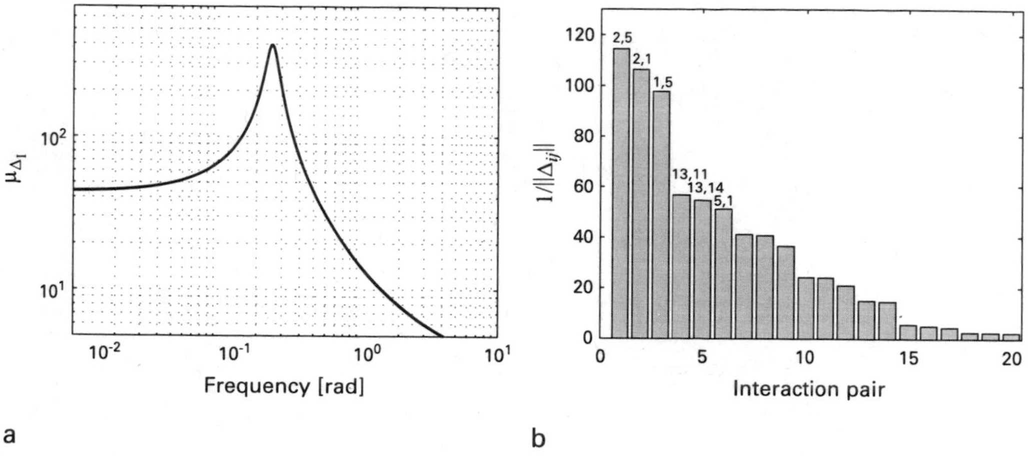

Figure 10.24
(a) Measure of the overall structural robustness of the neutrophil model using a complex diagonal perturbation $\mathbf{\Delta_I}$ corresponding to perturbing all metabolites simultaneously. (b) Robustness to perturbations of specific pairwise metabolite interactions.

are in the direct interaction between metabolites 2 and 5, and 2 and 1, all in the phagosome compartment, requiring perturbations of less than 1% to induce a Hopf bifurcation. The computed minimum perturbation for the effect of component 1 on component 2 is $\Delta_{2,1} = 0.0032 - j0.0089$ at $\omega = 0.26$ rad/s. We find that this is mainly a phase-lag perturbation and thus can be fitted to a pure time delay. That is, we can fit $1 + \Delta_{2,1}$ to a transfer function $e^{-\theta s}$ corresponding to a pure time delay θ in the effect of metabolite 1 on metabolite 2. The fitted value of the time delay is $\theta = 0.04$ s, which should be compared to the oscillation period of about 25 seconds.

Figure 10.25 shows the bifurcation diagram for the unperturbed and perturbed neutrophil model with a reaction rate constant k_{12} as bifurcation parameter. The nominal neutrophil model corresponds to $k_{12} = 30$. As shown in figure 10.25, a time delay $\theta = 0.04$ s in the effect of metabolite 1 on metabolite 2 induces a Hopf bifurcation point at the nominal steady state, thereby removing the oscillations in the nominal model. Also shown is the bifurcation diagram when the delay is increased to $\theta = 0.07$ s, with the result that the oscillations are removed for all values of the parameter k_{12}.

In summary, structural robustness analysis reveals the structure of the model proposed by Olsen et al. (2003) to be highly unrobust. The results serve to explain the difficulties in including spatial effects in the model, corresponding to a structural modification of the model. The main identified fragilities, involving interactions between metabolites in the phagosome, point to a need to refine this part of the model.

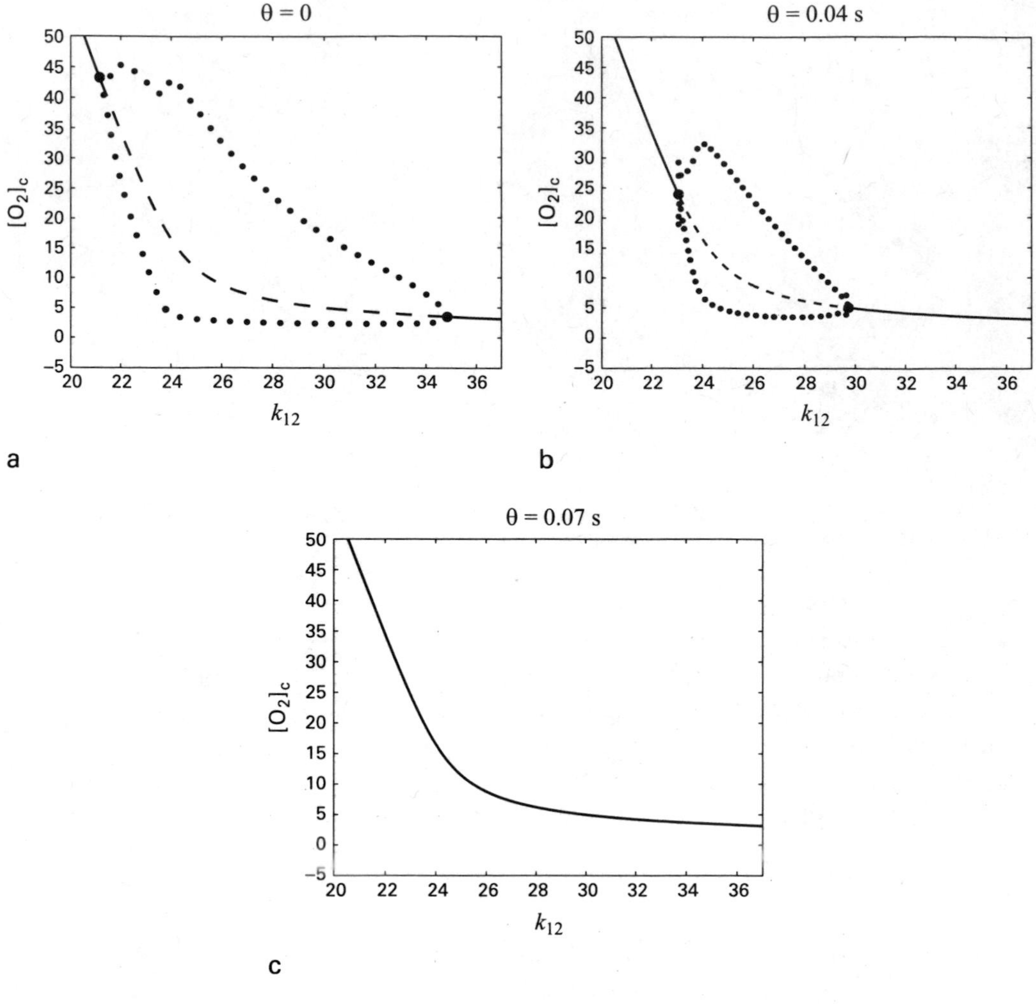

Figure 10.25
Bifurcation diagrams for neutrophil model with different perturbation delays θ in the effect of metabolite 1 on metabolite 2 in the phagosome. The nominal parameter value is $k_{12} = 30$.

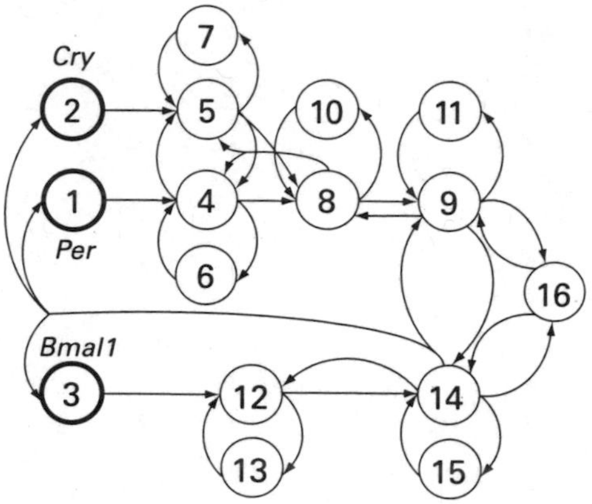

Figure 10.26
Schematic diagram of gene regulatory network proposed by Leloup and Goldbeter (2003). The numbers refer to the 16 states in the model, corresponding to mRNAs, proteins, and protein complexes in the cytoplasm and nucleus; the arrows indicate direct interactions between the components, resulting from biochemical reactions or transport mechanisms.

10.10 Case Study 3: The Mammalian Circadian Clock

To illustrate the use of our proposed method to identify key interactions underlying a given function, and to reduce models accordingly, we consider the model of circadian oscillations in mammals presented in Leloup and Goldbeter (2003). Based on experimental observations of intertwined positive and negative gene regulatory feedback loops in mice, the model consists of five genes and their products. Only three of the genes are considered to be directly involved in generating the circadian oscillations, while two genes are assumed to be constantly expressed. As shown by the network in figure 10.26, the directly involved genes correspond to three genes that are feedback regulated: *per*, *cry*, and *bmal1*. The model predicts autonomous sustained oscillations with a period of 23.8 hours in conditions corresponding to continuous darkness, which are entrained by 12-hour light/dark cycles to have a period of 24 hours (see figure 10.27). The entrainment is made possible by the fact that the *per* gene activity is influenced by light. Here we consider the structural robustness of the model in continuous darkness, that is, with autonomous oscillations.

Figure 10.28 shows the relative perturbations required in single components and specific pairwise interactions, respectively, to induce a Hopf bifurcation in the Leloup and Goldbeter model (2003). We first note that this model, compared to the MAPK and neutrophil models, is relatively robust in that perturbations of at least 42% in the concentrations of single components (and of a somewhat greater percentage for perturbations of specific interactions) are required to remove the circadian oscillations.

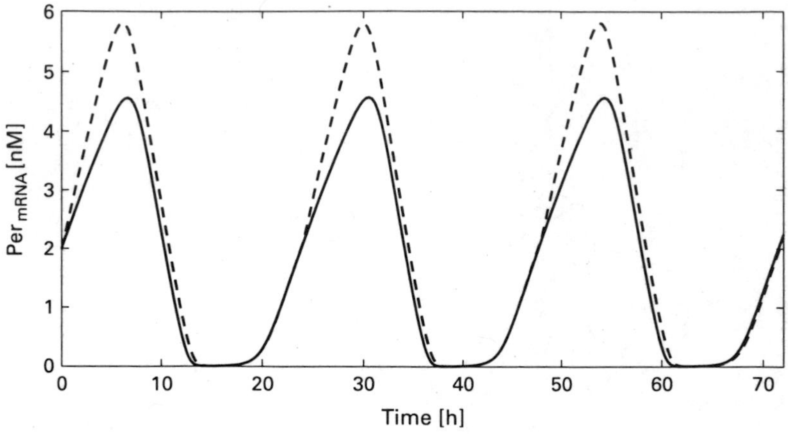

Figure 10.27
Circadian autonomous oscillations (*solid line*) and entrained oscillations to 12-hour light/dark cycles (*dashed line*) in the mRNA concentration of the *per* gene.

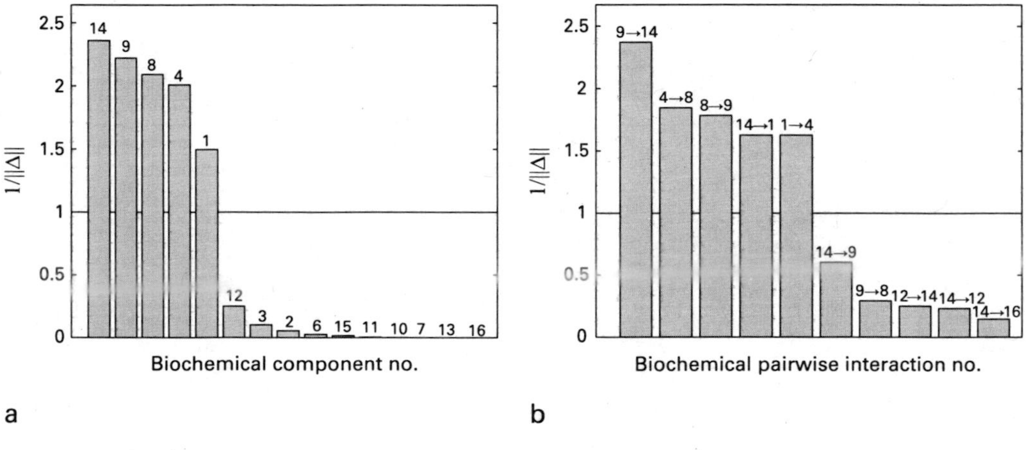

Figure 10.28
Minimum relative perturbations of (a) single components and (b) specific interactions required to induce a Hopf bifurcation in the Leloup and Goldbeter circadian model (2003).

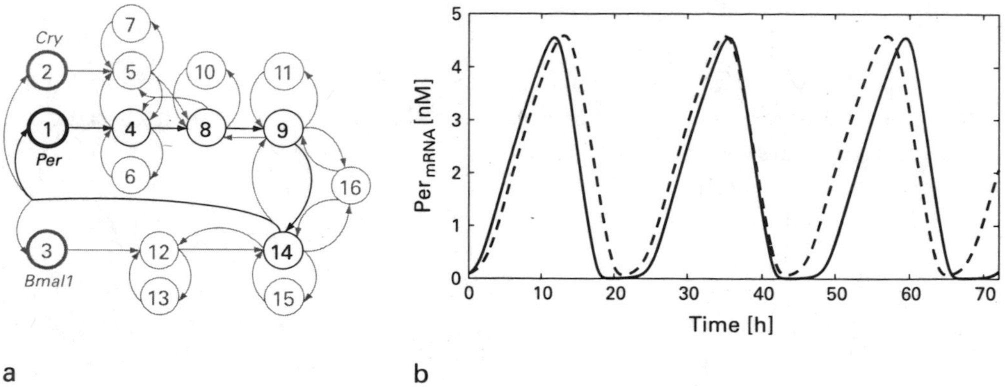

Figure 10.29
(a) Highlighted parts of the network correspond to the components and interactions identified as the mechanism underlying the circadian oscillations in the Leloup and Goldbeter model (2003). (b) Simulation of circadian oscillations with the full 16-state model (*solid line*) and the corresponding reduced 5-state model (*dashed line*).

We also note from the figure that only five components can be perturbed by less than 100% to remove the oscillations by inducing a Hopf point.

Because complete removal of the effect of changes in a single component corresponds to a perturbation $\Delta_i = -1$, if a component requires a perturbation $\|\Delta_i\| > 1$ to induce a bifurcation, this implies that the component can be removed without affecting the presence of the function, in this case, the circadian oscillations. Note that "removing a component" implies replacing it by a constant equal to its nominal concentration, that is, removing the dynamic interactions with the other components. The results in figure 10.28 thus indicate that all but five of the components can be removed from the model; we say "indicate" since the argument strictly only holds when single components are removed. The resulting subnetwork with five components is shown in figure 10.29, and as can be seen it involves a single feedback loop around the *per* gene (component 1). The oscillations predicted by the reduced 5-state model are shown together with the oscillations of the full 16-state model in figure 10.29; the 5-state model predictions are seen to be close to those of the full 16-state model. This final case study has demonstrated the usefulness of structural robustness analysis both for detecting key subnetworks underlying given functions and for performing nonlinear model reduction. Note that a significant advantage of the proposed reduced model is that it retains the biological interpretation of the full model's states.

10.11 Summary and Conclusions

Biochemical network models are characterized by a high level of parametric and structural uncertainty, reflecting incomplete knowledge of the underlying processes. The impact of parametric uncertainty is routinely analyzed to validate models. In contrast, structural uncertainty is rarely addressed, mainly due to a lack of readily available methods to quantify its impact. In this chapter, we have proposed a control-theoretic method for analyzing structural robustness of biochemical network models. As we have shown, our method can be used not only to quantify structural robustness, but also to identify specific network fragilities and determine key substructures underlying a given function. We have demonstrated the method by applying it to models of MAPK signaling, metabolic oscillations in activated neutrophils, and mammalian circadian clocks. For the MAPK model, it was found that small structural perturbations can induce bistability as well as sustained oscillations in the signaling. For the neutrophil model, a specific metabolic reaction was identified as a severe fragility and a small delay in this reaction was shown to remove the oscillatory behavior. For the relatively robust circadian clock model, structural robustness analysis was used to identify a 5-state subnetwork within the full 16-state network chiefly responsible for the circadian oscillations.

11 Robustness of Oscillations in Biological Systems

Jongrae Kim and Declan G. Bates

The oscillations observed in biological systems display a rich variety of dynamics and are typically generated by sophisticated multivariable feedback control mechanisms whose underlying design principles are obscure. It is hardly surprising, then, that the study of such systems using mathematical techniques, including methods from systems and control engineering, has a long and varied history (Goldbeter, 1996; Kholodenko et al., 1997; Linkens, 1979; Rapp, 1975). Until recent years, however, most such analyses of oscillatory biological systems have concentrated on establishing conditions for nominal stability or performance properties of the oscillations, while perhaps using phase-plane analysis to investigate the effect of varying individual parameters. Because many of these studies were undertaken long before the issue of robustness had come to dominate mainstream control theory, this is hardly surprising. And because oscillations in physical engineering systems are typically problems to be avoided, relatively few robust control theorists have been interested in analyzing their robustness. Recent recognition of the potential of robustness analysis methods to help in the development, refinement, and validation of mathematical models of biological systems has radically altered this situation, however. Intensive efforts are now under way to develop analytical techniques to quantify the robustness of models of oscillatory biological systems. This chapter describes a number of promising methods for the robustness analysis of such systems and shows how such analysis can shed new light on the underlying design principles of the systems concerned. We apply the proposed methods to analyze the robustness of oscillations in the concentration of adenosine $3', 5'$-cyclic monophosphate (cAMP) observed during the aggregation phase of starvation-induced development in *Dictyostelium discoideum*. We also highlight the important roles played by stochastic noise in ensuring the robustness of biological oscillations.

11.1 Robust cAMP Oscillations in Aggregating *Dictyostelium* Cells

The social amoeba *Dictyostelium discoideum* normally lives in forest soil, where it feeds on bacteria (Othmer and Schaap, 1998). Under conditions of starvation,

Dictyostelium cells begin a program of development during which they aggregate and eventually form spores atop a stalk of vacuolated cells. At the beginning of this process, the amoebae become chemotactically sensitive to cAMP and, by six to ten hours, almost all of them acquire competence to relay cAMP signals (Gingle and Robertson, 1976). After eight hours, a few pacemaker cells start to emit cAMP periodically (Raman et al., 1976). Surrounding cells move toward the cAMP source and relay the cAMP signal to more distant cells. Eventually, the entire population collects into mound-shaped aggregates containing up to 100,000 cells (Coates and Harwood, 2001). The processes involved in cAMP signaling in *Dictyostelium* are mediated by a family of cell surface cAMP receptors (cARs) that act on a specific heterotrimeric G protein to stimulate actin polymerization, and activation of adenylyl and guanylyl cyclases, among other responses (Parent and Devreotes, 1996). Most of the components of these pathways have mammalian counterparts, and much effort has been devoted in recent years to the study of signal transduction mechanisms in these simple microorganisms, with the eventual aim of improving understanding of defects in these pathways that may lead to disease in humans (Williams et al., 2006).

Laub and Loomis (1998) proposed a network model of interacting proteins that can account for the spontaneous oscillations in adenlylate cyclase activity that are observed in homogeneous populations of *Dictyostelium* cells four hours after the initiation of development (see figure 11.1). Analyses of the numerical solutions of the nonlinear differential equations making up the model suggest that it faithfully reproduces the observed periodic changes in adenosine $3', 5'$-cyclic monophosphate (cAMP). In particular, periods, amplitudes, and phase relations between oscillations in enzyme activities and internal and external cAMP concentrations were seen to agree well with experimental observations (Laub and Loomis, 1998).

The Laub and Loomis model for the cAMP oscillations is given by a set of seven nonlinear coupled ordinary differential equations as follows:

$$\frac{dx_1}{dt} = k_1 x_7 - k_2 x_1 x_2, \tag{11.1a}$$

$$\frac{dx_2}{dt} = k_3 x_5 - k_4 x_2, \tag{11.1b}$$

$$\frac{dx_3}{dt} = k_5 x_7 - k_6 x_2 x_3, \tag{11.1c}$$

$$\frac{dx_4}{dt} = k_7 - k_8 x_3 x_4, \tag{11.1d}$$

$$\frac{dx_5}{dt} = k_9 x_1 - k_{10} x_4 x_5, \tag{11.1e}$$

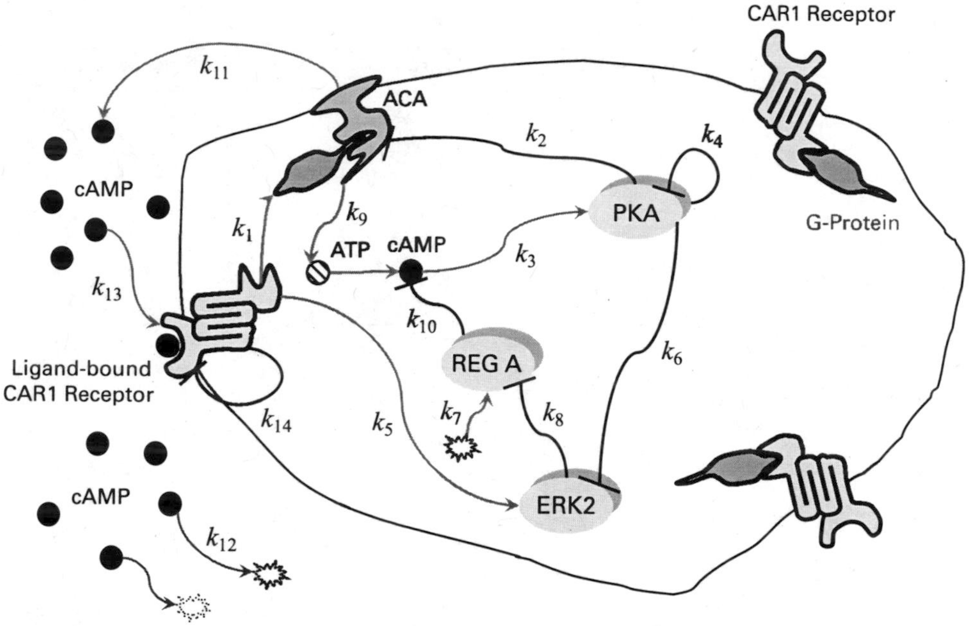

Figure 11.1
Dictyostelium discoideum cAMP oscillation network (Laub and Loomis, 1998), showing activation or self-degradation (*pointed arrows*) and inhibition (*blunted arrows*).

$$\frac{\mathrm{d}x_6}{\mathrm{d}t} = k_{11}x_1 - k_{12}x_6, \tag{11.1f}$$

$$\frac{\mathrm{d}x_7}{\mathrm{d}t} = k_{13}x_6 - k_{14}x_7, \tag{11.1g}$$

where t is time, x_1 is adenlylyl cyclase of aggregation (ACA), x_2 is the protein kinase A (PKA), x_3 is the MAP kinase ERK2 (extracellular signal-regulated kinase 2), x_4 is intracellular phosphodiesterase RegA, x_5 is internal cAMP, x_6 is external cAMP, and x_7 is the high-affinity cell surface cAMP receptor CAR1. The nominal values for the kinetic constants, k_i, are given in table 11.1. Note that all quantities are given as micromolar concentrations. As shown in figure 11.1, after external cAMP binds to the cell surface receptor CAR1, CAR1 activates adenylyl cyclase ACA and the mitogen-activated protein kinase ERK2. ACA then stimulates the production of intracellular cAMP, which in turn, activates the protein kinase PKA. PKA inhibits ACA and ERK2, forming two negative feedback loops controlling the level of internal cAMP. PKA activity is decreased as the internal cAMP is hydrolyzed by RegA. The internal cAMP is secreted to the outside of the cell and diffuses between cells. Thus, when external cAMP binds to CAR1, this forms a positive feedback loop.

Table 11.1
Nominal values of the kinetic constants

Parameter	Nominal value	Parameter	Nominal value
k_1 [1/min]	2.0	k_8 [1/(μM min)]	1.3
k_2 [1/(μM min)]	0.9	k_9 [1/min]	0.3
k_3 [1/min]	2.5	k_{10} [1/(μM min)]	0.8
k_4 [1/min]	1.5	k_{11} [1/min]	0.7
k_5 [1/min]	0.6	k_{12} [1/min]	4.9
k_6 [1/(μM min)]	0.8	k_{13} [1/min]	23.0
k_7 [μM/min]	1.0	k_{14} [1/min]	4.5

In the remainder of this chapter, three different approaches for the robustness analysis of uncertain systems are applied to the above network model. Here robustness is defined as the ability of the biochemical network model to reproduce the experimentally observed oscillations in cAMP, ACA, and so on in the presence of realistic levels of variation in multiple kinetic parameters. The first method measures local stability robustness properties of the oscillations using the structured singular value μ, a tool developed in the field of robust control theory to measure the robustness of feedback control systems to multiple forms of uncertainty. The second method uses a global optimization algorithm to search for the smallest variation in the nonlinear model parameters that drives the states of the system to a nonoscillatory behavior. Finally, a stochastic analysis is performed using Monte Carlo simulation to highlight the significant effect that intracellular noise can have on the robustness of biomolecular networks.

11.2 Deterministic Robustness Analysis

11.2.1 Local Robustness Analysis

The structured singular-value or μ-analysis method is a standard tool for the robustness analysis of linear systems in feedback control engineering (Balas et al., 2008). To make it easier to understand the basic concepts and formulations of μ-analysis, consider the following simple ordinary differential equation:

$$\frac{dx(t)}{dt} = kx(t), \tag{11.2}$$

where $x(t)$ is the concentration of some molecular species that degrades (for $k < 0$) with rate k. Let the rate be given by $k = -2(1 + \delta)$, where δ is the uncertainty in the estimate of the kinetic rate constant. The concentration of x converges to zero exponentially as t increases if k is strictly less than zero, as the solution to the differential equation is given by $x(t) = x_0 e^{-2(1+\delta)t}$, where x_0 is the initial concentration of x. The necessary and sufficient condition for $x(t)$ to converge to zero, that is, for the system to be stable, is that the exponent be strictly less than zero; in this case, that δ be greater than -1.

We now consider the vector case

$$\frac{d\boldsymbol{x}(t)}{dt} = K(\boldsymbol{\delta})\boldsymbol{x}(t), \tag{11.3}$$

where $\boldsymbol{x}(t)$ is an n-dimensional nonnegative real vector, whose elements represent the concentration of different molecular species, and $K(\boldsymbol{\delta})$ is a kinetic rate matrix whose dimension is $n \times n$ and whose value is a function of the uncertain vector $\boldsymbol{\delta}$ whose dimension is p. Again, the solution is given by

$$\boldsymbol{x}(t) = e^{K(\boldsymbol{\delta})t}\boldsymbol{x}_0,$$

where, for the vector case, the exponential is the matrix exponential (chapter 1). Similar to the scalar case, the necessary and sufficient condition for $\boldsymbol{x}(t)$ to converge to zero as t increases is that all the real parts of the eigenvalues of $K(\boldsymbol{\delta})$ be strictly less than zero (section 1.4). It is well known, however, that eigenvalues are often poor measures of robustness (Doyle and Stein, 1979). That is, there are many cases where a tiny perturbation could make the trajectories diverge to infinity for a specific uncertainty, even when the eigenvalues are far away from the positive real region. In robust control theory, the following alternative approach is usually adopted to avoid this problem (Balas et al., 2008). Consider the scalar example, equation (11.2), for k equal to $-2(1+\delta)$. It can be rewritten as follows by decoupling the known part and the uncertain part:

$$\frac{dx(t)}{dt} = -2x(t) + w(t), \tag{11.4a}$$

$$z(t) = -2x(t), \tag{11.4b}$$

where the uncertain part $w(t)$ is given by $\delta z(t)$. Hence the above system considers the effect of the uncertain part, $w(t)$, as an input to the system and $w(t)$ is given by the product of the system output, $z(t)$, and the uncertain gain, δ. This decoupling is always possible when the uncertain parameter, δ, appears as a rational polynomial function, and the resulting form is called a *linear fractional transformation* (LFT).

Using the Laplace transform, equation (11.4) may be transformed as follows:

$$Z(s) = M(s)W(s), \tag{11.5}$$

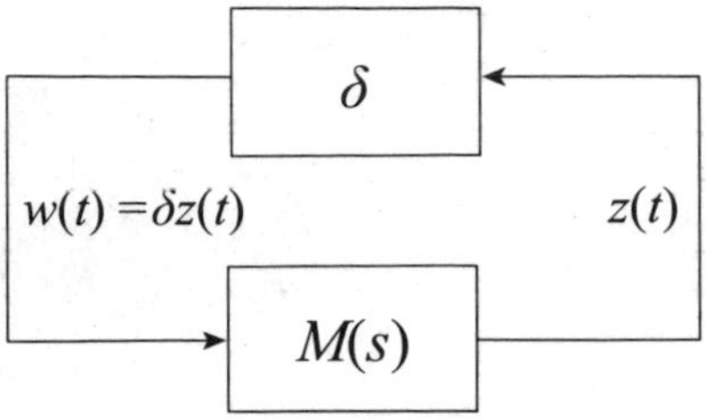

Figure 11.2
M-Δ structure for μ-analysis.

where s denotes the Laplace transform variable, $Z(s)$ and $W(s)$ are the Laplace transforms of $z(t)$ and $w(t)$, respectively, and

$$M(s) = -\frac{2}{s+2}. \tag{11.6}$$

The linear fractional transformation forms a feedback loop, as shown in figure 11.2. $M(s)$ can be considered as a system with input $w(t)$ and output $z(t)$. Because $M(s)$ is a linear system, it can be characterized completely by its frequency response. Suppose that we disconnect δ and $w(t)$, thus breaking the feedback loop of figure 11.2. Suppose, also, that we introduce the sinusoidal input $w(t) = e^{j\omega t}$, where $j = \sqrt{-1}$ and ω is the frequency which ranges from 0 to infinity. For each choice of frequency, $z(t)$ will eventually reach an oscillatory steady state given by

$$z(t) = M(j\omega)e^{j\omega t} = M(j\omega)w(t).$$

Equivalently,

$$Z(j\omega) = M(j\omega)W(j\omega).$$

We can now close the feedback loop by setting $W(j\omega) = \delta Z(j\omega)$ and observe how the internal signal changes. In particular, this leads to

$$[1 - M(j\omega)\delta]Z(j\omega) = 0. \tag{11.7}$$

From equation (11.7), if $1 - M(j\omega)\delta \neq 0$, then $Z(j\omega)$ must be equal to zero. On the other hand, if $1 - M(j\omega)\delta = 0$, then $Z(j\omega)$ is undefined. This represents the boundary between the stable and unstable regions of the feedback system.

For the vector case, where the uncertain parameters (δ_i) are real numbers, δ is replaced by a diagonal matrix, Δ, whose diagonal terms are given by the uncertain parameters. Moreover, the singularity condition of equation (11.7) is replaced by

$$\det[I - M(j\omega)\Delta] = 0, \tag{11.8}$$

where I is the identity matrix whose dimension is the same as $M(j\omega)\Delta$. Equation (11.8) characterizes perturbations Δ under which the system is unstable. Hence we want to calculate the uncertainty matrix, Δ, that makes the determinant equal to zero. Of course, in general, there will be an infinite number of matrices Δ that satisfy equation (11.8). We are interested in the uncertainty matrix whose magnitude is smallest among them because this defines the smallest variation in the model parameters for which the system loses stability. Although, in theory, this singularity condition must be checked at all frequencies $\omega \in [0, \infty)$, in practice, it is usually sufficient to check for a finite number of grid points over the frequency range.

The structured singular value, $\mu(\omega)$, is thus defined as

$$\frac{1}{\mu(\omega)} = \min_{\Delta}\{\bar{\sigma}(\Delta) \mid \det[I - M(j\omega)\Delta] = 0, \text{ for } \Delta \in B_\Delta\}, \tag{11.9}$$

for $\omega \in [0, \infty)$, where $\bar{\sigma}(\cdot)$ denotes the maximum singular value, and B_Δ is a set defined by

$$B_\Delta = \{\Delta \mid \Delta = \mathrm{diag}[\delta_1 I_1, \quad \delta_2 I_2, \quad \ldots, \quad \delta_p I_p]\}, \tag{11.10}$$

where diag[...] is a diagonal matrix, δ_i is the uncertain parameter, and I_i is the identity matrix whose dimension depends on how the uncertain parameter appears in the equations describing the system. Constructing a linear fractional transformation that has minimal dimensions for each I_i is sometimes challenging. Moreover, because calculating the value of μ exactly is most often prohibitively expensive from a computational point of view, in practice, lower and upper bounds on μ are generally computed (for more on μ-analysis, see Balas et al., 2001; Skogestad and Postlethwaite, 2005).

We now describe the process of converting the nonlinear oscillatory model (11.1) into a form that can be used for μ-analysis. Let the original nonlinear differential equations for the model (11.1) be written in compact form as

$$\frac{d\boldsymbol{x}}{dt} = \boldsymbol{f}(\boldsymbol{x}, \boldsymbol{k}), \tag{11.11}$$

where $\boldsymbol{x} = [x_1, \ x_2, \ \ldots, \ x_7]^T$, $\boldsymbol{k} = [k_1, \ k_2, \ \ldots, \ k_{14}]^T$, and $\boldsymbol{f}(\cdot, \cdot)$ is given by equation (11.1). Uncertainty is included in each kinetic parameter by setting:

$$k_i = \bar{k}_i(1 + \delta_i), \tag{11.12}$$

for $i = 1, 2, \ldots, 14$. With the nominal values of the k_i (corresponding to all δ_i being equal to zero) given in table 11.1, the model exhibits stable limit cycle trajectories in all states.

To obtain the limit cycle model, we use a harmonic balance method (Ma and Iglesias, 2002). First, we write the limit cycle, namely, the nominal trajectory, as

$$x_i^*(t) = a_{0,i} + \sum_{n=1}^{\infty} a_{n,i} \cos\left(\frac{2\pi n t}{\tau} + \phi_{n,i}\right), \tag{11.13}$$

for $i = 1, 2, \ldots, 7$, where τ is the period of the limit cycle. Though, in theory, the sum in equation (11.13) is infinite, in practice, the upper bound of the summation is limited to a finite number.

Next, we linearize the nonlinear differential equation about the nominal trajectory, $\boldsymbol{x}^*(t) = [x_1^*(t), x_2^*(t), \ldots, x_7^*(t)]^T$, writing the model as a linear periodically time-varying differential equation:

$$\frac{d\boldsymbol{x}_{\text{pert}}(t)}{dt} = \boldsymbol{A}(t)\boldsymbol{x}_{\text{pert}}(t) + \boldsymbol{B}(t)\boldsymbol{w}_s(t), \tag{11.14a}$$

$$\boldsymbol{z}_s(t) = \boldsymbol{C}(t)\boldsymbol{x}_{\text{pert}}(t) + \boldsymbol{D}(t)\boldsymbol{w}_s(t), \tag{11.14b}$$

where $\boldsymbol{A}(t) = \boldsymbol{A}(t+\tau)$, $\boldsymbol{B}(t) = \boldsymbol{B}(t+\tau)$, $\boldsymbol{C}(t) = \boldsymbol{C}(t+\tau)$, $\boldsymbol{D}(t) = \boldsymbol{D}(t+\tau)$, and each element of the vector $\boldsymbol{x}_{\text{pert}}$ is the displacement of the corresponding state from x_i^* for $i = 1, 2, \ldots, 7$. To transform this model to a linear, time-invariant system, the model is discretized. Using a technique called "lifting" (Ma and Iglesias, 2002), we then convert the resulting periodic state-space matrices to constant matrices.

For a fixed time, t_k, which is an element of $[t, t+\tau)$, equation (11.14) is discretized as follows:

$$\boldsymbol{x}_{\text{pert}}(t_{k+1}) = \boldsymbol{\Phi}(t_k)\boldsymbol{x}_{\text{pert}}(t_k) + \boldsymbol{\Gamma}(t_k)\boldsymbol{w}_s(t_k), \tag{11.15a}$$

$$\boldsymbol{z}_s(t_k) = \boldsymbol{H}(t_k)\boldsymbol{x}_{\text{pert}}(t_k) + \boldsymbol{J}(t_k)\boldsymbol{w}_s(t_k), \tag{11.15b}$$

where

$$\boldsymbol{\Phi}(t_k) = e^{\boldsymbol{A}(t_k)h}, \tag{11.16a}$$

$$\boldsymbol{\Gamma}(t_k) = \left(\int_0^h e^{\boldsymbol{A}(t_k)\eta}\, d\eta\right)\boldsymbol{B}(t_k), \tag{11.16b}$$

$$\boldsymbol{H}(t_k) = \boldsymbol{C}(t_k), \tag{11.16c}$$

$$\boldsymbol{J}(t_k) = \boldsymbol{D}(t_k), \tag{11.16d}$$

where $t_k = k\tau/n_{\text{dsc}}$, for $k = 0, 1, \ldots$, and t_0 is set to zero without loss of generality. The approximation error can be reduced by increasing the number of discretization points, n_{dsc}, but because this also increases the dimension of the problem and hence the computational burden of the μ bound calculations, in practice, a sensible trade-off is required. We now convert this to a discrete-time, time-invariant system through lifting. To demonstrate the lifting procedure, we first set k to zero. Then

$$\boldsymbol{x}_{\text{pert}}(t_1) = \boldsymbol{\Phi}(t_0)\boldsymbol{x}_{\text{pert}}(t_0) + \boldsymbol{\Gamma}(t_0)\boldsymbol{w}_s(t_0). \tag{11.17}$$

For $k = 1$,

$$\begin{aligned}\boldsymbol{x}_{\text{pert}}(t_2) &= \boldsymbol{\Phi}(t_1)\boldsymbol{x}_{\text{pert}}(t_1) + \boldsymbol{\Gamma}(t_1)\boldsymbol{w}_s(t_1)\\ &= \boldsymbol{\Phi}(t_1)(\boldsymbol{\Phi}(t_0)\boldsymbol{x}_{\text{pert}}(t_0) + \boldsymbol{\Gamma}(t_0)\boldsymbol{w}_s(t_0)) + \boldsymbol{\Gamma}(t_1)\boldsymbol{w}_s(t_1)\\ &= \boldsymbol{\Phi}(t_1)\boldsymbol{\Phi}(t_0)\boldsymbol{x}_{\text{pert}}(t_0) + [\boldsymbol{\Phi}(t_1)\boldsymbol{\Gamma}(t_0) \quad \boldsymbol{\Gamma}(t_1)]\begin{bmatrix}\boldsymbol{w}_s(t_0)\\ \boldsymbol{w}_s(t_1)\end{bmatrix}.\end{aligned}$$

Subsequently, accumulating all $\boldsymbol{w}_s(t_k)$ from t_0 to $t_{n_{\text{dsc}}-1}$, and propagating the state $\boldsymbol{x}_{\text{pert}}(t_0)$ to $\boldsymbol{x}_{\text{pert}}(t_{n_{\text{dsc}}})$, we obtain a time-invariant discrete-time system. A similar procedure is also applied for $\boldsymbol{z}_s(k)$ and the corresponding matrices.

Finally, using a zero-order hold or some other sampling method (Chen and Francis, 1996), we transform the linear, time-invariant discrete-time system back to the continuous-time domain to give

$$\frac{\mathrm{d}\boldsymbol{x}_{\text{pert}}(t)}{\mathrm{d}t} = \boldsymbol{A}\boldsymbol{x}_{\text{pert}}(t) + \boldsymbol{B}\boldsymbol{w}(t), \tag{11.18a}$$

$$\boldsymbol{z}(t) = \boldsymbol{C}\boldsymbol{x}(t) + \boldsymbol{D}\boldsymbol{w}(t), \tag{11.18b}$$

where $\boldsymbol{x}_{\text{pert}}(t)$ is a small displacement from $\boldsymbol{x}^*(t)$, $\boldsymbol{w}(t)$ is equal to $\Delta\boldsymbol{z}(t)$, and $\boldsymbol{w}(t)$ and $\boldsymbol{z}(t)$ are the accumulated vectors of $\boldsymbol{w}_s(t)$ and $\boldsymbol{z}_s(t)$ from the lifting procedure, respectively (for more on this transformation procedure, see Kim et al., 2006). The system is now in the standard form for application of μ-analysis techniques, and $\boldsymbol{M}(s)$ in figure 11.2 is given by

$$\boldsymbol{M}(s) = \boldsymbol{C}(s\boldsymbol{I} - \boldsymbol{A})^{-1}\boldsymbol{B} + \boldsymbol{D}. \tag{11.19}$$

The number of discretization points along the limit cycle, n_{dsc}, is set to 39, which is the minimum number of points guaranteeing that the eigenvalues of $\boldsymbol{A}$ do not change by more than 0.001 for subsequent increases in n_{dsc}. Note that, as a result of the transformations described above, the uncertainty matrix Δ is now made up of 39 repeated blocks of 13 real uncertain parameters. Because it is not multiplied by any x_i in equation (11.1), one of the 14 original uncertain parameters, k_7, does not appear. Hence the robustness analysis results derived from this approach could slightly underestimate the effects of uncertainty on the system. Application of the standard

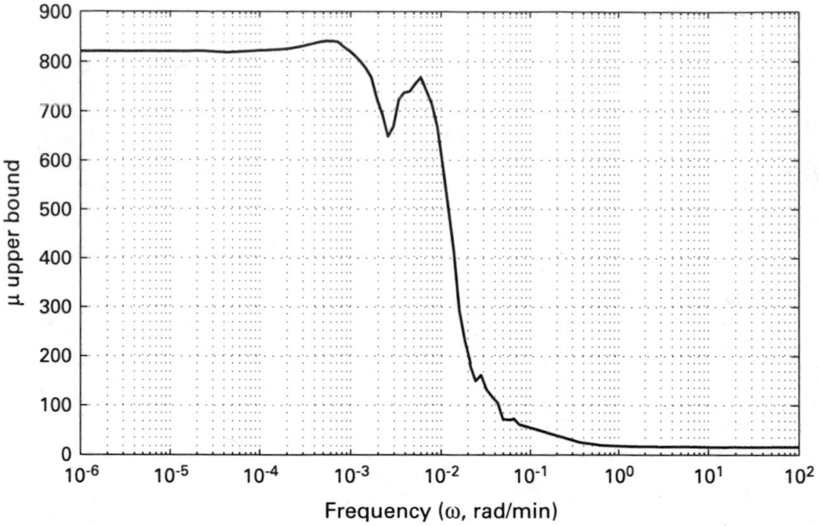

Figure 11.3
μ upper bound calculated using μ-toolbox at each ω (Balas et al., 2008).

algorithms for computing bounds on μ (Balas et al., 2001) to the system (11.18) produced the results shown in figure 11.3. The inverse of the peak of the upper bound on μ provides a maximum allowable level of uncertainty for which stable oscillations in the original nonlinear system are guaranteed to persist. From the figure this corresponds to a maximum allowable percentage variation in the parameters k_i of only $1/842 \approx 0.12\%$, which may indicate poor robustness indeed. Unfortunately, owing to the large number of repeated real parameters in the Δ matrix, the μ lower bound algorithms fail to converge; for the frequency range in the figure, they are all equal to zero. It is thus not possible to establish from this analysis whether the indicated lack of robustness is true (μ is close to its upper bound) or not (μ is much smaller than the computed upper bound, that is, the upper bound is conservative). In the next section, we resolve this issue by means of a global analysis. Note, however, that, despite its various limitations, μ-analysis provides a rigorous and elegant framework for the robustness analysis of highly complex systems subject to multiple sources of uncertainty. Its main advantage over other methods is that it can provide deterministic guaranteed robustness bounds, which may easily be used to compare the relative robustness of different systems or of different models of the same system.

11.2.2 Global Robustness Analysis

An alternative approach to robustness analysis is to employ optimization algorithms directly to search for particular combinations of parameters in the "uncertain parameter space" that maximize or minimize a particular cost function, whose value in some way reflects the level of robustness achieved by the model. Local optimization methods, for example, sequential quadratic programming (SQP; MathWorks, 2006), that use gradient information are computationally efficient but can easily get locked into local optima in the case of multimodal search spaces. On the other hand, global optimization methods such as genetic algorithms (GAs; Goldberg, 1989), use stochastic searches and evolutionary principles to approach the true global optimum, albeit at the cost of significantly increased computation. In the recent literature, several researchers have proposed combining the two approaches (Davis, 1991; Yen et al., 1995). Lobo and Goldberg (1996) provide some guidelines on designing hybrid genetic algorithms, along with experimental results and supporting mathematical analysis. In the present application, a probabilistic switching scheme based on that proposed by Lobo and Goldberg (1996) is used to switch dynamically between local (SQP) and global (GA) algorithms depending on which algorithm most effectively optimizes the cost function at each iteration (for full details of the algorithm, see Menon et al., 2006).

To apply the hybrid algorithm to test the robustness of the model's limit cycle the following cost function is minimized:

$$\min_{\delta \in \Delta} J = \min_{\delta \in \Delta} \int_{t_0}^{t_f} \dot{x}_1^2 \, \mathrm{d}t, \tag{11.20}$$

where

$$\Delta = \left\{ \boldsymbol{\delta} = [\delta_1, \ \delta_2, \ \ldots, \ \delta_{14}] \,\middle|\, \delta_i = \frac{p_\delta}{100} \times d_i, \text{ for } i = 1, 2, \ldots, 14 \right\},$$

where, in turn, $d_i \in [-1, 1]$, for $i = 1, 2, \ldots, 14$, p_δ is the percentage level of uncertainty, which will be specified for each optimization, and t_0 and t_f are chosen as 600 and 1,200 minutes, respectively. Note that, whereas δ_7 could not be included in the μ-analysis, $\boldsymbol{\delta}$ now includes all δ_i, from $i = 1$ to 14. This cost function was chosen because the state derivative will be zero whenever the states converge to a steady state and the limit cycle does not exist. The lower limit of integration, $t_0 = 600$ minutes, is chosen to reduce the effect of initial transient responses. The hybrid algorithm tries to find a $\boldsymbol{\delta}$ combination within the given boundary that minimizes the cost function. After the minimum is found, the nonlinear differential equations should be simulated over a range of initial conditions with the given values of k_i to check whether the state converges to an equilibrium point. Depending on the result of this test, p_δ is then increased or decreased until the minimum p_δ, denoted p_δ^*, is found. Results of the application of the hybrid optimization algorithm are displayed in figure 11.4, which shows ACA trajectories with the optimal combination of uncertainties in the

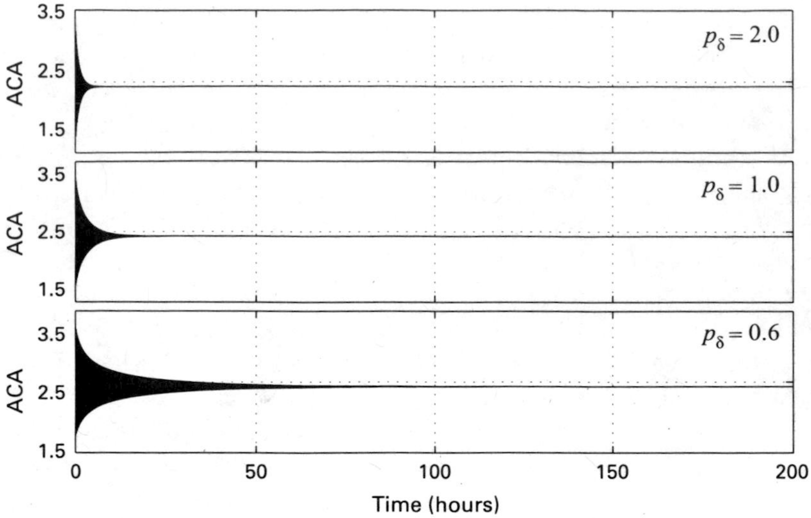

Figure 11.4
Effect of different levels of parameter variation on oscillatory behavior. The global optimization was performed on the interval from 10 to 20 hours. The longer simulations are shown to verify the results.

set Δ for three different values of p_δ. For all three cases, the optimal $\boldsymbol{\delta}$ minimizing the cost J occurs at the same boundary point, that is,

$$\delta^* = \frac{p_\delta}{100}[-1, -1, 1, 1, -1, 1, 1, -1, 1, 1, -1, 1, -1, 1].$$

Figure 11.4, clearly shows that, even for p_δ equal to 0.6 (corresponding to $\pm 0.6\%$ variations in the parameters), the optimization algorithm is able to find a parameter combination that destroys the limit cycle in the network model. As the allowable variation in the model parameters is increased, the rate of decay of the oscillations becomes even more rapid—for a $\pm 2\%$ variation the oscillations have completely ceased in less than 6 hours. Thus our results confirm the poor robustness findings of the previous μ-analysis; that is, extremely small changes in the values of the model's parameters can destroy the required oscillatory behavior.

11.3 Stochastic Robustness Analysis

The model for cAMP oscillations given in equation (11.1) in terms of ordinary differential equations corresponds to set of chemical reactions. Chemical reactions occur with probabilities proportional to the chances of collision of the molecules concerned.

Let a molecule C be a complex produced by the binding of molecules A and B:

$$\mathrm{A} + \mathrm{B} \xrightarrow{k} \mathrm{C}. \tag{11.21}$$

If this reaction is observed in a large volume, such as in a population of *Dictyostelium* cells, each of the species A, B, and C would be characterized by a concentration of molecules; if micromolar, the units of the rate constant k would be 1/μM/min. In that case, the following ordinary differential equation model for concentrations is valid:

$$\frac{\mathrm{d}C}{\mathrm{d}t} = kA \times B. \tag{11.22}$$

If, however, a small volume, say one cell as opposed to a population, is observed, then A, B, and C should be described by molecule number, and the rate constant converted into units of 1/(number of molecules)/min. Converting gives

$$k\left[\frac{1}{\mu\mathrm{M\ min}}\right] = \frac{k}{10^{-6}}\left[\frac{\ell}{\mathrm{mole\ min}}\right],$$

which leads to

$$\frac{k}{10^{-6}V}\left[\frac{1}{\mathrm{mole\ min}}\right] = \frac{k}{10^{-6}N_{\mathrm{av}}V}\left[\frac{1}{\#\ \mathrm{min}}\right], \tag{11.23}$$

where N_{av} is Avogadro's number, 6.023×10^{23}, and V is the volume where the reaction occurs, here the volume of the cell.

In a stochastic framework, the probability, P, that the reaction occurs within a time interval of length $\mathrm{d}t$ is given by the product of a propensity function a and $\mathrm{d}t$:

$$P = a\,\mathrm{d}t,$$

where the propensity function is given by

$$a = \frac{k}{10^{-6}N_{\mathrm{av}}V} A \times B,$$

where A and B now represent the number of molecules of the corresponding chemical species. The length of the time interval between reactions follows an exponential distribution, and which reaction occurs at that instant depends on the propensity functions. An exact simulation algorithm, Gillespie's direct method, can be used to realize the given chemical master equation (Gillespie, 1977; for details of the algorithm, see chapter 2).

Variations of equation (11.22) are treated as follows. If A is more abundant than both B and C, then the number of molecules of A is not significantly affected by the reaction. In this case, A could be considered as a constant and the reaction could be modified as follows:

$$\mathrm{B} \xrightarrow{A\times k/N_{\mathrm{av}}/V/10^{-6}} \mathrm{C}, \tag{11.24}$$

that is, C is produced directly from B with a reaction rate that depends on A. If, in addition, B is also abundant relative to C, then the reaction can be written as

$$\varnothing \xrightarrow{A\times B\times k/N_{\mathrm{av}}/V/10^{-6}} \mathrm{C}. \tag{11.25}$$

Similar rules can be applied for the reactants. Using these rules, we can reconstruct the chemical reaction network corresponding to the Laub and Loomis model as follows:

$$\begin{aligned}
&\mathrm{CAR1} \xrightarrow{k_1} \mathrm{ACA} + \mathrm{CAR1}\\
&\mathrm{ACA} + \mathrm{PKA} \xrightarrow{k_2/N_{av}/V/10^{-6}} \mathrm{PKA}\\
&\mathrm{cAMPi} \xrightarrow{k_3} \mathrm{PKA} + \mathrm{cAMPi}\\
&\mathrm{PKA} \xrightarrow{k_4} \varnothing\\
&\mathrm{CAR1} \xrightarrow{k_5} \mathrm{ERK2} + \mathrm{CAR1}\\
&\mathrm{PKA} + \mathrm{ERK2} \xrightarrow{k_6/N_{av}/V/10^{-6}} \mathrm{PKA}\\
&\varnothing \xrightarrow{k_7\times N_{av}\times V\times 10^{-6}} \mathrm{RegA}\\
&\mathrm{ERK2} + \mathrm{RegA} \xrightarrow{k_8/N_{av}/V/10^{-6}} \mathrm{ERK2}\\
&\mathrm{ACA} \xrightarrow{k_9} \mathrm{cAMPi} + \mathrm{ACA}\\
&\mathrm{RegA} + \mathrm{cAMPi} \xrightarrow{k_{10}/N_{av}/V/10^{-6}} \mathrm{RegA}\\
&\mathrm{ACA} \xrightarrow{k_{11}} \mathrm{cAMPe} + \mathrm{ACA}\\
&\mathrm{cAMPe} \xrightarrow{k_{12}} \varnothing\\
&\mathrm{cAMPe} \xrightarrow{k_{13}} \mathrm{CAR1} + \mathrm{cAMPe}\\
&\mathrm{CAR1} \xrightarrow{k_{14}} \varnothing,
\end{aligned} \tag{11.26}$$

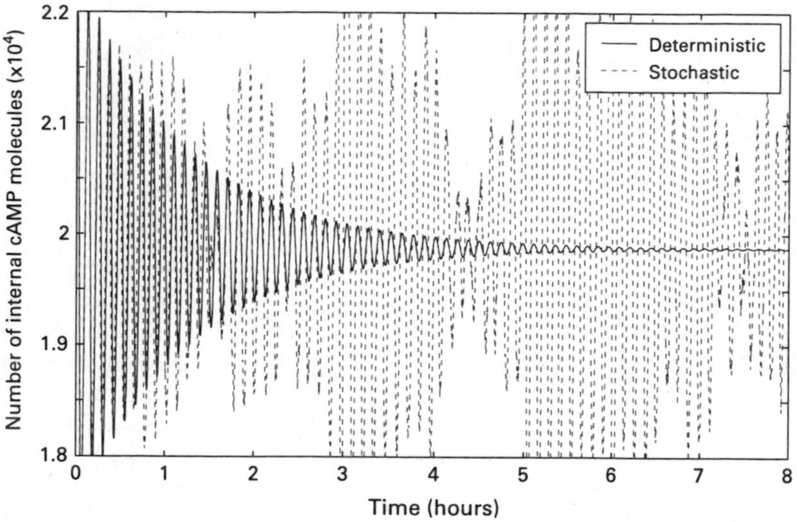

Figure 11.5
Internal cAMP oscillations with the worst-case perturbation for the deterministic and the stochastic simulations.

where V equal to $3.672 \times 10^{-14}\ell$ is obtained by adjusting the volume so that the variable representing ligand-bound CAR1 is approximately matched to the known number of cell receptors on the surface of a *Dictyostelium* cell (about 40,000; Laub and Loomis, 1998).

We can employ the above stochastic simulation approach to evaluate the effect of noise on the robustness of our model. As shown in figure 11.5, a 2% perturbation from the nominal values of the kinetic parameters in the original deterministic model is sufficient to destroy the stability of the oscillation and make the system converge to a steady state in about 6 hours (Kim et al., 2006). On the other hand, the figure shows that the stochastic model exhibits persistent oscillations under these perturbations to the nominal model parameters, when the stochastic simulation is performed using Gillespie's direct method (Gillespie, 1977). Although these results mirror those of Vilar et al. (2002) in revealing the qualitative differences in model dynamics that may result from consideration of noise, it is not yet clear whether the stochastic version of the model is actually more robust. It could be the case that a different worst-case parameter combination exists for this model. To clarify this issue, we proceed as follows. The kinetic parameters, the cell volume, and the initial conditions are all simultaneously perturbed according to

$$k_i = \bar{k}_i\left(1 + \frac{p_\delta}{100}\delta_i\right), \tag{11.27}$$

$$V = \overline{V}\left(1 + \frac{p_\delta}{100}\delta_v\right), \tag{11.28}$$

$$x_i(0) = \bar{x}_i(0)\left(1 + \frac{p_\delta}{100}\delta_{x_i}\right), \tag{11.29}$$

for $i = 1, 2, \ldots, 14$, where the kinetic parameters are perturbed in the same way as in the previous section, the nominal cell volume is equal to $3.672 \times 10^{-14}\ell$, the nominal initial conditions are as follows:

$$\bar{x}(0) = [7{,}290, \quad 7{,}100, \quad 2{,}500, \quad 3{,}000, \quad 4{,}110, \quad 1{,}100, \quad 5{,}960]^T, \tag{11.30}$$

and δ_i, δ_v, and δ_{x_i} are uniform random number in the range of $[-1, 1]$. As shown in figure 11.5, the number of molecules considered here is relatively large. Consequently, the time between individual reactions is small, thus the progress of the stochastic simulations will be slow. To overcome this problem, we used a variation of Gillespie's direct method called the τ-leap algorithm (Gillespie, 2001; described in section 2.3.1). The accuracy of the algorithm is highly dependent on the chosen local error tolerance. To compare the simulation results from the τ-leap and direct methods, we chose the maximum allowed relative error to be 5×10^{-5}. For simulating the cAMP network with the τ-leap algorithm, we used a software package called Dizzy, version 1.11.4 (CompBio Group, Institute for Systems Biology, 2006), which is freely available.

Finally, 100 Monte Carlo simulations were performed for random variations in the uncertain parameters for $p_\delta = 10\%$ and the time history of each internal cAMP variable was inspected. The time samples were obtained from 0 to 200 min, with a sample time of 0.01 min. Then the Fourier transform of the samples was taken using the fast Fourier transform (FFT) algorithm in Matlab (MathWorks, 2003). Trajectories for which more than 70% of the frequency content appeared outside of the peak amplitude were identified as nonoscillatory. Exactly the same scenario was applied for the deterministic model. The final results are shown in figure 11.6. The bars at the 20-minute mark on the horizontal axis represent the proportion of nonoscillating cases among the 100 simulations. For the deterministic model, this is around 13%; for the stochastic model, it is only 3%. Thus the number of nonoscillating cases is significantly reduced by including the effects of stochasticity in the robustness analysis. The standard deviation of the period is also significantly reduced for the stochastic case, which is to say, the changes in the period due to the effects of the uncertain parameters are smaller for the stochastic model than for the deterministic one. Similar improvements in the robustness of both the period and amplitude of the oscillations in the stochastic model were observed for a range of different uncertainty levels in the kinetic parameters (Kim et al., 2007). It is thus quite clear that stochastic noise

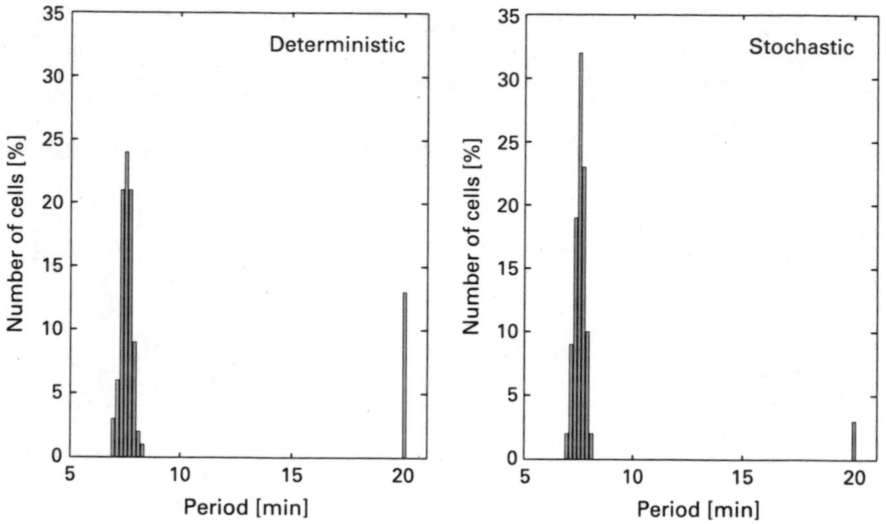

Figure 11.6
Period distributions for the deterministic and the stochastic simulations. The bars at 20 minutes represent the proportion of nonoscillating cases.

in the *Dictyostelium* cAMP network represents an important source of robustness to variations between different cells and to changes in their environment.

11.4 Conclusions

In the recent systems biology literature, the concept of robustness has been proposed as a key validator for models of many types of biological systems. This chapter has shown how analysis tools from the field of control engineering can be used to provide insight into the robustness of models of oscillatory biochemical networks. μ-analysis techniques provide allowable levels of parameter variations for which model robustness is guaranteed (something that optimization-based search or statistical methods can never do). On the other hand, hybrid optimization methods that combine global and local approaches can overcome the computational complexity of certain robustness analysis problems, and thus compute actual worst-case parameter combinations that can be used to check the theoretical robustness levels predicted by μ-analysis. Finally, statistical methods based on Monte Carlo simulation allow a rigorous comparison of the robustness of deterministic and stochastic models of oscillatory biological systems. Such analysis reveals the crucial role played by intracellular noise in ensuring the robustness of the resulting oscillations. Although in this chapter, we have focused our analyses on individual *Dictyostelium* cells, synchronization effects between neighboring cells can also have a significant impact on the robustness of the overall biological system, as shown by Kim et al. (2007). Stochastic simulations of large numbers of interacting cells require huge amounts of computing power, however, strongly motivating the development of analytical tools for evaluating the stability and robustness of stochastic systems (for promising new results in this area, see Scott et al., 2007; and Kim et al., 2008).

12 A Theory of Approximation for Stochastic Biochemical Processes

David Thorsley and Eric Klavins

The difficulties of modeling stochastic processes inside the cell are twofold. A stochastic model must be complex enough to explain and predict the biologically interesting features of a process's behavior, but not be so large and complex as to make analysis of the model intractable. The control-theoretic approach to modeling seeks not to faithfully describe every possible interaction that could affect the process behavior but to produce a model that approximates the true system closely enough to accurately describe key aspects. In this chapter, we propose a method for producing such approximate models to describe the behavior of stochastic biochemical processes. The method uses the mathematical construct known as the Wasserstein pseudometric and is general enough to address many modeling problems of interest, such as model comparison, parameter estimation, and model invalidation.

12.1 Modeling Stochastic Phenomena

The processes that govern behavior at the cellular level are inherently stochastic (Mettetal and van Oudenaarden, 2007). In the traditional view of cellular biology, it was believed that the observed variation in the behavior of cells was due to differences in their genetic material and in their environment (extrinsic noise, differences in inputs to the system) and that genetically identical cells placed in identical environments would behave identically. Recent studies have suggested that this is not the case, however, that the observed variation is also the result of noise intrinsic to all chemical reactions. Although the large numbers of reacting molecules on a macroscale result in noise playing a small enough role that such reactions can be modeled as deterministic systems, on a nanoscale, where there may only be a few copies of a particular reacting species, intrinsic noise can cause significant fluctuations (McAdams and Arkin, 1999) and the reactions must be modeled as stochastic systems. Moreover, regulatory mechanisms inside the cell do not necessarily attempt to minimize or eliminate this noisy behavior. Exploiting fluctuations can have beneficial effects for the cell; for example, fluctuations can be used to increase the strength of a molecular signal (Paulsson et al., 2000). The recognition that stochastic effects play an important role in cellular behavior has motivated the development of new experimental methods to observe and quantify noisy behavior inside the cell (Elowitz et al., 2002) and promoted a new appreciation of mathematical techniques to simulate and analyze cellular processes stochastically (Gillespie, 2007).

This chapter investigates problems that arise when trying to produce a realistic mathematical model of a stochastic biochemical process. The model of the phage λ lysis-lysogeny decision proposed by Arkin et al. (1998), though an early success of stochastic modeling and simulation, suffers from several limitations. Perhaps the most significant of these is the size and complexity of the mathematical model, which includes hundreds of species and hundreds of reactions. Formal analysis of such enormous models is intractable, and studying such systems through stochastic simulation is time consuming and computationally intensive. Avoiding the difficulties associated with enormous models is the objective of *model reduction.*

Another important task in modeling biochemical processes is *parameter estimation*, which determines the rates at which the reactions in the system occur. A typical model of a biochemical process relies on reaction rates that have been collected independently by dozens of prior experiments over a period of years (see, for example, Arkin et al., 1998; de Atauri et al., 2004). When constructing a large model of a process, it is often the case that some reaction rates have not been experimentally validated; in such situations, these parameters are estimated to produce models that best match the expected behavior of the process in an ad hoc approach to parameter estimation.

There are many approaches to modeling processes based on both the level of detail desired by the modeler and the amount of a priori knowledge that is available. This chapter will concentrate on *stochastic reaction networks*, mechanistic models that are closely related to the underlying physical and chemical laws that govern the processes. Even at this level, there is room for abstraction and approximation: whereas Arkin et al. (1998) propose a model of stochastic gene expression with hundreds of defined species and reactions, Swain et al. (2002) consider only 8 species and 11 reactions, and Singh and Hespanha (2008) consider only 2 species and 4 reactions.

Reaction network models are most useful when we have a good idea of the underlying biochemistry. Alternative approaches based on the analysis of high-throughput data produce alternative models such as probabilistic Boolean networks and Bayesian networks (Price and Shmulevich, 2007) and stochastic differential equations (van Kampen, 1981). When confronted with different models of the same process that are based on different formalisms, rely on different levels of abstraction, and are constructed from different first principles (mechanism-driven or data-driven), we would like to verify whether these different models are behaviorally equivalent. We call the task of verifying the equivalence of models *model comparison*. If it turns out that the different models are not equivalent to each other, we would then want to determine which candidate models are poor representations of the real process, a task called *model invalidation*.

Given the complexity of the biochemical processes that we aim to model, finding compact models that exactly describe the behavior of these systems is likely to be computationally intractable at best. From a control-theoretic perspective, there is no need to find an exact model that faithfully explains and predicts every aspect of a process's behavior; what matters is to find a model that is "good enough" to solve a given problem. A less detailed model may be more useful than a more detailed model if it is much easier to analyze the simpler model and use it to make predictions. The tools we develop in this chapter for model reduction, comparison, and invalidation are based on the control-theoretic idea that the simplest model sufficient for the task at hand is usually the best. Because the simplest model of a stochastic process may be a reaction network, a Bayesian network, a stochastic differential equation, or even just a data set, we propose a general method for addressing these modeling issues that can, in principle, be applied to all of these different classes of models.

12.2 The Need for Approximation Metrics

The models used to describe stochastic biochemical processes are based on simplifying assumptions and approximations known not to be fully accurate. Mechanistic models like reaction networks are based on stochastic chemical kinetics (Gillespie, 2007). The mathematical formulation of stochastic chemical kinetics is based on several assumptions—the chemical reaction chamber must be a fixed volume, the reactants must be well mixed within that volume, the system must be at thermal equilibrium, and most molecular collisions must be nonreactive. These assumptions allow us to approximate the molecular positions as uniformly distributed throughout the reaction chamber and the molecular velocities as governed according to the Maxwell-Boltzmann distribution; they simplify the analysis and make it easier to understand the model, thereby making it easier to use it to make predictions about the system's behavior. Despite the apparent limitations of the simplifying assumptions (we know, for example, that processes are affected by spatial effects and by temperature gradients), models that use the stochastic chemical kinetics formalism are accurate enough to have both explanatory and predictive power.

Our goal for the chapter is to present a theory of approximation that facilitates the development of simpler models of stochastic biochemical processes and that is applicable to all classes of stochastic models, be they based on mechanistic principles or empirical observation. To convey what we mean by quality of an approximation, we introduce and define a notion of distance between stochastic processes. Because different problems have different essential features, it must be possible to alter the precise formulation of the distance based on which aspects of a system's behavior are the most important for the problem under consideration. Using this distance, we approach common tasks such as model reduction, parameter estimation, model comparison, and model invalidation as problems of finding how "close" a model is—in some relevant sense—to a set of data or to some other model.

Engineers and mathematicians have developed different metrics for quantifying the differences between systems. For example, information theorists use the *Kullback-Leibler divergence* to quantify the difference between the distributions defining the probabilities of various sequences of data. Control theorists have defined the *gap metric* to measure the difference in performance that results when unstable systems are stabilized using feedback; two systems that are close in the gap metric have similar performance when placed in a feedback loop. The gap metric is thus often used to evaluate the performance of model reduction and parameter estimation techniques when the final objective is to design a feedback control system. But when considering stochastic biochemical processes, the final objective is not as clear. Different decisions as to what aspects of the system's behavior are relevant call for different metrics. Accordingly, to quantify the differences between stochastic biochemical processes, we use the class of *Wasserstein pseudometrics*, a class broad enough to incorporate many different notions of distance between systems without requiring a fundamental reworking of the algorithms needed to calculate the distance whenever a new notion is required.

12.3 Stochastic Processes and Wasserstein Pseudometrics

The stochastic chemical kinetics formalism, described in chapter 2, is commonly used for modeling stochastic phenomena inside the cell. We use this framework to describe a stochastic biochemical process and explain how to construct a probability measure on the set of all trajectories from the reaction network formalism. The approach we propose here is general enough that it can be applied to other classes of stochastic models, such as Bayesian networks and stochastic differential equations (e.g., linear noise approximations and chemical Langevin equations). We begin this section by expanding on the material presented in section 2.2.1 and specializing the notation to the problem under consideration.

12.3.1 Stochastic Reaction Networks

A *reaction network* $RN = (\mathcal{S}, \mathcal{R})$ consists of a set of species $\mathcal{S}$ and a set of reactions $\mathcal{R}$. Each reaction takes the form

$$n_{1,L}S_1 + \cdots + n_{p,L}S_k \xrightarrow{c} n_{1,R}S_1 + \cdots + n_{q,R}S_k, \tag{12.1}$$

where for all $i = 1, \dots, k$, S_i is a species, and each $n_{i,L}$ and $n_{i,R}$ is a nonnegative integer. Each reaction has a rate constant c, whose role will be described shortly.

Reaction networks can be interpreted either deterministically or stochastically (McQuarrie, 1976). To interpret a reaction network stochastically, we first define the state of RN as a k-dimensional vector $\boldsymbol{x}(t) = [N_1(t), \dots, N_k(t)]^T$, where $N_i(t)$ denotes the number of the species S_i at time t. The probability distribution of the initial state (at $t = 0$) is denoted by π_0.

The firing of a reaction in $\mathcal{R}$ produces a state transition

$$\boldsymbol{x} \mapsto \boldsymbol{x} - \begin{bmatrix} n_{1,L} \\ \vdots \\ n_{k,L} \end{bmatrix} + \begin{bmatrix} n_{1,R} \\ \vdots \\ n_{k,R} \end{bmatrix}, \tag{12.2}$$

which corresponds to the consumption of reactants and the creation of products. Reactions in stochastic reaction networks fire at distinct, discrete, instances in time. If the state of RN is $\boldsymbol{x}$ at time t, the probability that a reaction R will fire in the interval $[t, t + \mathrm{d}t)$ is defined as $a_R(\boldsymbol{x})\,\mathrm{d}t$, where

$$a_R(\boldsymbol{x}) = c \prod_{i=1}^{k} \binom{N_i}{n_{i,L}} \tag{12.3}$$

is the *propensity function* for a reaction $R \in \mathcal{R}$. The precise form of the propensity function relies on kinetic arguments that are described in detail in chapter 2 and in Gillespie (2007).

The propensity function for each reaction depends only on the current state of the network and not on any previous states. Stochastic reaction networks therefore satisfy the Markov property. Indeed, a stochastic reaction network can be described and analyzed using an equivalent continuous-time Markov process (CTMP), which consists of a countable set of states X, a transition rate matrix $\mathbf{Q}$, and an initial probability distribution π_0. If $\mathbf{Q}_{ij}$ is a nondiagonal element of $\mathbf{Q}$, then the probability of a transition from $\boldsymbol{x}_i$ to $\boldsymbol{x}_j$ in the interval $[t, t + \mathrm{d}t)$ is defined as $\mathbf{Q}_{ij}\,\mathrm{d}t$; if $\mathbf{Q}_{ii}$ is a diagonal element of $\mathbf{Q}$, we require that $\mathbf{Q}_{ii} = -\sum_{j \neq i} \mathbf{Q}_{ij}$.

To construct a continuous-time Markov process from a reaction network, we define the state space X as the k-dimensional lattice of nonnegative integers $(\mathbf{Z}^{\geq 0})^k$. Each nondiagonal element of $\mathbf{Q}$ is related to the propensity functions according to the equation

$$\mathbf{Q}_{ij} = \sum_{R \in \mathcal{R}: \boldsymbol{x}_j = \boldsymbol{x}_i - [n_{1,L} \dots n_{k,L}]^T + [n_{1,R} \dots n_{k,R}]^T} a_R(\boldsymbol{x}_i).$$

The initial distribution of the CTMP, π_0, is the same as that of the reaction network.

In practice, because it is not possible to observe the state of a biochemical system modeled by a stochastic reaction network completely, we introduce a set of outputs Y and a state output function $h : X \to Y$ to describe our incomplete knowledge of the full system state. For example, if we are capable of only observing the quantity of a species S_i (e.g., because it is tagged with green fluorescent protein), we would define the output set Y as the nonnegative integers and the state output function as $h([N_1, \ldots, N_k]^T) = N_i$.

Example The following reaction network is a simple model of gene expression (Singh and Hespanha, 2008):

$$R_1: \quad \varnothing \underset{k_{-1}}{\overset{k_1}{\rightleftharpoons}} \text{mRNA},$$

$$R_2: \quad \text{mRNA} \xrightarrow{k_2} \text{mRNA} + \text{protein},$$

$$R_3: \quad \text{protein} \xrightarrow{k_3} \varnothing.$$

The set of species is $\mathcal{S} = \{\text{mRNA}, \text{protein}\}$; the symbol $\varnothing$ is a shorthand that denotes that no species is present on one side of the reaction. Reaction networks are often displayed graphically as shown in figure 12.1a. Because there are two species,

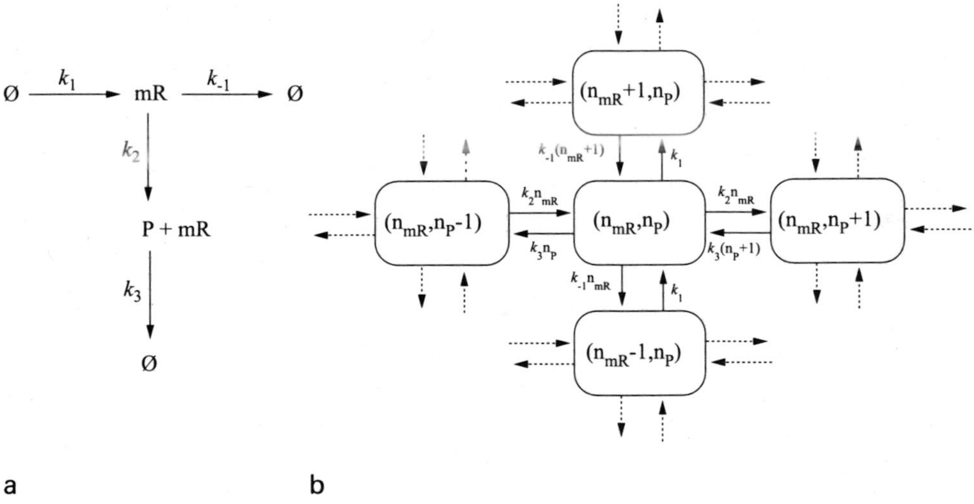

Figure 12.1
(a) Graphical version of a reaction network modeling gene expression. (b) Typical state in the continuous-time Markov process (CTMP) generated by the stochastic interpretation of the reaction network in a. Each state is indicated by a pair (n_{mR}, n_P), and the labels on each arrow indicate the rate propensity for each transition in or out of (n_{mR}, n_P).

the continuous-time Markov process generated by the reaction network above has the state space $X = \mathbf{Z}^{\geq 0} \times \mathbf{Z}^{\geq 0}$, that is, the state space consists of pairs of nonnegative integers. Each state of the CTMP is of the form $\boldsymbol{x} = [n_{mR}\ n_P]$. A typical state of the CTMP is shown in figure 12.1b. The output function h is defined as $h([n_{mR}\ n_P]) = n_P$; this output function describes the situation where we can observe protein number in the system, which can be accomplished by incorporating a fluorescent marker into the protein. ■

12.3.2 Constructing Probability Measures

Stochastic reaction networks and continuous-time Markov processes comprise specific classes of stochastic processes, which are collections of random variables defined on a probability space $(\Omega, \mathcal{F}, \mathcal{P})$, where Ω is a sample space, $\mathcal{P}$ is a probability measure, and $\mathcal{F}$ is a σ-field. The σ-field $\mathcal{F}$ is the domain of the probability measure and, intuitively, describes the sets of trajectories that can be distinguished by the available measurement apparatus. Stochastic reaction networks are used to model dynamic processes whose behaviors are time-varying trajectories. A trajectory of a stochastic reaction network associates with each time $t \geq 0$ a state $\boldsymbol{x} \in X$ and an output $h(\boldsymbol{x}) \in Y$. A trajectory generated by a reaction network is thus a function $\omega : \mathbf{R}^{\geq 0} \to Y$ that assigns an output to each time instant $t \geq 0$. The sample space Ω is the set of all such trajectories.

It is straightforward to define pseudometrics on the sample space Ω to capture a notion of distance between trajectories. A *pseudometric* d is a function $d : \Omega \times \Omega \to \mathbf{R}^{\geq 0}$ that is always nonnegative ($d(\omega, \eta) \geq 0$ for all $\omega, \eta \in \Omega$) and satisfies the "triangle inequality" ($d(\omega, \varphi) + d(\varphi, \eta) \geq d(\omega, \eta)$ for all $\omega, \eta, \varphi \in \Omega$). (We say that d is a "pseudometric" because it only satisfies two of the three conditions necessary to be a true metric: we do not require d to satisfy the property $d(\omega, \eta) = 0$ only if $\omega = \eta$. A technical restriction on d is that it must be measurable with respect to the σ-field $\mathcal{F}$.) The σ-field $\mathcal{F}$ is the domain of the probability measure $\mathcal{P}$. A stochastic process is, fundamentally, a method for defining this probability measure on the space of trajectories Ω. The procedure by which a probability measure on Ω is constructed from a given continuous-time Markov process is described in Breiman (1992, chapter 15). Because there exists a CTMP corresponding to any stochastic reaction network, it follows that each stochastic reaction network defines a measure $\mathcal{P}$ on the space of trajectories. Therefore, to define a distance between stochastic reaction networks, it suffices to define a distance between probability measures.

12.3.3 Wasserstein Pseudometrics

For simplicity, we restrict ourselves to pseudometrics of the form

$$d(\omega, \eta) = |Z(\omega) - Z(\eta)|, \tag{12.4}$$

where $Z : \Omega \to \mathbf{R}$ is an arbitrary random variable. We call Z a *reporter random variable* because it describes, or reports, one interesting aspect of a trajectory. For the network shown in figure 12.1a, some potential reporter random variables of interest include

- Z can represent the amount of a protein at a given time t:

$Z(\omega) = \omega(t)$.

- Z can represent the first time that n proteins are present:

$Z(\omega) = \min\{\omega^{-1}(n)\}$.

- Z can represent the average amount of protein over an interval (t_s, t_f):

$$Z(\omega) = \frac{1}{t_f - t_s} \int_{t_s}^{t_f} \omega(t)\,\mathrm{d}t.$$

- Z can indicate if more than n proteins are ever present:

$Z(\omega) = 1$ if $n_P(t) > n$ for some $t \in \mathbf{R}^{\geq 0}$ and $Z(\omega) = 0$ otherwise.

Each probability measure $\mathcal{P}$ defines a cumulative distribution function (CDF) of Z:

$$F_{\mathcal{P},Z}(z) = \mathcal{P}(Z < z). \tag{12.5}$$

The inverse cumulative distribution function of Z is

$$F^{-1}_{\mathcal{P},Z}(y) = \inf\{z : F_{\mathcal{P},Z}(z) \geq y\}. \tag{12.6}$$

Let $\mathcal{P}_1$ and $\mathcal{P}_2$ denote two probability measures on Ω. Using $F^{-1}_{\mathcal{P}_1,Z}$ and $F^{-1}_{\mathcal{P}_2,Z}$, the CDFs of Z with respect to the two probability measures, we can define the following pseudometric to quantify the difference between them.

Definition 12.1 For any $p > 0$, the Wasserstein pseudometric W^p_d between two probability measures $\mathcal{P}_1$ and $\mathcal{P}_2$ on a sample space Ω equipped with a pseudometric d defined according to equation (12.4) is

$$W^p_d(\mathcal{P}_1, \mathcal{P}_2) = \left(\int_0^1 |F^{-1}_{\mathcal{P}_1,Z}(y) - F^{-1}_{\mathcal{P}_2,Z}(y)|^p \,\mathrm{d}y \right)^{1/p}. \tag{12.7}$$

■

Intuitively, the Wasserstein pseudometric W^1_d is the area between the inverse CDFs generated by the two probability measures, as shown in figure 12.2a (the Was-

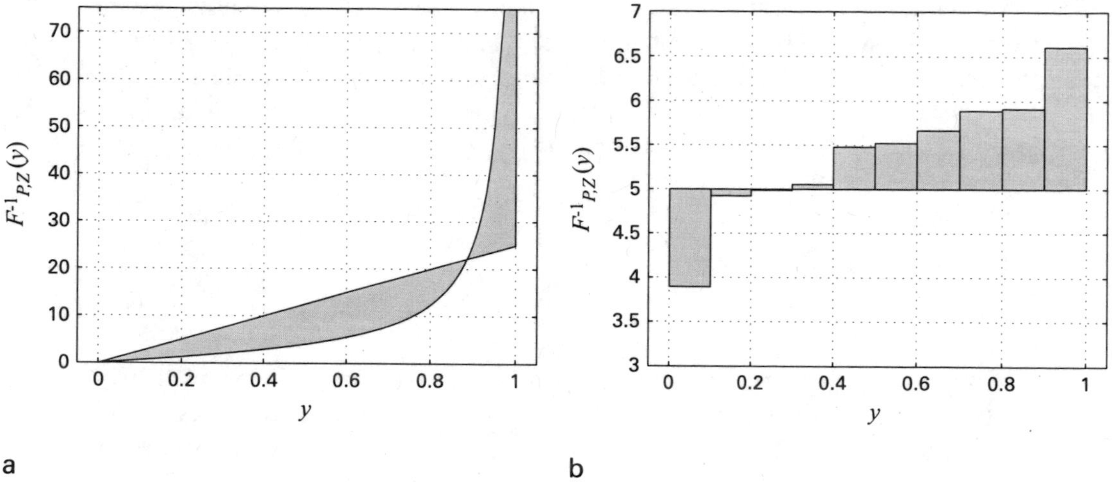

Figure 12.2
(a) Graphical interpretation of the Wasserstein pseudometric as the area between two inverse cumulative distribution functions (CDFs), shown here as those of a uniform distribution and a truncated Cauchy distribution. (b) Interpretation of the root-mean-square (RMS) distance as a Wasserstein pseudometric. The area between the two inverse CDFs is shaded. By squaring the area of each rectangle and summing, we can find the value of the integral in equation (12.8), which is the square of the RMS distance.

serstein pseudometric is defined more generally with respect to an arbitrary distance d in Dudley, 2002).

12.3.4 Related Concepts

In the Monge-Kantorovich transportation problem, a classical problem of mathematical economics (Vershik, 2006), the two probability distributions $\mathcal{P}_1$ and $\mathcal{P}_2$ represent the distributions of locations where a type of good is produced and where it is consumed, respectively. The goal of the transportation problem is to find an optimal transportation plan that minimizes the average cost of transporting goods from where they are produced to where they are consumed. If the locations of the producers and the consumers are along a straight line, the integral in equation (12.7) is the exact solution to this problem. The Wasserstein pseudometric is one of many similarly defined "transportation metrics" that have been proposed as solutions (for a general definition of the Wasserstein pseudometric, see Dudley, 2002, chap. 11). The equivalence of the general definition and definition 12.1 above is proven for $p = 1$ in Vallender (1974); the proof can be easily generalized to the case where $p > 1$. The Wasserstein pseudometric is used in many different application areas and thus has many different names: the "Kantorovich metric," the "Mallows distance," and the "earth mover's distance," among others.

Wasserstein pseudometrics are generalizations of techniques for quantifying errors between a noisy data set and a deterministic system or fixed parameter. In particular, the Wasserstein pseudometric W_d^2 is a generalization of the root-mean-square distance between a parameter θ and a set of data $\{\hat{\theta}_1, \hat{\theta}_2, \dots, \hat{\theta}_n\}$, where each data point is an estimate of the parameter. We can define the two probability distributions as $\mathcal{P}_1(\hat{\theta}_i) = 1/n$, for all i, to describe the data set, and $\mathcal{P}_2(\theta) = 1$ to describe the fixed parameter. The Wasserstein pseudometric W_d^2 between these two distributions is

$$W_d^2(\mathcal{P}_1, \mathcal{P}_2) = \left(\int_0^1 |F_{\mathcal{P}_1,Z}^{-1}(y) - F_{\mathcal{P}_2,Z}^{-1}(y)|^2 \, dy \right)^{1/2}. \tag{12.8}$$

If we plot the two inverse cumulative distribution functions as in figure 12.2b, we see that the integral can be calculated by finding the area of n rectangles, each of which has width $1/n$. The height of the ith rectangle is $|\hat{\theta}_i - \theta|$. The value of the integral is therefore

$$W_d^2(\mathcal{P}_1, \mathcal{P}_2) = \left(\sum_{i=1}^{n} \frac{1}{n} |\hat{\theta}_i - \theta|^2 \right)^{1/2}, \tag{12.9}$$

which is the root-mean-square distance criterion. Similarly, the Wasserstein pseudometric W_d^1 is a generalization of the mean absolute error.

There is a rich literature on metrics between probability distributions; there are many different possible methods for defining such metrics (Gibbs and Su, 2002). Another commonly used metric is the *total variation distance*, defined for discrete sample spaces as

$$TV(\mathcal{P}_1, \mathcal{P}_2) = \frac{1}{2} \sum_{\boldsymbol{x} \in X} |\mathcal{P}_1(\boldsymbol{x}) - \mathcal{P}_2(\boldsymbol{x})|. \tag{12.10}$$

For example, Munsky and Khammash (2006) demonstrate the effectiveness of the finite-state projection method by showing that the total variation distance between the full chemical master equation model and a reduced master equation model is small. The total variation distance is equal to the Wasserstein distance W_d^1 when the distance function d is equal to the *discrete metric*, defined as $d(\omega, \eta) = 1$ if $\omega \neq \eta$ and $d(\omega, \eta) = 0$ otherwise. However, it is possible for the total variation distance between two stochastic reaction networks to become arbitrarily small while a Wasserstein pseudometric defined with respect to a different d remains large. If this new distance function d represents a biologically interesting feature of the system's behavior, a total variation distance criterion may not be sufficient to decide that two systems are close. The Wasserstein pseudometric W_d^1 has a dual representation, often used in computer science to define "bisimulation metrics," which are also used to measure the difference between a pair of systems. Desharnais et al. (2004) extend the Wasserstein pseudometric to define a bisimulation metric on the space of probabilistic transition systems; van Breugel and Worrell (2006) take a similar approach and develop a polynomial-time algorithm for computing the distances between such systems.

12.4 Algorithms for Calculating Wasserstein Pseudometrics

Like van Breugel and Worrell (2006), we are interested in developing efficient algorithms for calculating Wasserstein pseudometrics. Because the continuous-time Markov processes generated from reaction networks may have infinite state spaces, algorithms that are polynomial in the number of states in the system are of no help. Calculation of Wasserstein pseudometrics with respect to general distance functions d is difficult; the restriction we make in equation (12.4) reduces the complexity of the algorithms.

The primary reason for the difficulty in calculation is that the probability measure $\mathcal{P}$ on $(\Omega, \mathcal{F})$ generated by a continuous-time Markov process is usually too complex to determine exactly and thus must remain unknown. Moreover, the CTMP is merely an approximation of a physical process that is also an unknown probability measure on $(\Omega, \mathcal{F})$.

We work around this difficulty by approximating the unknown probability measure $\mathcal{P}$ with an empirical probability measure $\hat{\mathcal{P}}_n$, generated by taking n independent samples of Ω according to $\mathcal{P}$. The empirical probability measure created from these n samples $\{\omega_1, \omega_2, \ldots, \omega_n\}$ is defined as

$$\hat{\mathcal{P}}_n(\omega_i) = \frac{1}{n}, \qquad \text{for } i = 1, \ldots, n.$$

We calculate a Wasserstein pseudometric between empirical probability distributions to approximate the pseudometric between the unknown underlying distributions. The sample data used to make this approximation may be obtained either from physical processes by performing experiments or from reaction network models by using the stochastic simulation algorithm (SSA; Gillespie, 1977).

If we specialize equation (12.7) to the case of empirical probability distributions, we obtain an algorithm for estimating Wasserstein pseudometrics between distributions. Consider two unknown probability distributions $\mathcal{P}_1$ and $\mathcal{P}_2$, and take n independent samples $\{\omega_1, \ldots, \omega_n\}$ from $\mathcal{P}_1$ and ℓn independent samples $\{\eta_1, \ldots, \eta_{\ell n}\}$ from $\mathcal{P}_2$, where $\ell \in \mathbf{N}$. The empirical cumulative probability distribution functions of Z with respect to $\mathcal{P}_1$ and $\mathcal{P}_2$ are

$$F_{\hat{\mathcal{P}}_{1,n}, Z}(z) = \frac{|\{\omega : Z(\omega) < z\}|}{n} \quad \text{and} \quad F_{\hat{\mathcal{P}}_{2,\ell n}, Z}(z) = \frac{|\{\omega : Z(\omega) < z\}|}{\ell n},$$

Input n samples $\{\omega_1, \ldots, \omega_n\}$ generated according to P_1. **Input** ℓn samples $\{\eta_1, \ldots, \eta_{\ell n}\}$ generated according to P_2.

1. **Calculate** $Z(\omega_i)$ for each ω_i, $i = 1 \ldots n$.
2. **Sort** and re-index $\{\omega_1, \ldots \omega_n\}$ so that $Z(\omega_1) \leq Z(\omega_2) \leq \cdots \leq Z(\omega_n)$.
3. **Calculate** $Z(\eta_i)$ for each η_i, $i = 1 \ldots \ell n$.
4. **Sort** and re-index $\{\eta_1, \ldots \eta_{\ell n}\}$ so that $Z(\eta_1) \leq Z(\eta_2) \leq \cdots \leq Z(\eta_{\ell n})$.
5. **Calculate** $|Z(\omega_{\lceil \frac{i}{\ell} \rceil}) - Z(\eta_i)|$ for $i = 1 \ldots \ell n$.
6. **Calculate** $W_d^1(\hat{P}_{1,n}, \hat{P}_{2,\ell n}) = \frac{1}{n} \sum_{i=1}^{n} |Z(\omega_{\lceil \frac{i}{\ell} \rceil}) - Z(\eta_i)|$.

Figure 12.3
Algorithm for estimating the Wasserstein pseudometrics between two probability distributions.

where $\hat{\mathcal{P}}_{1,n}$ denotes our sampled estimate of $\mathcal{P}_1$ and $\hat{\mathcal{P}}_{2,\ell n}$ denotes our sampled estimate of $\mathcal{P}_2$.

Without loss of generality, we can sort these sets of samples so that $Z(\omega_1) \leq Z(\omega_2) \leq \cdots \leq Z(\omega_n)$ and $Z(\eta_1) \leq Z(\eta_2) \leq \cdots \leq Z(\eta_{\ell n})$. The inverse empirical cumulative distribution functions of $\mathcal{P}_1$ and $\mathcal{P}_2$ are then of the form

$$F^{-1}_{\hat{\mathcal{P}}_n, Z}(y) = Z(\omega_i), \quad \text{where } \frac{i-1}{n} < y \leq \frac{i}{n}. \tag{12.11}$$

The Wasserstein pseudometric between the underlying probability distributions is expressed in terms of inverse empirical probability distributions according to the following theorem.

Theorem 12.1 Suppose $E_{\mathcal{P}_i}(|Z|) < \infty$ for $i = 1, 2$. The Wasserstein pseudometric $W_d^1(\mathcal{P}_1, \mathcal{P}_2)$ between two probability distributions $\mathcal{P}_1$ and $\mathcal{P}_2$ with respect to a pseudodistance function $d(\omega, \eta) = |Z(\omega) - Z(\eta)|$ on Ω is equal to

$$W_d^1(\mathcal{P}_1, \mathcal{P}_2) = \lim_{n \to \infty} \frac{1}{\ell n} \sum_{i=1}^{\ell n} |Z(\omega_{\lceil i/\ell \rceil}) - Z(\eta_i)| \tag{12.12}$$

almost surely. ■

Theorem 12.1 suggests an algorithm for approximating a Wasserstein pseudometric between two probability measures $\mathcal{P}_1$ and $\mathcal{P}_2$ (this algorithm is shown in figure 12.3; for a proof see Thorsley and Klavins, 2008).

To calculate the integral between the two empirical probability measures, we choose the number of samples taken from $\mathcal{P}_2$ to be an integer multiple of the number of samples taken for $\mathcal{P}_1$. The reason for this restriction is that it reduces the complexity of the calculation; to compute the area between the two empirical cumulative distribution functions, we must ensure that the edges of the rectangles whose areas we are adding together line up as evenly as possible.

The complexity of this algorithm is $O(\ell n \log \ell n)$, as the rate determining step is the sorting of the outcomes $\{\eta_1, \ldots \eta_{\ell n}\}$. Since generating the samples $\{\omega_1, \ldots \omega_n\}$ and $\{\eta_1, \ldots \eta_{\ell n}\}$ involves either performing a significant number of experiments or running the stochastic simulation algorithm a large number of times, the complexity of estimating a Wasserstein pseudometric from empirical probability measures is much less than the complexity of generating the data used to construct those measures.

12.5 Applications to Stochastic Gene Expression

The theoretical results of the previous sections provide tools for calculating and estimating Wasserstein pseudometrics between sample data sets generated by arbitrary stochastic processes. In this section, we illustrate how these tools are used to address various problems related to the modeling of biochemical reaction networks.

12.5.1 Model Comparison

Figure 12.4a is the reaction network model of gene expression found in Swain et al., (2002). In this model, RNA polymerase binds to a promoter (D), creating a complex (C) that in turn produces a transcribing polymerase (T). From the transcribing polymerase, unbound messenger RNA strands (mR^u) are produced. The mRNA can bind to either a degradasome (mC^1) or a ribosome (mC^2). If the mRNA strand binds to a ribosome, it produces a transcribing complex (mT) and then a protein (P). The mRNA strand is preserved until it binds to a degradasome. In contrast to Swain et al. (2002), we simplify the model by assuming that the rates are fixed throughout the process and do not vary with changes in cell size.

Suppose the output of this reaction network is the protein number, which could be estimated if the protein were, for example, fluorescent. We define a reporter random variable $Z(\omega) = \omega(1{,}440)$, the number of proteins present in the system at $t = 1{,}440$ seconds. This is the time at which cell division first occurs in the model proposed by Swain et al. (2002) and is a reasonable value for the length of the cell cycle in *E. coli*.

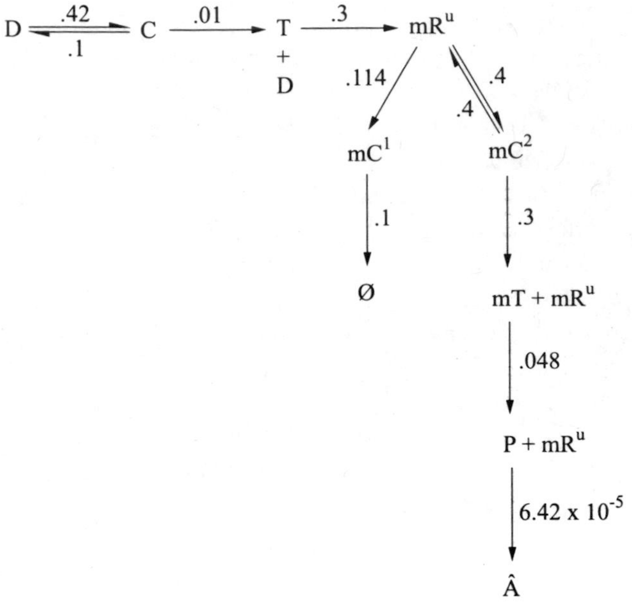

a

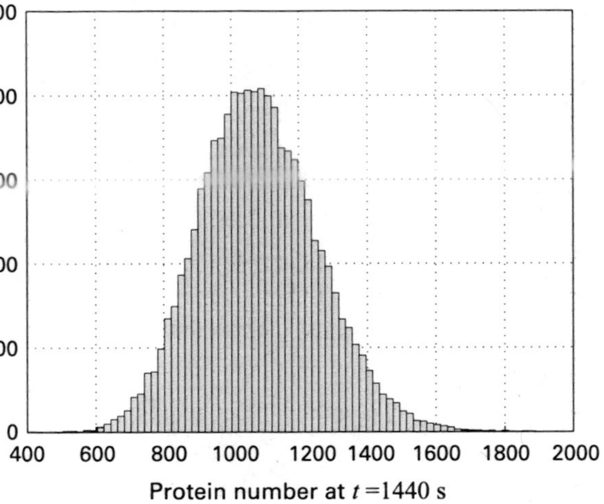

b

Figure 12.4
(a) Gene expression model from Swain et al., 2002. (b) Histogram showing the values of $Z(\omega)$ for 45,000 stochastic simulations ω of the reaction network in a.

Using the stochastic simulation algorithm, we generated $n = 45{,}000$ samples from the reaction network and computed the value $Z(\omega)$ for each sample. A histogram generated from the data is shown in figure 12.4b.

A reduced model of gene expression, shown in figure 12.5a is also considered in Swain et al. (2002). The translation process has been simplified. The transcribing polymerase T produces an mRNA strand (mR) that either decays or produces a protein (P). The complexes containing ribosomes and degradasomes are abstracted away in this model. Again using the stochastic simulation algorithm, we generated $n = 45{,}000$ samples from this reaction network and computed $Z(\eta)$ for each sample; a histogram of the data for this reduced system is shown in figure 12.5b.

Using the algorithm shown in figure 12.3, we approximated a Wasserstein pseudometric between $\mathcal{P}_1$, the probability measure generated by the full model, and $\mathcal{P}_2$, the probability measure generated by the reduced model. The value of this approximation is

$$W_d^1(\hat{\mathcal{P}}_{1,n}, \hat{\mathcal{P}}_{2,n}) = 5.44 \text{ proteins.}$$

As a heuristic, we define the percentage error between the reduced model and the full model as

$$\varepsilon(\mathcal{P}_1 \| \mathcal{P}_2) = \frac{W_d^1(\mathcal{P}_1, \mathcal{P}_2)}{E_{\mathcal{P}_1}(Z)}.$$

Because the average protein number in the full system is $E_{\hat{\mathcal{P}}_{1,n}}(Z) = 1{,}080.2$, the error introduced by the reduced model is 5.44/1,080.2, or $\varepsilon(\hat{\mathcal{P}}_{1,n} \| \hat{\mathcal{P}}_{2,n}) = 0.50\%$.

Using the bootstrap percentile method (Shao and Tu, 1995), we estimated a 95% confidence interval for $W_d^1(\mathcal{P}_1, \mathcal{P}_2)$. By resampling from the probability measures $\hat{\mathcal{P}}_{1,n}$ and $\hat{\mathcal{P}}_{2,n}$, we took 5,000 estimates of $W_d^1(\mathcal{P}_1, \mathcal{P}_2)$ and arrived at a 95% confidence interval of $(3.63, 7.50)$. Thus there is approximately a 97.5% probability that the difference between the full model and the reduced model is $\varepsilon < 7.50/1{,}080.2 = 0.69\%$. These numbers demonstrate a close agreement between the full and reduced models (for more on the bootstrap percentile method, see Thorsley and Klavins, 2008).

12.5.2 Parameter Estimation

The simple model in figure 12.1a, with an appropriate choice of parameters, serves as a reduced model for the reaction network in figure 12.4a. Following Swain et al. (2002) we assume that the protein decay rate $k_3 = 6.42 \times 10^{-5}\ \text{s}^{-1}$ and that $k_2 = 15k_{-1}$. Under these constraints, we performed an approximate stochastic gradient descent using the finite-difference method to estimate the values of k_1 and k_2 to

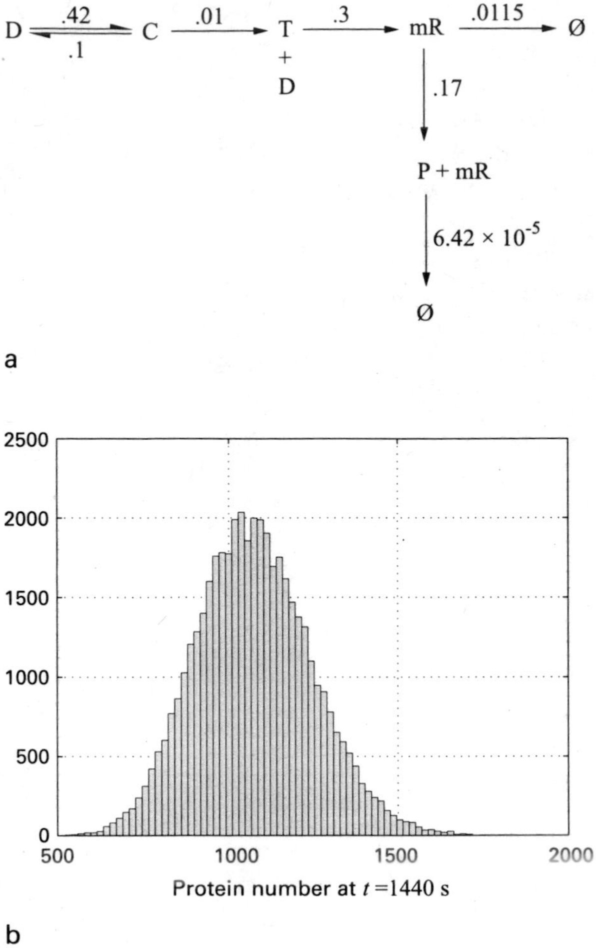

Figure 12.5
(a) Reduced gene expression model from Swain et al., 2002. (b) Histogram showing the values of $Z(\eta)$ of 45,000 stochastic simulations η of the reaction network in a.

minimize the Wasserstein pseudometric $W_d^1(\mathcal{P}_1, \mathcal{P}_3(k_1, k_2))$, where $\mathcal{P}_3(k_1, k_2)$ is the probability measure generated by the simple reaction network in figure 12.1a. This algorithm is an extension of the standard gradient descent method that is used when the gradient cannot be calculated directly and the objective function can only be observed with noise. The algorithm is initialized by taking an initial guess as to the optimal values of k_1 and k_2. At each iteration, the gradient is estimated by calculating the Wasserstein pseudometric with the parameters $k_1 \pm \epsilon$ and $k_2 \pm \epsilon$. Using the noisy values of the Wasserstein pseudometric obtained at these points, one estimates the value of the gradient at (k_1, k_2) by taking a finite difference estimate of the derivative (for a thorough treatment of the procedure, see Spall, 2003, chapter 6).

After running the approximate stochastic gradient descent algorithm, we estimated the optimal values for the parameters k_1 and k_2 to be

$$k_1^* = 0.0554 \text{ mRNA/s} \quad \text{and} \quad k_2^* = 0.17 \text{ protein/(mRNA.s)},$$

and the Wasserstein pseudometric between the full model and the optimized reduced model to be

$$W_d^1(\hat{\mathcal{P}}_{1,n}, \hat{\mathcal{P}}_{3,n}(k_1^*, k_2^*)) = 1.08 \text{ proteins.}$$

The difference between the optimized reduced model and the full model is $\varepsilon(\hat{\mathcal{P}}_{1,n} \| \hat{\mathcal{P}}_{3,n}(k_1^*, k_2^*)) = 1.08/1{,}080.2 = 0.10\%$. Thus not only does the reaction network in figure 12.1a have fewer species and reactions than the model in figure 12.5a; it is also a closer approximation to the model in figure 12.4a with respect to the Wasserstein pseudometric W_d^1.

The optimal value of this Wasserstein pseudometric is sensitive to changes in k_1, as shown in figure 12.6. A 10% change in the value of k_1 increases the value of $W_d^1(\hat{\mathcal{P}}_{1,n}, \hat{\mathcal{P}}_{3,n}(k_1, k_2))$ to about 100, resulting in an error of about 10%. This example illustrates the difficulty in finding optimal reduced models by hand, as small changes in the parameter k_1 quickly diminish the accuracy of the reduced model.

12.5.3 Model Invalidation

The values of the transcription and translation rates k_1^* and k_2^* found by using approximate stochastic gradient descent have been demonstrated to produce a close agreement between the reduced model of gene expression and the full model. The convergence properties and numerical performance of the algorithm do not guarantee, however, that our choice of parameters is optimal; by varying the transcription and translation rates, we may find a model that is an even better predictor of the network's behavior. Because the estimates of the Wasserstein pseudometric are noisy, repeated runs of the algorithm give slightly different values for the optimal parameters.

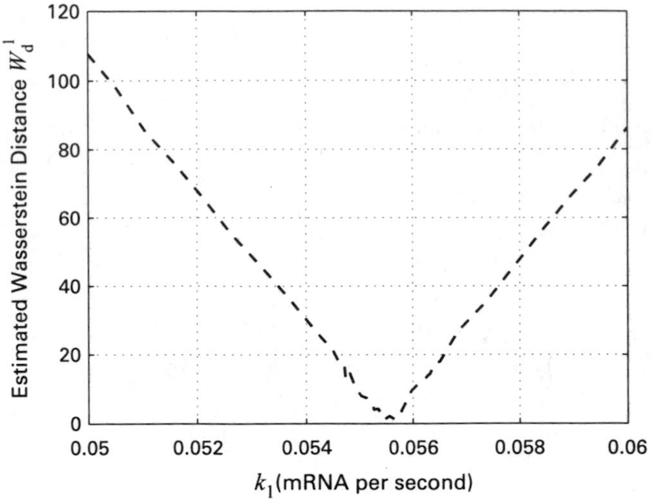

Figure 12.6
Sensitivity of the Wasserstein pseudometric $W_d^1(\hat{\mathcal{P}}_{1,n}, \hat{\mathcal{P}}_{3,n}(k_1, k_2))$ to change in transcription parameter k_1. The translation parameter is held constant at $k_2 = 0.17$ protein/(mRNA.s).

The Wasserstein pseudometric we have been using as an objective function captures only one feature of the system's behavior—the distribution of the protein number at $t = 1{,}440$ seconds. By considering different pseudometrics, we can distinguish between different estimates of the parameter values that are equally good at predicting this one feature of the behavior by determining which parameter values best predict other aspects of the behavior.

We considered two variant hypotheses: either the translation rate in the reduced model could be faster ($k_2 = 0.3$ protein/(mRNA.s)) or it could be slower ($k_2 = 0.1$ protein/(mRNA.s)) than the value k_2^*. Using these choices for the translation rate, we optimized the transcription rates so as to minimize W_d^1:

$$\left.\begin{aligned} k_{1,\text{fast}} &= 0.0541 \text{ mRNA/s}, \\ k_{2,\text{fast}} &= 0.3 \text{ protein/(mRNA.s)} \end{aligned}\right\} \quad \text{(fast translation)}$$

and

$$\left.\begin{aligned} k_{1,\text{slow}} &= 0.0581 \text{ mRNA/s}, \\ k_{2,\text{slow}} &= 0.1 \text{ protein/(mRNA.s)} \end{aligned}\right\} \quad \text{(slow translation).}$$

We denoted the previously obtained optimal values as the "medium translation" case and set $k_{1,\text{medium}} = k_1^* = 0.0554$ mRNA/s and $k_{2,\text{medium}} = 0.17$ protein/(mRNA.s). We generated 1,000 samples by simulating each reduced model and observing the protein number every 10 minutes from $t = 0$ to $t = 1{,}440$. We esti-

Table 12.1
Wasserstein pseudometrics between the full model of figure 12.4 and the reduced model

k_1	k_2	$d(\omega,\eta)$	$W_d^1(\mathcal{P}_{\text{full}}, \mathcal{P}(k_1,k_2))$	$\varepsilon_d(\mathcal{P}_1 \| \mathcal{P}_2)$
0.0554	0.1734	$\lvert\omega(1{,}440) - \eta(1{,}440)\rvert$	6.961	0.065%
0.0581	0.1	$\lvert\omega(1{,}440) - \eta(1{,}440)\rvert$	7.960	0.074%
0.0541	0.3	$\lvert\omega(1{,}440) - \eta(1{,}440)\rvert$	9.608	0.089%
0.0554	0.1734	$\lvert\min \omega^{-1}(50) - \min \eta^{-1}(50)\rvert$	1.259	7.67%
0.0581	0.1	$\lvert\min \omega^{-1}(50) - \min \eta^{-1}(50)\rvert$	1.484	9.05%
0.0541	0.3	$\lvert\min \omega^{-1}(50) - \min \eta^{-1}(50)\rvert$	2.973	18.12%
0.0554	0.1734	$\lvert\min \omega^{-1}(500) - \min \eta^{-1}(500)\rvert$	0.728	0.99%
0.0581	0.1	$\lvert\min \omega^{-1}(500) - \min \eta^{-1}(500)\rvert$	3.330	4.54%
0.0541	0.3	$\lvert\min \omega^{-1}(500) - \min \eta^{-1}(500)\rvert$	2.298	3.13%

Note: The transcription rates were optimized to minimize the Wasserstein pseudometric with respect to $d(\omega,\eta) = |\omega(1{,}440) - \eta(1{,}440)|$. Computing the Wasserstein pseudometric with respect to $d(\omega,\eta) = |\min \omega^{-1}(50) - \min \eta^{-1}(50)|$ indicates that the fast translation model does a much poorer job of predicting the distribution of times when 50 proteins are first present. Similarly, the distribution of times when 500 proteins are first present is better predicted by the medium translation model than either of the other models.

mated the values of the Wasserstein pseudometrics taken from these data sets with respect to $d(\omega,\eta)$ to be

$$W_d^1(\mathcal{P}_{\text{full}}, \mathcal{P}_{\text{fast}}) = 9.608 \text{ proteins},$$

$$W_d^1(\mathcal{P}_{\text{full}}, \mathcal{P}_{\text{medium}}) = 6.961 \text{ proteins},$$

$$W_d^1(\mathcal{P}_{\text{full}}, \mathcal{P}_{\text{slow}}) = 7.960 \text{ proteins}.$$

These values are sufficiently close to each other that we cannot conclude that the medium translation model is the best model; the differences in the values may be due to noise from the sampling process. The value of $W_d^1(\mathcal{P}_{\text{full}}, \mathcal{P}_{\text{medium}})$ is larger here than it was in the parameter-fitting sample because we are working with fewer samples.

We have observed in experimental simulations that this sample impoverishment tends to bias our estimates of Wasserstein pseudometrics upward. To determine which of these models is most likely the best description of the process, we tested their ability to predict other aspects of the full system's behavior by evaluating the Wasserstein pseudometrics between these models and the full model with respect to other distances on Ω.

The values of the Wasserstein pseudometrics computed with respect to different distances are shown in table 12.1 and figure 12.7. We first evaluated how well each reduced model predicts the "start-up" time of the process by defining a reporter random variable $Z_2(\omega) = \min \omega^{-1}(50)$, the first time along a trajectory at which

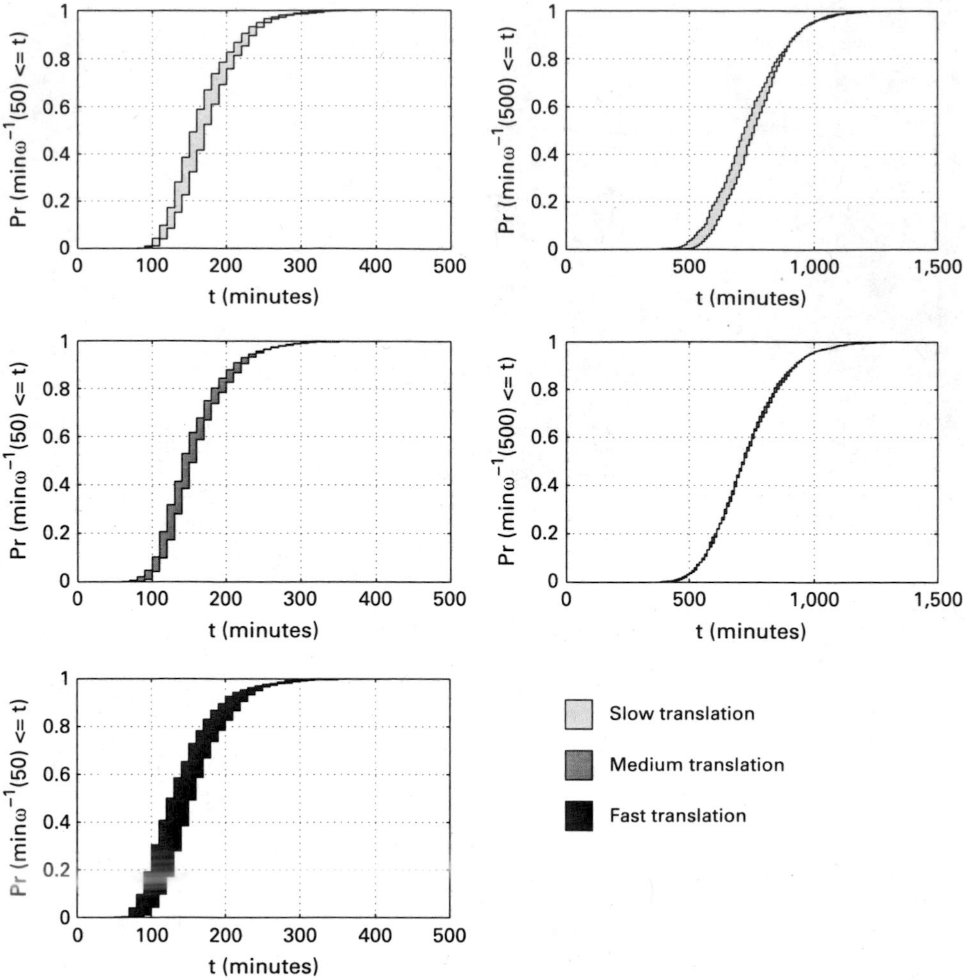

Figure 12.7
Wasserstein distances between each of the candidate reduced models and the full model are equal to the areas between the two cumulative distribution functions in each of the panels. In the left column, each reduced model is compared to the full model using $W^1_{d_2}$, the Wasserstein pseudometric with respect to the "start-up" distance $d_2(\omega, \eta) = |Z_2(\omega) - Z_2(\eta)|$. The fast translation model does very poorly with respect to this distance. In the right column, the slow and medium translation models are compared to the full model using $W^1_{d_3}$, the Wasserstein pseudometric with respect to the "midpoint" distance. The medium translation model outperforms the slow translation model with respect to this distance.

50 proteins are present. We then defined a one-dimensional distance $d_2(\omega, \eta) = |Z_2(\omega) - Z_2(\eta)|$ and evaluated $W^1_{d_2}(\mathcal{P}_{\text{full}}, \mathcal{P})$ for each reduced model $\mathcal{P}$. The value of $W^1_{d_2}(\mathcal{P}_{\text{full}}, \mathcal{P}_{\text{fast}})$ was twice as large as the Wasserstein pseudometric between the full-system model and either of the other two reduced models, indicating that the fast translation model is a poor predictor of the network's start-up behavior.

We then evaluated each reduced model's performance according to a similar "midpoint" criterion by defining a reporter random variable $Z_3(\omega) = \min \omega^{-1}(500)$ and a corresponding one-dimensional distance d_3. According to this criterion, the medium translation rate model is superior to the other models as $W^1_{d_2}(\mathcal{P}_{\text{medium}}, \mathcal{P}_{\text{full}})$ is much smaller than either $W^1_{d_2}(\mathcal{P}_{\text{full}}, \mathcal{P}_{\text{fast}})$ or $W^1_{d_2}(\mathcal{P}_{\text{slow}}, \mathcal{P}_{\text{fast}})$. Because the medium translation model was a better predictor of both the start-up and midpoint behavior, we consider it the most likely "best" model of the three possibilities.

Finally, we defined a set of Wasserstein pseudometrics $d_t(\omega, \eta) = |\omega(t) - \eta(t)|$ for each $t \in \{0, 10, \ldots, 1{,}440\}$. The values of these Wasserstein pseudometrics between the full model and each reduced model are shown in figure 12.8. These criteria provide further evidence that $\mathcal{P}_{\text{medium}}$ is the best model as it generally produces the lowest values of $W^1_{d_t}$, although the slow translation rate performs slightly better when t is

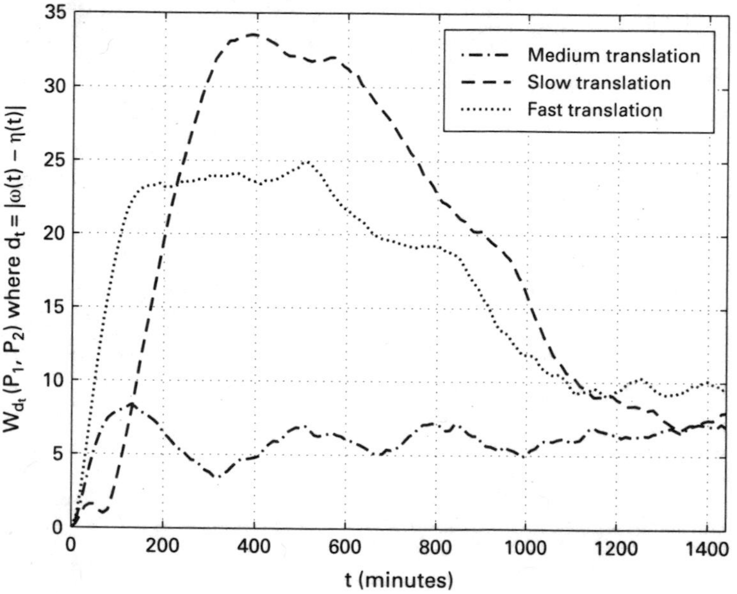

Figure 12.8
Wasserstein pseudometrics between the full model and the candidate reduced models with respect to the distance $d_t(\omega, \eta) = \min|\omega(t) - \eta(t)|$ for $t = \{0, 10, \ldots, 1{,}440\}$. The medium translation rate model (*dashed-dotted line*) produces the smallest Wasserstein pseudometric with respect to most of the distances d_t, although the slow translation model performs better when t is close to zero.

small. By expanding the class of candidate models, we may be able to produce an even better model; for example, by making the translation rate time varying (as a result of the amount of transcribing polymerase T in the full model going from zero toward its steady-state value), it may be possible to produce an even more precise and predictive model.

12.6 Future Directions

The overwhelming complexity of biochemical processes requires a method of approximation that will allow scientists and engineers to develop simple models that describe with sufficient accuracy the most relevant aspects of a behavior of the system. Depending on the design goals of the engineer or the hypotheses being tested by the scientist, the notion of "relevant aspects" may vary not only when different systems are being studied, but also when the same system is being studied by investigators with differing objectives. Wasserstein pseudometrics provide a method for comparing stochastic processes that effectively captures the idea that the relevant aspects of a system's behavior may change depending on the problem under consideration. By defining and calculating Wasserstein pseudometrics with respect to different underlying pseudodistances d, we can produce problem-specific approximate models of complex processes. Each choice of a distance d defines a different performance criterion. We can use a fixed performance criterion to compare a model to a data set or to a reference model for model comparison, model reduction, or parameter estimation, or we can use a set of performance criteria to perform model invalidation.

The Wasserstein pseudometric method proposed in this chapter has been developed with an eye toward maximum generality; in principle, it can be used with any class of stochastic model given enough time and enough sample data. This generality comes at a price. To take a unified approach for different classes of models, we must use sampling and simulation; the values of Wasserstein pseudometric that we calculate using this procedure are not as precise as those we would obtain using more analytical methods. An open area of research is to derive more efficient algorithms for approximating specific classes of stochastic processes.

13 System-Theoretic Approaches to Network Reconstruction

Jorge Gonçalves and Sean Warnick

One of the fundamental aims of systems biology is the discovery of the specific biochemical mechanisms that explain the observed behavior of a particular system. These mechanisms are composed of complex networks of reactions between various chemical species; detailing the species and interactions forming such a network can be an overwhelming task, even for the simplest of biochemical systems.

Our research outlines an experimental process that makes such discovery possible despite the intrinsic difficulty of network reconstruction. If nothing is known about the network connections between measured species, then, for a given network composed of p measured species, p experiments must be performed, with each experiment independently controlling a measured species, that is, control input i must first affect measured species i. If something is known about the network, then these conditions may be relaxed. The network representation resulting from this process yields a predictive model commensurate with the data used to create it: steady-state data yield static network information, whereas time-series data generate a dynamic representation of the system suitable for simulation.

We demonstrate that, in the absence of this essential experimental design, the network cannot be reconstructed and every conceivable network structure, from a fully decoupled to a fully connected network, can be equally descriptive of any particular set of input-output data. If this correct methodology is not followed, any best-fit measure for network reconstruction can yield arbitrarily poor and misleading results.

13.1 Motivation: Biochemical Complexity

There are a number of characteristics of biochemical systems that make them complex and difficult to understand. One is the sheer number of distinct chemical species present in a given system. For example, although there are only 20 amino acids from which all proteins are built, the number of possible proteins composed of n amino acids grows exponentially, as 20^n. Many proteins are composed of hundreds or thousands of amino acids. Moreover, proteins themselves combine to form chemical complexes leading to intricate and sophisticated chemical machinery. Also, proteins form complexes with other molecules, such as RNA in the case of ribosomes, yielding a nearly unbounded potential for generating distinct chemical species within a system.

Another characteristic of biochemical systems that leads to their complexity is their dynamic nature. Chemical species react, forming new compounds that, in turn, interact and affect a changing chemical environment. In such an environment, fluctuating concentrations of one species can critically affect the chemical behavior of others, thus leading to an ongoing cycle of assembly and disassembly that is associated with the resulting phenotype and chemical behavior of the entire biological system.

In the face of this daunting complexity, there is nevertheless a hope of developing a real understanding of the mechanisms responsible for the observed system behavior, since most chemical species interact with a relatively small number of reactants in the system, so that the torrent of chemical activity characterizing the system is actually quite structured. This structure is captured in the way chemical concentrations or molecular counts of one species directly depend on the concentrations or counts of a limited number of other species. Such dependencies describe the biochemical network of the system. Once the structure of this network becomes apparent, it is possible to eliminate numerous interactions and relationships between species that would otherwise be considered. In this way, knowledge of the structure of the biochemical network can facilitate an understanding of the entire system.

Our work describes fundamental limitations associated with learning network structure and uses knowledge about these limitations to outline practical methods for reconstructing biochemical networks from data. Section 13.2 introduces the biochemical network, emphasizing its role in managing the complexity of biochemical systems and its relationship to system dynamics. Section 13.3 then introduces network reconstruction and discusses its fundamental difficulties and limitations and the effect of noise and nonlinearities in a biochemical system. Section 13.4 considers the case where nothing is known about a network and explains why in this case it is necessary to design experiments as noted above. It shows how network reconstruction can be achieved using two common types of experiments based on gene silencing and overexpression, discusses trade-offs between steady-state and time-series data, and concludes with an illustrative example. Section 13.5 sets forth the chapter's conclusions and section 13.6 its mathematical details.

13.2 Biochemical Networks

Every biological system is rooted in the chemistry that governs the particular sequences of reactions that lead to its specific phenotype. This chemistry is described by the complete list of chemical species in the system and the kinetic processes that govern reactions between them. This section describes how these interactions lead to the notion of a biochemical network. First, it discusses the complete biochemical network and explains the network's fundamental role of encoding the essential chemistry driving the biochemical system. Next, it introduces an approximation of the complete network that we describe as network structure, a simplified network that illustrates the relationships between a subset of the chemical species in the system. Finally, it discusses the relationship between network structure and the dynamics of the biochemical system.

13.2.1 The Complete Biochemical Network

Consider the hypothetical situation in which every protein, carbohydrate, lipid, nucleic acid, and other biomolecule capable of being produced by a particular biochemical system is known and enumerated. Chemical kinetics then define the relationship between each of these chemical species and the rest of the system. There are various models that have been developed to describe such kinetics, but all of these models reflect the same underlying causal relationship between species that would define the complete biochemical network if such an enormous model could be developed. For example, stochastic models (chapter 2) generally attempt to capture a refined view of chemical processes by modeling the molecular count of the relevant species. Such models are important for describing processes that are very sensitive to the presence of small numbers of particular types of molecules. Probability theory is used to represent the impact of various dynamic effects, such as thermal noise, on the fluctuations in the numbers of these molecules, and these effects then combine into one large, coupled, differential equation called the master equation. In this equation, the structure of the biochemical network is encoded in the algebraic coupling between variables.

Deterministic models, on the other hand, represent species by their chemical concentrations and typically use mass-action assumptions and Michaelis-Menten and Hill (sigmoidal) equations to represent the subsequent chemical kinetics (chapter 1). Combining the effects of these dynamics over all species leads to a set of coupled ordinary differential equations in which, once again, the structure of the biochemical network is encoded in the algebraic coupling of these equations. Gillespie (2000) gives a detailed description of some differences between stochastic and deterministic models for biochemical systems. In either case, however, the structure of the biochemical network is a fundamental characteristic of the system, encoded in the algebraic structure of a dynamic model that considers every species over the entire system.

One way to conceptualize this network is as a graph, where every node represents a particular chemical species whose abundance at any particular time is measured by either concentration or molecular count. The quantity of the species corresponding to any particular node fluctuates over time, depending on the quantities of other species and on environmental variables such as temperature. This model allows us to add nodes to our graph for each relevant environmental variable and then draw a directed edge from every node of influence to the impacted node. The connections between nodes, known as *edges*, characterize the dependencies between quantities that influence the time behavior of species in the system, which is to say, dynamic relationships between species are defined over the edges in this graph. Since most biomolecules are highly specific and target a few specialized reactions, our graph is both tightly structured and sparse, significantly reducing the number of possible connections among species.

Figure 13.1 illustrates the graph structure for a network with ten species. Despite this enormous reduction in complexity, however, the complete biochemical network is still too large to handle, even for relatively simple biochemical systems. Even with the advent of remarkable measurement technologies, such as high-throughput DNA microarrays and mass spectrometry techniques, we still have not been able to measure every species or collect enough information to build a complete model of an entire system. As a result, scientists have necessarily focused on subsets of the species in the system and worked toward an in-depth understanding of various pieces of the complete network.

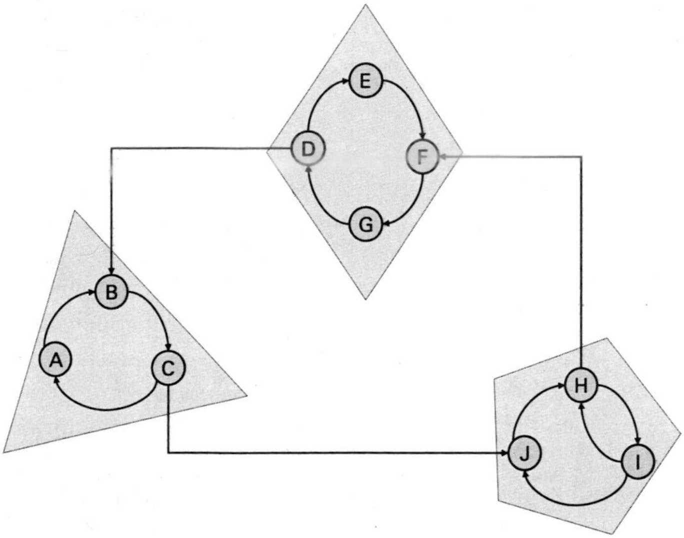

Figure 13.1
Complete biochemical network for a system with ten distinct species. The network can be viewed as the interaction of three specific mechanisms, represented by a shaded triangle, diamond, and pentagon. Edges here indicate direct reactions between species.

13.2.2 Network Structure: A Simplified Representation

The complete biochemical network is a sparse graph that characterizes dynamic relationships among all molecules in a biochemical system. Despite its sparsity, the sheer number of potential species makes even this characterization unwieldy. A simplified representation is necessary, one that can incrementally reveal more about the complete biochemical network as additional information becomes available.

The traditional approach to dealing with such large, complex problems is to break them apart and study them in pieces. For example, scientists have targeted specific mechanisms such as signal transduction, protein synthesis, and cell metabolism and have identified many of the species and reactions involved in these mechanisms. Although such efforts have provided important insights into the local behavior of these mechanisms, the global behavior is often obscured by a lack of understanding about the network interactions.

Here we offer an alternative to modeling a specific piece of the system in full detail. Instead, we consider a subset of species in the system, possibly species that have no direct chemical relationship, and develop a coarse model of the entire system by considering the causal relationships among these measured species. These causal relationships may include excitation, inhibition, or some combination; the critical factor is whether the presence of one species impacts the presence of another. We refer to this coarse representation of the system as the *network structure*, as opposed to the complete biochemical network, and note that it implicitly references a particular set of measured species.

Like the complete biochemical network, network structure can be represented as a graph, with nodes of the graph indicating measured species and edges in the graph indicating a causal relationship between two measured species. We refer to the edge structure of this graph as the *Boolean network structure*. Figure 13.2 demonstrates the Boolean structure of four different network structures for the complete biochemical network shown in figure 13.1.

In the network structure, unlike the complete biochemical network, an edge between two measured species does not necessarily indicate a direct chemical interaction. Since the number of measured species may be considerably smaller than the total number of species in the system, a large number of unmeasured species may be present. An edge between measured species may therefore represent a complex chain of reactions through an entire network of intermediate species so long as none of these intermediate species is part of the set of measured species that constitute the nodes of the graph. The presence of an edge in the network structure thus indicates a direct relationship between the measured species, but may well indicate an indirect relationship when considering the complete set of chemical species in the system. If every species in the system were measured, then network structure would coincide

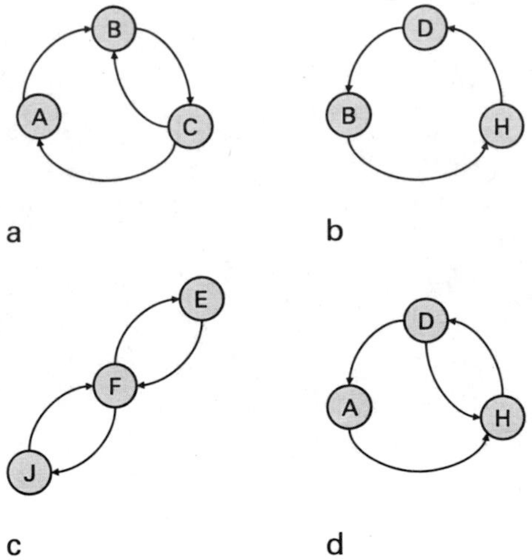

Figure 13.2
Four possible network structures as simplified representations of the complete biochemical network from figure 13.1. The network structure differs depending on which species are measured. Unlike the complete biochemical network, some edges here may represent direct chemical reactions (e.g., A to B in panel a, or D to B in panel b), whereas others may represent indirect ones, so long as the intermediate reactions only involve unmeasured species (e.g., C to B, in panel a.)

with the complete biochemical network for the system, but in all other situations (which we will generally assume to be the case), network structure is a simplification of the complete network.

This difference between the meaning of edges is apparent in figure 13.2, where some edges indicate a direct reaction between species, while others may represent an indirect relationship, so long as the intermediate reactions do not involve any measured species. This is the reason that a traditional approach to network discovery, focusing on understanding each mechanism in detail, may have difficulty obtaining the network structure. Consider, for example, the mechanism represented by the triangle in figure 13.1. If we examine the network structure obtained by measuring only the species (A,B,C) describing this mechanism (figure 13.2a), we note that a traditional approach to the problem may easily miss the feedback from C to B since this link is not a direct reaction. This flexible interpretation of an edge as being either a direct reaction or an indirect chain of unmeasured reactions enables the network structure to offer a coarse view of the entire system rather than a detailed view of a particular component.

Since network structure offers a view of the whole system, small changes can yield different structures in surprising ways. For example, figure 13.2b, d demonstrates two distinct structures between representative species from all three mechanisms (triangle, diamond, and pentagon) of the system in figure 13.1. The structure in figure 13.2b uses species B as a representative for the triangle mechanism, and this choice results in the simple ring structure one might expect from the relationship between mechanisms in the complete biochemical network. The structure in figure 13.2d, however, replaces B with A as the representative species for the triangle mechanism, and now a pathway from D to H appears in the network structure. This change illustrates that network structure is providing a view of the whole system, as differing views of the relationship between D and H appear depending on whether A or B is measured. This is particularly interesting since the change in structure resulting from the choice between A or B does not even involve the mechanism (triangle) to which A and B belong. This global information about the network structure indicates that investigation of one mechanism can offer understanding of other mechanisms in the system.

On the other hand, the network structure of a system makes no assumptions about the relationship or structure among unmeasured species in the system. For example, we note in the network structure for (A,B,C) illustrated in figure 13.2a that C affects both A and B. The pathways from C to A and C to B, however, are entirely distinct, and share no common mechanisms or reactions. This is in contrast with the network structure for (A,D,H) in the same figure (panel d), where D affects both A and H, but the pathways $D \to B \to C \to A$ and $D \to B \to C \to J \to H$ share a number of intermediate reactions, as seen in figure 13.1. The unmeasured species B and C contribute to both edges in the network structure diagram, whereas the unmeasured species J only contributes to the edge between D and H. Thus network structure offers a global view of the relationship between measured species without saying anything about the structure or relationship between unmeasured species.

This lack of detail, while retaining the correct relationship between measured species, is the critical feature of network structure. It enables an accurate, if incomplete, view of the entire interconnected system, rather than providing a detailed partial understanding of one component of the system, disconnected from the rest of the network. This coarse view of the system remains consistent with more refined representations that are generated as more species become available for measurement, leading to an incremental but effective discovery process (figure 13.3). The incremental character of this discovery process ensures that the information load necessary to obtain accurate, if less precise, representations of the system remains bounded by practical constraints on data availability. Network structure thus offers the simplified representation necessary to enable a developing understanding of the complete biochemical system despite technological limitations or the network's formidable complexity.

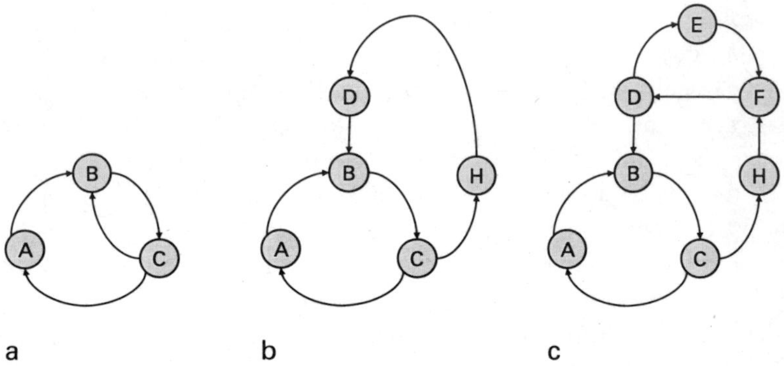

Figure 13.3
Network structure offers a simplified, but accurate, view of the complete biochemical network. This accuracy is reflected in the fact that a coarse network structure (panel a) of the system from figure 13.1 remains consistent with refinements generated as additional species are measured (panels b and c). Notice that refinements simply add detail to edges of the coarser structure.

13.2.3 Network Structure and System Dynamics

Network structure offers a simplified view of the complete biochemical network, specifying the structural relationships among measured species without defining those among unmeasured species. It has a Boolean component represented by the presence or absence of edges between measured species; this Boolean structure remains consistent with refinements as more species in the system become available for measurement. This section formalizes the Boolean representation of network structure and extends it to include the dynamic properties of the system through the use of dynamical structure functions.

The Boolean component of network structure can be formally represented by the transpose of its adjacency matrix. For a network structure with p measured species, this is a $p \times p$ matrix $\boldsymbol{M}$ with entries $M_{ij} = 1$ if species i depends on species j, and zero otherwise. That is, $M_{ij} = 1$ if there is an edge from species j to species i. For example, the Boolean components of the networks in figure 13.3 can be written as

$$\begin{bmatrix} 0 & 0 & 1 \\ 1 & 0 & 1 \\ 0 & 1 & 0 \end{bmatrix}, \quad \left[\begin{array}{ccc|cc} 0 & 0 & 1 & 0 & 0 \\ 1 & 0 & 0 & 1 & 0 \\ 0 & 1 & 0 & 0 & 0 \\ \hline 0 & 0 & 0 & 0 & 1 \\ 0 & 0 & 1 & 0 & 0 \end{array}\right], \quad \left[\begin{array}{ccccc|cc} 0 & 0 & 1 & 0 & 0 & 0 & 0 \\ 1 & 0 & 0 & 1 & 0 & 0 & 0 \\ 0 & 1 & 0 & 0 & 0 & 0 & 0 \\ 0 & 0 & 0 & 0 & 0 & 0 & 1 \\ 0 & 0 & 1 & 0 & 0 & 0 & 0 \\ \hline 0 & 0 & 0 & 1 & 0 & 0 & 0 \\ 0 & 0 & 0 & 0 & 1 & 1 & 0 \end{array}\right], \qquad (13.1)$$

where these matrices implicitly reference nodes (A,B,C), left; (A,B,C,D,H), middle; and (A,B,C,D,H,E,F), right. The partitioning of the second and third matrices is provided to facilitate an easy comparison with each previous matrix, demonstrating that zeros in the Boolean structure remain present in all refinements. This preservation of zeros over refinements characterizes the consistency of coarse network structures with respect to their own refinements.

Nevertheless, the 1s in the matrix representation do not necessarily remain under a refinement. For example, M_{23} in the left matrix, representing an edge from C to B, is not present in either of the next two refinements. Compare this with any of the other 1s in the same matrix, all of which remain present in all refinements. This discrepancy in how edges are handled by this Boolean representation of network structure occurs because some edges represent indirect reactions, and thus will disappear at some level of refinement. Other edges represent direct reactions and thus remain in every refinement down to the complete biochemical network. This inconsistency toward edges, despite the consistency toward zeros, highlights a drawback of the Boolean representation: it uses the same symbol for every edge, even though the nature of these edges can vary considerably.

Another drawback of the Boolean representation is that it is insufficient to characterize the system dynamics. Although the edge structure of a system can reveal something about its behavior, in and of itself, Boolean structure cannot provide the details of these dynamics. Moreover, specific details about behavior, such as stoichiometry and reaction rates, are not captured in the Boolean representation of network structure.

To overcome these drawbacks, consider an alternative representation of network structure which, like the Boolean representation $\boldsymbol{M}$, captures the edge structure between measured species through the zero structure of a $p \times p$ matrix. As a result, it retains the consistency properties of $\boldsymbol{M}$ over refinements, in that zeros are preserved, although nonzero terms may not be. This new representation replaces each 1 in $\boldsymbol{M}$ with a characterization of the specific reaction, or chain of reactions, symbolized by the presence of that edge between measured species in the network structure. In this way, the dynamics between measured species are specified for each edge; we symbolize these dynamics, which we will describe later, from species j to species i with the notation Q_{ij}. Taken together, the dynamics of the entire system can then be encoded in this matrix structure, $\boldsymbol{Q}$, called the *internal structure* of the system. When external control variables are present, we encode the network structure between these m variables and the p measured species with an additional $p \times m$ matrix $\boldsymbol{P}$, called the *control structure*, defined similarly to $\boldsymbol{Q}$. These two matrices $(\boldsymbol{Q}, \boldsymbol{P})$ are called the *dynamical structure function* of the system; they characterize the network structure, at the resolution consistent with the number of measured species, as well as the dynamics of the system—thereby enabling simulation. As an example, consider the complete biochemical network in figure 13.1. Suppose, in addition, that an externally controlled variable is introduced that directly targets species A in the system. If only species A, B, and C, along with the external control variable, are measured, then the dynamical structure function of the system becomes:

$$(\boldsymbol{Q}, \boldsymbol{P}) = \left(\begin{bmatrix} 0 & 0 & Q_{AC} \\ Q_{BA} & 0 & Q_{BC} \\ 0 & Q_{CB} & 0 \end{bmatrix}, \begin{bmatrix} P_A \\ 0 \\ 0 \end{bmatrix} \right). \tag{13.2}$$

Notice that the zero structure of $(\boldsymbol{Q}, \boldsymbol{P})$ describes the Boolean network structure of the system. This raises the question as to how Q_{ij} and P_{kl} encode the dynamics between species i and j or between species k and l.

Although a complete derivation can be found in section 13.6 and in Gonçalves and Warnick (2008), the short answer to this question is that we consider the system to be near equilibrium and analyze the relationship between species for small deviations from this steady state. This analysis yields a *linearization* (as discussed in chapter 1) of the true chemical dynamics around one particular set of equilibrium values of the system variables, characterized by the pairwise interactions among measured species and control variables. Each of these interactions is either a direct reaction or an indirect reaction through a number of unmeasured intermediate species (figure 13.4). In either case, the dynamics of any pairwise interaction are then ideally described by a linear, time-invariant (LTI) system with an order that is one greater than the number of unmeasured intermediate species involved in that particular reaction chain. Each

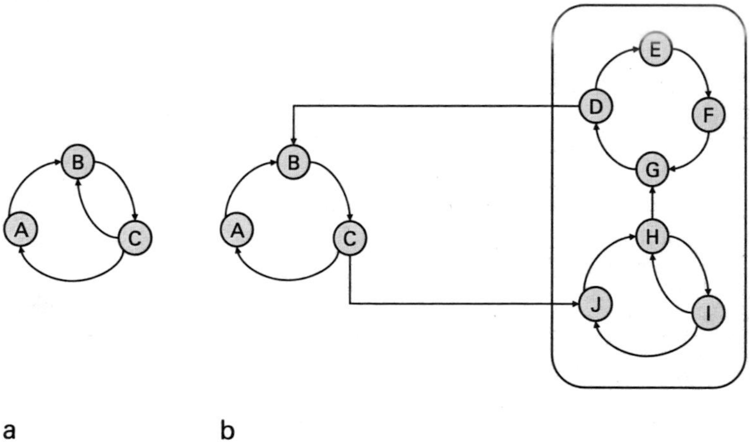

Figure 13.4
Edge C → B in the network structure (panel a) of the complete biochemical network in figure 13.1 is composed of the subnetwork including all of the intermediate unmeasured nodes (panel b). The impact of C on B near equilibrium is then given by the transfer function Q_{BC}, associated with this subnetwork.

of these single-input, single-output LTI systems can then be represented by its transfer function, Q_{ij} or P_{kl}. Thus different edges in the dynamical structure function will exhibit different levels of complexity, reflected by the order of their transfer functions, according to the intricacy of the unmeasured components and mechanisms embedded along that particular pathway. As a result, we will sometimes distinguish this richer view of network structure from its Boolean component by calling it *dynamical structure*. This flexibility suggests that $(\boldsymbol{Q}, \boldsymbol{P})$ is a powerful way to describe both the network structure and the dynamic behavior of the system without attempting to specify every detail or understand every mechanism of its complete biochemical network.

13.3 The Network Reconstruction Problem

The establishment of dynamical structure as a simplification of both the complete biochemical network and its inherent dynamics introduces a new paradigm for the reverse engineering of the biochemical system. Instead of attempting to reconstruct the complete biochemical network one piece at a time and to somehow correctly connect these pieces, we explore the recovery of the dynamical structure at a particular resolution. This dynamical structure can then be incrementally refined toward a comprehensive understanding of the complete biochemical system as the necessary measurement technology and data become available.

This section describes the nature of this new problem, distinguishing the reconstruction of dynamical structure from other, well-established reverse engineering problems: system identification and realization. It then leverages this new perspective on reconstruction to demonstrate fundamental limitations on network discovery, even for the partially descriptive, idealized representations captured by dynamical structure. Finally, it introduces the mechanisms needed to extend our idealized analysis to the nonlinear, noisy setting that characterizes real biochemical systems.

13.3.1 Reconstruction versus Identification and Realization

There are a number of different ways to represent a linear, time-invariant dynamical system. Each representation reveals different information about the system, and thus each also requires different amounts of data—possibly different kinds of data—as well as different computational procedures to be specified from these data. Here we discuss transfer functions, state-space realizations, and dynamical structure and use them to contrast the problems of identifying, realizing, and reconstructing systems. The transfer function of an LTI system is a description of the input-output behavior of the system. It provides a model capable of simulation; knowing the transfer function allows one to predict the response of the system to a new input. *System identification* uses the time-series data characterizing stimulus and response to determine the input-output dynamics of the system. Developing reliable algorithms capable of operating on real (i.e., noisy and limited) data is a research topic in its own right (see chapter 14). The system identification process reads time-series data to produce a mathematical expression characterizing the dynamic input-output behavior of the system; for linear, time-invariant systems, this expression is called the transfer function. For a system with m inputs and p outputs, the input-output map is a $p \times m$ matrix of transfer functions that is also called the transfer function, with any ambiguity resolved by context. In this work, we denote the transfer function matrix $\boldsymbol{G}(s)$ and note that each individual element of $\boldsymbol{G}$ is a proper rational complex function of the complex Laplace variable s.

In contrast to the transfer function, which describes only the input-output behavior of an LTI system, a state-space model of the system describes the precise network architecture used to realize a particular input-output behavior. This representation of the system is a set of coupled ordinary differential equations of the form

$$\dot{\boldsymbol{x}}(t) = \boldsymbol{A}\boldsymbol{x}(t) + \boldsymbol{B}\boldsymbol{u}(t),$$

for an $n \times n$ matrix $\boldsymbol{A}$, an $n \times m$ matrix $\boldsymbol{B}$, and a vector of m inputs $\boldsymbol{u}$. In this work, we assume that $p < n$ states are measured, leading to the additional output equation $\boldsymbol{y}(t) = \boldsymbol{C}\boldsymbol{x}$, where $\boldsymbol{C} = [\boldsymbol{I}\ \boldsymbol{0}]\boldsymbol{x}$ (where $\boldsymbol{I}$ is the $p \times p$ identity matrix, and $\boldsymbol{0}$ is the $p \times (n - p)$ matrix of zeros). This output equation thus indicates that the first p elements of the state vector $\boldsymbol{x}$ are exactly the measured variables in the system; the remaining $(n - p)$ state variables are unmeasured "hidden" states. A state-space model that includes every species in a given chemical system would describe the complete biochemical network; the zero structure of the $\boldsymbol{A}$ and $\boldsymbol{B}$ matrices would exactly describe the Boolean structure of the network, and the values of these matrices would encode the dynamics of the system.

In general, there are many state-space realizations that generate the same input-output behavior, or transfer function. *Realization* is the process of finding a state-space model that produces a given system; a state-space model derived from a given transfer function is called a realization of the transfer function. There is a well-developed realization theory for linear systems that answers questions such as what is the minimum number of states, n, needed to describe a given transfer function, $\boldsymbol{G}(s)$; how to find state-space realizations of particular canonical forms; and how to relate $(\boldsymbol{A}, \boldsymbol{B}, \boldsymbol{C})$ to $\boldsymbol{G}(s)$. This theory establishes that beyond the input-output data used to generate a transfer function, more information is needed to distinguish one state-space realization from another as a description of a particular system.

One exception to this principle is when all states are measured, that is, when there are no hidden states. In this situation, we know $\boldsymbol{C}$ is the identity, and there is a unique state-space realization associated with every transfer function $\boldsymbol{G}$. This knowledge enables the discovery of a system's complete network structure from data by first solving a system identification problem to obtain $\boldsymbol{G}$, and then solving the system realization problem to find $(\boldsymbol{A}, \boldsymbol{B})$. The network structure is then encoded in the zero structure of $\boldsymbol{A}$ and $\boldsymbol{B}$. Nevertheless, given the complexity of biochemical systems, it is unreasonable to assume that all states are measured.

Even with just one hidden state, the realization problem becomes ill posed; a transfer function will have many state-space realizations, and each of these may suggest an entirely different network structure for the system. This is true even if it is known that the true system is, in fact, a minimal realization of the identified transfer function. As a result, failure to explicitly acknowledge the presence of hidden states and the ambiguity in network structure that results can lead to a deceptive and erroneous process for network discovery. Consider the following example.

Example: Assuming Full-State Measurements Can Be Deceptive Here we demonstrate how assuming full-state measurements can lead to erroneous and deceptive structure results. When full-state measurements are assumed, one approach to network discovery is to first solve an identification problem yielding a transfer function that fits the observed data well. Then, since full-state measurements are assumed, the minimal realization corresponding to this transfer function can be found, and this realization will specify a network structure for the system. Nevertheless, if even one hidden state is present, this predicted structure can be completely different from the true system structure.

What is particularly deceptive about this approach is that one may gain confidence in the accuracy of the predicted network structure based on the ability of the transfer function to fit the measured data. Moreover, an excellent fit may be achieved with a transfer function of an order equal to the number of its outputs, thereby suggesting that a full-state measurement assumption is reasonable. Thus a high-quality fit can lead one to think that the resulting network structure is correct when, in fact, it is not.

Suppose the concentrations of two species are measured at regular intervals during a particular dynamic transition until the system reequilibrates. The objective of the experiment is to discover the network structure between the two compounds.

Although one may use various identification techniques to find a model that explains the measured data, our purpose is not to compare them or even to comment on the identification step at all. Instead, we are interested in how assuming there are no hidden states obscures the network reconstruction problem.

Suppose the measured data are as illustrated in figure 13.5. We observe that the following system fits the data well:

$$\boldsymbol{G}(s) = \begin{bmatrix} \dfrac{1.65}{s+1.65} \\ \dfrac{2.77}{s+2.77} \end{bmatrix}.$$

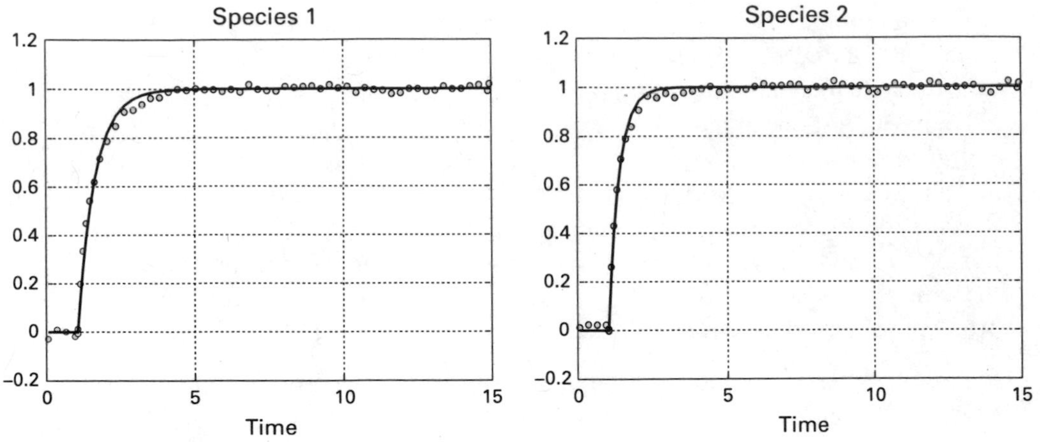

Figure 13.5
Concentration data for two compounds measured during a particular dynamic transition, showing the fit of a decoupled second-order system (*solid line*).

Although this model may have been derived in a number of different ways, the important observation is that it is second order. Since two species are measured and a transfer function that fits the data well is second order, we might assume that we have full-state measurements. Moreover, a system with full-state measurements has a unique state-space realization. The transfer function $\boldsymbol{G}(s)$ corresponds to the state-space model

$$\boldsymbol{A} = \begin{bmatrix} -1.65 & 0 \\ 0 & -2.77 \end{bmatrix} \quad \text{and} \quad \boldsymbol{B} = \begin{bmatrix} 1.65 \\ 2.77 \end{bmatrix}.$$

The state-space realization of a system exposes its network structure. Here the two measured species are completely decoupled. Nevertheless, the actual system generating these data is given by

$$\boldsymbol{A} = \begin{bmatrix} -0.1 & -0.1 & -2.6 \\ 1.3 & -2.7 & -2.2 \\ 2.3 & 1.3 & -9.2 \end{bmatrix} \quad \text{and} \quad \boldsymbol{B} = \begin{bmatrix} 2.8 \\ 3.6 \\ 5.6 \end{bmatrix},$$

where only the first two states are measured, and the third state is hidden. The actual network structure between the measured states is thus fully connected. Thus we see that the network structure determined by assuming no hidden states can be incorrect, even when the model's fit to data is good.

This example illustrates that, even for the simplest of systems, with only two measured outputs and with knowledge of linear dynamics, failing to acknowledge the possibility of hidden states can lead to a highly convincing, yet erroneous, estimate of the network structure between measured states.

Such an error in network reconstruction results from making a full-state measurement assumption when hidden states are present, which fixes a unique realization and network structure corresponding to the transfer function fit to the data. Because the presence of even a single hidden state enables many realizations to be compatible with the same input-output model, if one erroneously assumes full-state measurements when in fact unmeasured states are present, then the actual structure might well be different from the one determined by the full-state assumption.

In this way, assuming full-state measurements hides the network reconstruction problem within the system identification and realization problems; once a model is identified from data, the network structure is fixed by assumption. Nevertheless, to find a system's true structure from data, one must develop a theory of reconstruction independent from the identification and realization processes, one that explicitly addresses the presence of hidden states and their subsequent dynamic effects. Such an approach is developed through the use of dynamical structure and the formulation of the reconstruction problem.

Dynamical structure is a new representation for linear, time-invariant systems. Like both the state-space representation and the transfer function, dynamical structure encodes the input-output dynamics of the system. Nevertheless, unlike either the transfer function, which encodes no information about network structure, or the state-space realization, which encodes all the information about network structure, dynamical structure encodes only part of the information about the structure of the network. In particular, it encodes the structure between measured states while offering no information about the structure among hidden states. Dynamical structure is a pair of matrices, $(\boldsymbol{Q}(s), \boldsymbol{P}(s))$, like the state-space realization $(\boldsymbol{A}, \boldsymbol{B})$, but these are matrices of transfer functions, much like $\boldsymbol{G}(s)$.

Reconstruction is the process of finding $(\boldsymbol{Q}(s), \boldsymbol{P}(s))$ from input-output data or, if identification has already been performed, from the transfer function of the system. It is distinct from both system identification and realization, and its formulation enables an alternative approach to network discovery. This alternative approach solves the reconstruction problem at a given resolution, and then incrementally refines the model as more measurements become available. In doing so, even though no assumptions about full-state measurements are made, the data requirements necessary to reconstruct part of the network structure can be fully controlled through the choice of resolution.

13.3.2 Fundamental Limitations of Reconstruction

Using dynamical structure functions to formulate precisely the network reconstruction problem enables us to carefully analyze the nature of its solutions. In particular, we are interested in understanding what data, and hence what kinds of experiments, are necessary to solve this problem. This section describes simplifications that enable clear analysis, then summarizes the limitations for network reconstruction, even for these idealized situations. From these limitations, we obtain a clear characterization of the nature of the experiments necessary to solve the problem.

Formulating the network reconstruction problem hinges on essential simplifications that expose the central relationships between input-output behavior and network structure. In particular, recall that we have restricted our attention to linear, time-invariant systems, presumably derived as linearizations of the nonlinear dynamics of the biochemical system developed near particular equilibrium values. Moreover, we have assumed that the noise reflected in the data is sufficiently small that a system identification routine is able to deliver the correct transfer function of the system. Further, recall that reconstruction itself is a less ambitious objective than realization, in that only the structural relationships between measured states are desired. With all of these simplifications, we have formulated a problem with the best chance for a solution. As a result, we can expect that any challenges with the solution of this simplified problem are fundamental, in that they will also hinder a more ambitious approach.

Given these simplifications, reconstruction becomes the derivation of $(\boldsymbol{Q}, \boldsymbol{P})$ given $\boldsymbol{G}$. With this formulation, it is possible to show that, in general, not even Boolean reconstruction is possible without more information about the system. In fact, for any transfer function $\boldsymbol{G}(s)$, it is possible to find a viable structure $(\boldsymbol{Q}(s), \boldsymbol{P}(s))$, consistent with $\boldsymbol{G}$, for *any* internal structure $\boldsymbol{Q}(s)$. That is, any nontrivial input-output behavior can be realized by a completely decoupled internal structure, a completely coupled internal structure, or anything in between. This implies that the information from input-output data is not sufficient to determine the network structure of the system, even in the idealized scenario of a linear network with perfect knowledge of the transfer function. This fact is illustrated by the following example.

Example: Minimal Realizations Do Not Specify Unique Structure Consider a system with the following transfer function:

$$\boldsymbol{G} = \frac{1}{s+3}\begin{bmatrix} \dfrac{1}{s+1} \\ \dfrac{1}{s+2} \end{bmatrix}.$$

It can be shown that this transfer function is consistent with two systems having quite different internal structures, of the form $\dot{\boldsymbol{x}} = \boldsymbol{A}\boldsymbol{x} + \boldsymbol{B}u$, $\boldsymbol{y} = \boldsymbol{C}\boldsymbol{x}$ given by

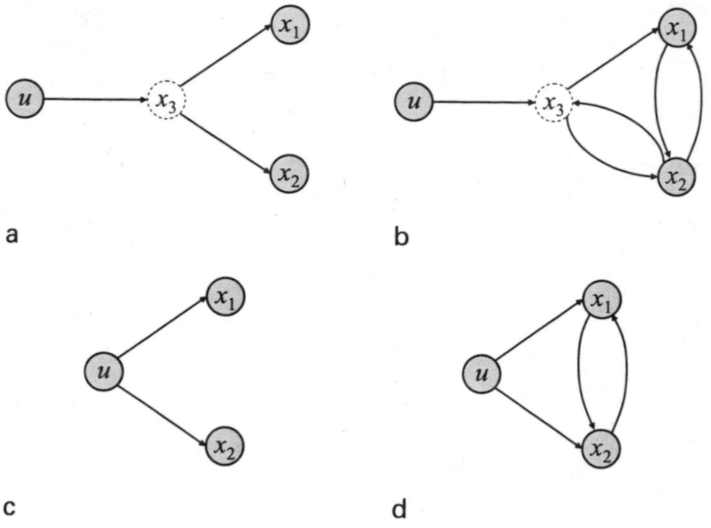

Figure 13.6
The same transfer function yields two different minimal realizations and two different network structures: decoupled internal structure (a,c) and coupled internal structure (b,d). The complete biochemical network corresponding to each state space realization is shown in panels a and b; measured species are shaded. The corresponding network structure, characterized by the dynamical structure function, is shown in panels c and d.

$$A_1 = \begin{bmatrix} -1 & 0 & 1 \\ 0 & -2 & 1 \\ 0 & 0 & -3 \end{bmatrix}, \quad A_2 = \begin{bmatrix} -2 & -1 & 1 \\ -1 & -3 & 1 \\ 0 & -1 & -1 \end{bmatrix},$$

$$\boldsymbol{B}_1 = \boldsymbol{B}_2 = [0 \quad 0 \quad 1]^T \quad \text{and} \quad \boldsymbol{C}_1 = \boldsymbol{C}_2 = [\boldsymbol{I} \quad \boldsymbol{0}],$$

where $\boldsymbol{C}_1$, and $\boldsymbol{C}_2$ are 2×3 matrices. The networks in figure 13.6 correspond to these two realizations of $\boldsymbol{G}$. Note that both realizations are minimal. This demonstrates that knowing that the true system is minimal is not enough to specify even the Boolean structure of the system; every transfer function $\boldsymbol{G}(s)$ admits a consistent dynamical structure for *any* internal structure $\boldsymbol{Q}(s)$, whether it be perfectly decoupled or fully connected, as shown here.

Even in this idealized setting, reconstruction is not possible without additional information about the system. The same analysis that conclusively determines that input-output data yield no information about internal structure also demonstrates exactly what additional information is necessary for the transfer function to uniquely determine the structure. In particular, exactly p elements in each row of the matrix $[\boldsymbol{Q} \ \boldsymbol{P}]$, including the zero on the diagonal of $\boldsymbol{Q}$, must be known a priori for a transfer function to uniquely determine the rest of the dynamical structure function. This partial structure information is the basis for the experimental design that solves the reconstruction problem; discovering the fundamental limitations of reconstruction also sheds light on the solution to this problem.

13.3.3 Robustness: Noise and Nonlinearities

Although the fundamental limitations of network reconstruction are distilled by focusing on an idealized context for the problem, in practice, we want to ensure that any reconstruction methodology we develop will work on real biochemical systems. This implies that the results developed for linear systems, with no noise and perfect system identification, must be extended to include situations with noise and unmodeled dynamics, including nonlinearities.

In real biochemical networks, the data are typically very noisy, and some of the dynamic effects, such as diffusion and transport, may not be accounted for in a kinetic model. As a result, system identification will typically not produce the actual transfer function of the linearized real system; it will instead generate an approximation that captures the essence of the dynamics and fits the data well. For a properly designed experiment with the appropriate partial structure information, this transfer function will then uniquely specify a particular dynamical structure. Nevertheless, since the transfer function is an approximation of the true transfer function, the resulting dynamical structure is also only an approximation of the true structure. Moreover, this structure will typically be fully connected since it is very unlikely that noisy data will exactly produce zeros in the appropriate locations to eliminate edges in the network structure.

To discover the true network structure in this situation, then, one must solve the network robustness problem, computing the distance from a given transfer function to all possible Boolean structures. This is done by discovering the smallest perturbation to the transfer function that subsequently yields the desired Boolean structure when combined with the essential partial structure information. Computing this distance from the specified transfer function to each Boolean structure enables us to discover the Boolean structures that are "closest" to the given transfer function derived from noisy data. This drives the selection of a realistic structure in spite of noisy data and unmodeled dynamics, and provides a practical reconstruction process that allows measured data to discover the network structure most descriptive of its particular dynamic.

13.4 Experimental Design for Reconstruction

Analyzing the dynamical structure function of a system and its relationship to its transfer function reveals the fundamental limitations of network reconstruction.

In particular, we discover that perfect knowledge of the input-output dynamics, reflected by the system's true transfer function, is insufficient to solve the reconstruction problem, even if only the Boolean structure is desired. Further analysis, however, reveals that knowledge of exactly p elements in each row of the matrix $[\boldsymbol{Q}\ \boldsymbol{P}]$ is sufficient additional information to enable us to reconstruct a network from a system's transfer function. This section describes two examples of experiments that can satisfy this condition and enable network reconstruction.

The critical feature of this experimental process is the design of a set of perturbations that independently target each of the measured chemical species modeled in the network. Such perturbations are inputs to the system with a special structure to their control network, $\boldsymbol{P}$. In particular, if we know that each input only affects the measured species through one particular targeted species, then we know that exactly one element in each column of $\boldsymbol{P}$ is nonzero. This implies that exactly p such inputs, one for each measured species, will result in exactly $p - 1$ zeros in every row of $\boldsymbol{P}$. This partial structure information, along with the knowledge that each diagonal element of $\boldsymbol{Q}$ is zero, satisfies the condition of knowing exactly p elements in each row of the matrix $[\boldsymbol{Q}\ \boldsymbol{P}]$. Reconstruction of the rest of the entries of the dynamical structure function is then possible from perfect knowledge of the system's transfer function.

In situations where the linear approximation of the actual biochemical system is sufficiently descriptive, and where the signal-to-noise ratios enable confidence in the identified transfer function, this procedure will deliver the network structure of the system. Moreover, an experimental design that fails to augment knowledge of the transfer function with knowledge of p elements of each row of $[\boldsymbol{Q}\ \boldsymbol{P}]$ cannot deliver knowledge of the network structure. This experimental design process is essential in that every conceivable internal structure between species, $\boldsymbol{Q}$, can be equally descriptive of any particular transfer function. Unless this methodology is followed, and the essential partial-structure information is obtained through an appropriate experimental design, any best-fit measure for estimating a possible network structure can yield poor and misleading results. The results can be poor because the network estimate can be incorrect, and misleading because the resulting transfer function can accurately fit the measured data.

In the presence of unmodeled dynamics or significant noise, more work is necessary to recover the network structure. In particular, the network robustness problem must be solved, which is described in detail in section 13.6. Since the noise and unmodeled dynamics create a discrepancy between the transfer function identified from data and the true input-output behavior of the system, one would like to measure the distance from the identified transfer function to various Boolean network structures, selecting one that is most parsimonious in a particular sense. Current research indicates that doing so can lead to recovering the actual network structure in many situations. To make this experimental design more concrete, we discuss specific situations where the necessary conditions for reconstruction are satisfied. We explore the use of microarray techniques for measuring both messenger RNA and protein concentrations. Finally, we discuss the implications of measuring only steady-state values. We demonstrate that the network representation resulting from this experimental process yields a predictive model commensurate with the data used to create it: steady-state data yield static network information, whereas time-series data generate a dynamic representation of the system suitable for simulation.

13.4.1 Gene Silencing

A relatively easy way to interact directly with a specific gene is to perturb a system with RNA interference (RNAi), a mechanism that either inhibits gene expression at the stage of translation or hinders the transcription of specific genes. This leads to *gene silencing*, a term generally used to describe the "switching off" of a gene by a mechanism other than genetic modification. Gene-silencing experiments are now commonplace and can be interpreted as modifying the structure of the biological system.

We first assume that our network has p observed states consisting of mRNA concentrations. We model gene silencing by RNAi as an "operator" that adds an additional state variable and an input to the system. The input represents an RNAi molecule, and the additional state is the complex resulting from the binding of the RNAi with the corresponding mRNA molecule. Silencing an mRNA results in temporarily driving its concentration to zero.

Performing a silencing experiment on species y_i and measuring the dynamical behavior of this modified system yields a column vector transfer function, $\boldsymbol{G}_i$. The control structure function, $\boldsymbol{P}_i$, for this particular experiment is a column vector with only one nonzero entry, P_{ii}. We know this because the control input, u_i, affects only the new hidden state, which in turn only directly affects the measured state, y_i. Note that y_i may then affect other measured states, but the input is specifically designed to target y_i. Thus there is only one unknown transfer function in $\boldsymbol{P}_i$.

Repeating the gene-silencing experiments on all other p measured mRNA species gives a transfer function $\boldsymbol{G} = [\boldsymbol{G}_1 \ \boldsymbol{G}_2 \ \cdots \ \boldsymbol{G}_p]$ and a diagonal $\boldsymbol{P} = \text{diag}(P_{11}, P_{22}, \ldots, P_{pp})$. Because we know that the $p - 1$ nondiagonal terms on each row of $\boldsymbol{P}$ are zero, there are enough equations in $(\boldsymbol{I} - \boldsymbol{Q})\boldsymbol{G} = \boldsymbol{P}$ to solve all the unknowns in $\boldsymbol{Q}$, as explained in section 13.6. Exactly p independent RNAi experiments are required to reconstruct the network.

Measuring protein concentrations rather than mRNA yields the same dynamics, except that now all mRNA amd RNAi are hidden states. In this case, RNAi lowers the concentration of mRNA, and so, in time, eliminates the associated protein. It follows that p of these experiments will also enable as to reconstruct the system's dynamical structure since, with a similar argument, the control structure $\boldsymbol{P}$ is diagonal.

13.4.2 Inducible Overexpression

Gene overexpression may be constitutive, where the gene is always being transcribed, or inducible, where the gene may be externally activated. In a second set of structural perturbation experiments, we consider the latter. External activation occurs through the introduction of a transgene into the host specifically designed to temporarily increase the abundance of the desired transcript. The target specificity of these methods allows the control of the expression of specific genes without directly affecting other genes in the network.

We can posit a simple model of inducible overexpression in much the same way we did the model of gene silencing described above. In this case, the input, u_i, can be either a transcription factor or an activator of a transcription factor for gene i. The new hidden states can represent one or more states. One example of this is a transcription factor or a complex composed of the transcription factor bound to specific parts of DNA that increases the affinity for RNA polymerase. The concentration of mRNA corresponding to y_i will then increase (if it is not already saturated) and then possibly affect other measured species. These changes result in a control structure, $\boldsymbol{P}_i$, that has, again, only one nonzero entry. Thus inducibly overexpressing all measured states in a similar way will lead to a diagonal $\boldsymbol{P}$, making it possible to reconstruct the dynamical structure of the network. A similar argument follows if protein concentrations are measured instead of mRNA.

13.4.3 Steady-State versus Time-Series Data

Thus far, we have assumed that time-series data are available to perform system identification and obtain a transfer function. Frequently, however, experimental costs permit only steady-state measurements. As shown below, most of the connectivity of the network can still be reconstructed, but with no dynamical information. The advantage, however, is that we can perform a greater number of experiments for the same time and effort.

Assume that, after some time, the control input concentrations are maintained at a constant value giving us access to $\boldsymbol{y}(t)$, where t is large. This is equivalent (provided the system is stable) to knowledge of $\boldsymbol{G}_0 = \boldsymbol{G}(0)$, in other words, $\boldsymbol{G}(s)$ evaluated at $s = 0$. This follows from the final value theorem (see, for example, Chen, 1999). If $\boldsymbol{Q}_0 = \boldsymbol{Q}(0)$ and $\boldsymbol{P}_0 = \boldsymbol{P}(0)$, then $(\boldsymbol{I} - \boldsymbol{Q}(s))\boldsymbol{G}(s) = \boldsymbol{P}(s)$ evaluated at $s = 0$ becomes $(\boldsymbol{I} - \boldsymbol{Q}_0)\boldsymbol{G}_0 = \boldsymbol{P}_0$. With this equation, all of the above results follow, except in the case that $\boldsymbol{G}(s)$ has a zero at $s = 0$.

13.4.4 A Nonlinear System with Noise

Here we consider an example where we measure the concentration of three mRNA molecules. The full network is shown in figure 13.7a. Each mRNA is translated into

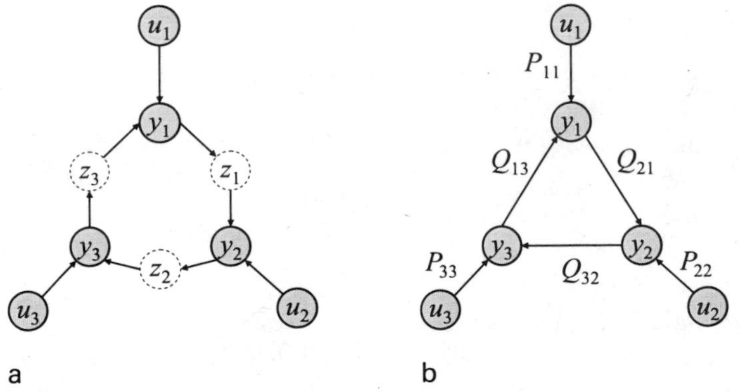

Figure 13.7
(a) Complete network, which includes hidden states (*dashed circles*). (b) Network structure with only measured states.

a protein that, together with an external transcription factor (the input), regulates the transcription of another mRNA. Because they can be administered in large dosages or replenished often enough, we assume that the inputs are steps. To obtain the network structure, three separate experiments are performed, one for each input, and mRNA measurements are taken.

The dynamics of the system are linear except for $\dot{y}_1$:

$$\dot{y}_1 = -y_1 + \frac{0.5}{0.1 + z_3^3} + u_1,$$

$$\dot{y}_2 = -2y_2 + 1.5z_1 + u_2,$$

$$\dot{y}_3 = -1.5y_3 + 0.5z_2 + u_3,$$

$$\dot{z}_1 = 0.8y_1 - 0.5z_1,$$

$$\dot{z}_2 = 1.2y_2 - 0.8z_2,$$

$$\dot{z}_3 = 1.1y_3 - 1.3z_3.$$

To each simulation, we add independent Gaussian noise with mean zero and standard deviation 0.1, as explained in section 13.6.4. The simulations are repeated three times and only steady-state data are used. The results can be found in table 13.1.

In an attempted reconstruction from the data, candidate networks are compared by associating a cost to each structure. Most network structures have a very high cost, which suggests they are not correct. When fit to this data, eight networks have low and similar costs. These are further differentiated by penalizing connections, again, as explained in section 13.6.4, and the true network is recovered.

Table 13.1
Result of steady state for three different experiments, from each input to the measured outputs

	Experiment 1	Experiment 2	Experiment 3
$u_1 \rightarrow y_1$	0.4382	0.4504	0.4353
$u_1 \rightarrow y_2$	0.4372	0.4285	0.4253
$u_1 \rightarrow y_3$	0.2349	0.2411	0.2264
$u_2 \rightarrow y_1$	−0.2639	−0.2803	−0.2975
$u_2 \rightarrow y_2$	0.1455	0.1653	0.1283
$u_2 \rightarrow y_3$	0.1076	0.0852	0.1077
$u_3 \rightarrow y_1$	−0.6850	−0.7267	−0.7225
$u_3 \rightarrow y_2$	−0.7307	−0.6862	−0.7642
$u_3 \rightarrow y_3$	0.3397	0.2787	0.2055

Moreover, a reconstruction from time-series data would recover $\boldsymbol{Q}(s)$ and $\boldsymbol{P}(s)$, which contain all of the dynamical behavior between measured states. This reconstruction allows for simulation, analysis, and modeling of structural perturbations. For example, a gene knockout, which turns a gene off constitutively can be modeled simply by removing the corresponding rows and columns of $\boldsymbol{Q}$. Alternatively, the effect of a mutation that causes gene i to lose its sensitivity to transcription factor j is obtained by making $Q_{ij} = 0$. Such flexibility is not available when working with a model consisting solely of the transfer function $\boldsymbol{G}(s)$.

13.5 Conclusions

Biological networks play a fundamental role relating a biochemical system's physical structure to its dynamic behavior. This chapter introduced network reconstruction as a problem distinct from system identification or realization and demonstrated its role in identifying the structure of a networked system. Dynamical structure functions were presented as a new representation of linear, time-invariant systems that incorporates exactly the structural information between measured states without fixing structural relationships between hidden states of the system. This tool enabled a precise formulation of the reconstruction problem that explicitly acknowledged the presence of unmeasured states, and thus differentiated reconstruction from system realization. Moreover, the use of dynamic structure functions revealed the dynamic properties of structurally perturbed systems, enabling convenient analysis of modified structures.

Through the use of these tools, fundamental limitations of network reconstruction became apparent. Not even Boolean reconstruction of linear, time-invariant systems is possible from input-output data alone. To discover a system's network structure from data, one must perform structural perturbation experiments, where the experimental design reveals certain structural elements. Subsequent analysis of the nature of a system's dynamical structure function revealed how to design these experiments; gene silencing and inducible overexpression were shown to be effective mechanisms yielding the information necessary for network reconstruction for linear, time-invariant systems. These results were then extended to develop reconstruction methods for nonlinear systems through the solution of the network robustness problem. These techniques thus yield an effective mechanism for obtaining network structure, and this structural understanding can be iteratively refined as new measurement capabilities become available.

13.6 Appendix: Mathematical Details

Consider a nonlinear system $\dot{\bar{\boldsymbol{x}}} = \boldsymbol{f}(\bar{\boldsymbol{x}}, \boldsymbol{u}, \boldsymbol{w})$, $\bar{\boldsymbol{y}} = \boldsymbol{h}(\bar{\boldsymbol{x}}, \boldsymbol{w})$ with p measured states $\bar{\boldsymbol{y}}$, hidden states $\bar{\boldsymbol{z}}$ (potentially a large number of them), m inputs $\boldsymbol{u}$, and noise $\boldsymbol{w}$. The system is linearized around an equilibrium point (a point such that $\boldsymbol{f}(\bar{\boldsymbol{x}}^*, \boldsymbol{0}, \boldsymbol{0}) = \boldsymbol{0}$) such that the inputs and noise do not move the states too far from the equilibrium and the linearized system remains close to the original nonlinear system. The linearized system can be written as $\dot{\boldsymbol{x}} = \boldsymbol{A}\boldsymbol{x} + \boldsymbol{B}\boldsymbol{u}$, $\boldsymbol{y} = \boldsymbol{C}\boldsymbol{x}$, where $\boldsymbol{x} = \bar{\boldsymbol{x}} - \bar{\boldsymbol{x}}^*$ and $\boldsymbol{y} = \boldsymbol{h}(\bar{\boldsymbol{x}}, \boldsymbol{0}) - \boldsymbol{h}(\bar{\boldsymbol{x}}^*, \boldsymbol{0})$. The transfer function of the system is given by $\boldsymbol{G}(s) = \boldsymbol{C}(s\boldsymbol{I} - \boldsymbol{A})^{-1}\boldsymbol{B}$. Typically, we can use data to find a transfer function $\boldsymbol{G}$ with standard system identification tools (Ljung, 1987).

13.6.1 Dynamical Structure Functions

Like system realization, network reconstruction begins with a transfer function, but it attempts to determine the network structure between measured states without imposing any additional structure on the hidden states. To accomplish this, a new representation of linear, time-invariant systems is required. Partition the linear system as

$$\begin{bmatrix} \dot{y} \\ \dot{z} \end{bmatrix} = \begin{bmatrix} A_{11} & A_{12} \\ A_{21} & A_{22} \end{bmatrix} \begin{bmatrix} y \\ z \end{bmatrix} + \begin{bmatrix} B_1 \\ B_2 \end{bmatrix} \boldsymbol{u}, \tag{13.3a}$$

$$y = [\boldsymbol{I} \quad \boldsymbol{0}] \begin{bmatrix} y \\ z \end{bmatrix}, \tag{13.3b}$$

where $\boldsymbol{x} = (y, z) \in \mathbf{R}^n$ is the full state vector, $\boldsymbol{y} \in \mathbf{R}^p$ is a partial measurement of the state, z are the $n - p$ "hidden" states, and $\boldsymbol{u} \in \mathbf{R}^m$ is the control input. In this work, we restrict our attention to situations where output measurements constitute partial-state information, that is, where $p < n$. We consider only systems with full-rank transfer function that do not have entire rows or columns of zeros because such "disconnected" systems are somewhat pathological and only serve to complicate the exposition without fundamentally altering the conclusions of this work.

Consider the Laplace transform of the signals in (13.3), yielding

$$\begin{bmatrix} sY \\ sZ \end{bmatrix} = \begin{bmatrix} A_{11} & A_{12} \\ A_{21} & A_{22} \end{bmatrix} \begin{bmatrix} Y \\ Z \end{bmatrix} + \begin{bmatrix} B_1 \\ B_2 \end{bmatrix} U, \tag{13.4}$$

where Y, Z and U are the Laplace transforms of y, z, and u, respectively. Solving for Z gives

$$Z = (sI - A_{22})^{-1} A_{21} Y + (sI - A_{22})^{-1} B_2 U.$$

Substituting into equation (13.4) then yields

$$sY = WY + VU, \tag{13.5}$$

where $W = A_{11} + A_{12}(sI - A_{22})^{-1} A_{21}$ and $V = A_{12}(sI - A_{22})^{-1} B_2 + B_1$. Let D be a matrix with the diagonal term of W, i.e. $D = \mathrm{diag}(W_{11}, W_{22}, \dots, W_{pp})$. Then

$$(sI - D)Y = (W - D)Y + VU.$$

Note that $W - D$ is a matrix with zeros on its diagonal. We then have

$$Y = QY + PU, \tag{13.6}$$

where

$$Q = (sI - D)^{-1}(W - D) \tag{13.7}$$

and

$$P = (sI - D)^{-1} V. \tag{13.8}$$

Note that Q is zero on the diagonal.

Definition 13.1 Given the system (13.3), we define the *dynamical structure function* of the system to be (Q, P), where Q and P are the *internal structure* and *control structure*, respectively, and are given as in equations (13.7) and (13.8). ■

Note that there are two main reasons we choose to work with (Q, P) instead of (W, V). The first is that by imposing zeros on the diagonal of Q, the matrix Q contains p variables less than the matrix W, which is important in the reconstruction problem (discussed later). The second reason is that, if all the measured states are removed from the system except for Y_i and Y_j, then the transfer function Q_{ij} (Q_{ji}) corresponds to the exact transfer function between the states when Y_j (Y_i) is the input and Y_i (Y_j) is the output. The same holds for P in terms of U_j (U_i) and Y_i (Y_j). In addition, $Q(s)$ and $P(s)$ have important predictive properties that are missing in

$G(s)$. Having the dynamics in (Q, P) and not just the Boolean structure allows us to modify or mutate the system at the measured states and obtain the new $(\bar{Q}, \bar{P})$ and the corresponding new transfer function $\bar{G}(s)$ (lemma 13.1). For example, if a measured state is eliminated, $(\bar{Q}, \bar{P})$ is obtained by eliminating the associated row and column in Q and the associated row in P. New data can then be obtained by simulation without the need for new experiments. Note that this could not have been obtained from just $G(s)$.

Given any system (13.3), its dynamical structure function, like its transfer function, is uniquely specified. This implies that the state-space description of a system completely specifies both the dynamical and Boolean structures of the system at *every* resolution level of measurements. We next consider the situation where the state-space description of the system is not known, and only the transfer function of the system is available.

Definition 13.2 A dynamical structure function, (Q, P), is *consistent* with a particular transfer function, G, if there exists a realization of G, of some order, of the form of system (13.3), such that (Q, P) are specified by equations (13.7) and (13.8). ■

Definition 13.3 Consider a system characterized by a transfer function G. The dynamical structure of the system can be *reconstructed* if there is only one admissible dynamical structure function, (Q, P), that is consistent with G. Likewise, the Boolean structure of the system can be reconstructed if all admissible dynamical structure functions that are consistent with G have the same Boolean structure. Here "admissible" refers to the entries of Q and P being strictly proper rational functions, that Q is zero on the diagonal, and that (Q, P) satisfy any additional constraint characterizing the information about the system that is available a priori. ■

13.6.2 Fundamental Limitations

We begin by characterizing the situations when reconstruction is possible knowing only the transfer function of a system. We consider both dynamical and Boolean reconstruction and show that neither is possible except in the degenerate case where there is only a single measurement, $p = 1$, thus trivializing the reconstruction problem. In the next section, we consider what "extra information" would be necessary and sufficient to reconstruct the dynamical structure of a system with a known transfer function.

For this section, we assume that no additional information about the system is available other than its transfer function G. We begin by characterizing the relationship between a system's dynamical structure function and its transfer function. The proofs for the results in this appendix can be found in Gonçalves and Warnick (2008).

Lemma 13.1 A dynamical structure function $(\boldsymbol{Q}, \boldsymbol{P})$ is consistent with a transfer function $\boldsymbol{G}$ if and only if the following relationship holds:

$$[\boldsymbol{G}^T \quad \boldsymbol{I}]\begin{bmatrix} \boldsymbol{Q}^T \\ \boldsymbol{P}^T \end{bmatrix} = \boldsymbol{G}^T,$$

where $(\cdot)^T$ indicates conjugate transpose. ■

We see, then, that the mapping from a system's dynamical structure function to its transfer function is a linear transformation, leading immediately to the following observations.

Lemma 13.2 Given a $p \times m$ transfer function $\boldsymbol{G}$, the operator $[\boldsymbol{G}^T \quad \boldsymbol{I}]$:

1. has dimension $m \times (m + p)$;
2. has full row rank; and
3. has a nullspace of dimension p.

Lemma 13.3 Given a $p \times m$ transfer function $\boldsymbol{G}$, every dynamical structure $[\tilde{\boldsymbol{Q}} \ \tilde{\boldsymbol{P}}]^T$ in the nullspace of the operator $[\boldsymbol{G}^T \quad \boldsymbol{I}]$ satisfies

$\tilde{\boldsymbol{P}} = -\tilde{\boldsymbol{Q}}G.$ ■

Lemma 13.4 Given a transfer function $\boldsymbol{G}$, the set $\mathcal{S}_G$ of all dynamical structure functions consistent with $\boldsymbol{G}$ can be parameterized by a $p \times p$ internal structure function $\tilde{\boldsymbol{Q}}$ and is given by

$$\mathcal{S}_G = \left\{ (\boldsymbol{Q}, \boldsymbol{P}) : \begin{bmatrix} \boldsymbol{Q}^T \\ \boldsymbol{P}^T \end{bmatrix} = \begin{bmatrix} \boldsymbol{0} \\ \boldsymbol{G}^T \end{bmatrix} + \begin{bmatrix} \boldsymbol{I} \\ -\boldsymbol{G}^T \end{bmatrix} \tilde{\boldsymbol{Q}}^T, \tilde{\boldsymbol{Q}} \in \mathcal{Q} \right\},$$

where $\mathcal{Q}$ is the set of internal structure functions. Moreover, the set $\mathcal{S}_G$ has $p^2 - p$ degrees of freedom. ■

This characterization of dynamical structure functions consistent with a given transfer function reveals the solution to the reconstruction problem. Note that, when $p = 1$, both dynamical and Boolean reconstructions are trivial since the only admissible $\boldsymbol{Q}$ is $\boldsymbol{Q} = \boldsymbol{0}$, thus fixing $\boldsymbol{P} = \boldsymbol{G}$. Aside from this rather pathological situation, we characterize the solution to the reconstruction problem with the following theorem:

Theorem 13.1: Reconstruction from $\boldsymbol{G}$ Given any $p \times m$ transfer function $\boldsymbol{G}$, with $p > 1$ and no other information about the system, dynamical and Boolean reconstruction is not possible. Moreover, for *any* internal structure $\boldsymbol{Q}$, there is a dynamical structure function $(\boldsymbol{Q}, \boldsymbol{P})$ that is consistent with $\boldsymbol{G}$. ■

In particular, this shows that both a completely decoupled ($\boldsymbol{Q} = \boldsymbol{0}$) and a fully connected internal structure (among others) are consistent with any given $\boldsymbol{G}$. This fact highlights the point that a selection criterion for reconstruction, such as sparsity, must be justified independently of the input-output data from the system.

13.6.3 Reconstruction with Extra Information

Theorem 13.1 makes it clear that not even Boolean reconstruction is generally possible from $\boldsymbol{G}$ without additional information. We know from lemma 13.4 that the set of all dynamical structure functions consistent with a given transfer function has $p^2 - p$ degrees of freedom. The role of additional information, then, is to add enough constraints to the set of admissible dynamical structure functions so that the intersection of admissible dynamical structure functions with the set of dynamical structure functions $\mathcal{S}_G$ consistent with a given transfer function contains a unique element, $(\boldsymbol{Q}, \boldsymbol{P})$. The following theorem indicates when partial-structure information, which refers to knowledge of some of the elements of $\boldsymbol{Q}$ or $\boldsymbol{P}$, is necessary and sufficient for dynamical reconstruction.

Theorem 13.2: Reconstruction with Partial Structure Given a $p \times m$ transfer function $\boldsymbol{G}$, dynamical structure reconstruction is possible from partial structure information if and only if $p - 1$ elements in each column of $[\boldsymbol{Q} \;\; \boldsymbol{P}]^T$ are known that uniquely specify the component of $(\boldsymbol{Q}, \boldsymbol{P})$ in the nullspace of $[\boldsymbol{G}^T \;\; \boldsymbol{I}]$. ■

Theorem 13.2 identifies exactly what information about a system's structure, beyond knowledge of its transfer function, must be obtained to recover the rest of its structure, enabling the design of experiments targeting precisely the extra information needed for reconstruction.

In many situations, we have no information on the internal structure $\boldsymbol{Q}$, but we may partially know $\boldsymbol{P}$ since we are designing the mechanisms actuating the system. When there is precisely one more measured state than inputs, $m = p - 1$, then we observe that knowing the $mp = p^2 - p$ elements of $\boldsymbol{P}$ is sufficient to fully recover $\boldsymbol{Q}$, as the conditions of theorem 13.2 are met. When $m < p - 1$, however, some knowledge of $\boldsymbol{Q}$ will be essential for reconstruction.

A special case of theorem 13.2 occurs when $p = m$ and $\boldsymbol{G}$ is full rank. In this situation, we observe that simply knowing that $\boldsymbol{P}$ is diagonal, that is, that the $p^2 - p$ off-diagonal elements of $\boldsymbol{P}$ are zero, is sufficient for reconstruction. Moreover, we demonstrate a particularly simple formula for $(\boldsymbol{Q}, \boldsymbol{P})$ in the following corollary:

Corollary 13.2.1 If $m = p$, $\boldsymbol{G}$ is full rank, and there is no information about the internal structure of the system, $\boldsymbol{Q}$, then the dynamical structure can be reconstructed if each input controls a measured state independently, that is, without loss of generality, the inputs can be indexed such that $\boldsymbol{P}$ is diagonal. Moreover, $\boldsymbol{H} = \boldsymbol{G}^{-1}$ characterizes the dynamical structure as follows:

$$Q_{ij} = -\frac{H_{ij}}{H_{ii}} \quad \text{and} \quad P_{ii} = \frac{1}{H_{ii}}. \qquad \blacksquare$$

Finally, we observe that knowing that full-state measurements are available, that $p = n$, is equivalent to knowing the structure of the system. This is because the realization of $\boldsymbol{G}$ with $\boldsymbol{C} = \boldsymbol{I}$ is unique, thereby generating a unique structure $(\boldsymbol{Q}, \boldsymbol{P})$ consistent with $\boldsymbol{G}$. Nevertheless, such a situation seems impractical, because the presence of noise in a system may demand hidden states to model the impact of this noise. Moreover, as discussed earlier, assuming full-state measurements when unmodeled dynamics are present can be disastrous for reconstruction.

13.6.4 Robustness

This section gives the mathematical details relating to section 13.3.3. We recall that data are derived from a noisy nonlinear system, generated according to corollary 13.2.1: that is, each input u_i controls first the measured state y_i so that $\boldsymbol{P}$ is diagonal. We use linear system identification to obtain $\boldsymbol{G}(s)$, the best possible fit, in some sense, to the data. We have to assume we remain close enough to the equilibrium so that the linearization is still valid. To average out the noise, the experiments are repeated N times. We want to quantify the *distance* from all possible Boolean structures to $\boldsymbol{G}$. There are several possible ways to quantify such distances, but most lead to nonconvex problems and thus are difficult to solve for large systems. With $\boldsymbol{\Delta}$ as a perturbation to the nominal system, the problem would be formulated as the minimization of $\|\boldsymbol{\Delta}\|$ such that some of the $Q_{ij} = 0$ and all other Q_{ij} and P_{ii} are free but strictly causal and $(\boldsymbol{I} - \boldsymbol{Q})(\boldsymbol{G}(\boldsymbol{\Delta})) = \boldsymbol{P}$. There are several uncertainty models and norms to choose from, and choosing the correct one is key to obtaining a convex problem. For example, uncertainty that is additive, multiplicative, or in the coprime factors all lead to nonconvex problems.

Quantifying Distances between Boolean Networks

Let $\boldsymbol{G}$ be the nominal plant obtained from our noisy data using some nonperfect system identification algorithm. $\boldsymbol{G}$ also misses unmodeled dynamics such as nonlinearities. Assume we model uncertainty with output (or input) feedback uncertainty, that is, the true system is given by $(\boldsymbol{I} + \boldsymbol{\Delta})^{-1}\boldsymbol{G}$. This choice of model uncertainty plays an important role in reaching a convex problem.

The problem is as follows: given a Boolean structure with several $Q_{ij} = 0$ and all other Q_{ij} and P_{ii} free, find the smallest $\boldsymbol{\Delta}$ (in some norm) such that the $\boldsymbol{Q}$ obtained from $(\boldsymbol{I} + \boldsymbol{\Delta})^{-1}\boldsymbol{G}$ has the desired Boolean structure. Basically, we want $(\boldsymbol{I} + \boldsymbol{\Delta})^{-1}\boldsymbol{G} = (\boldsymbol{I} - \boldsymbol{Q})^{-1}\boldsymbol{P}$ to have the desired set of $Q_{ij} = 0$ while minimizing $\|\boldsymbol{\Delta}\|^2$. We can rewrite the above equation as $\boldsymbol{\Delta} = \boldsymbol{G}\boldsymbol{P}^{-1}(\boldsymbol{I} - \boldsymbol{Q}) - \boldsymbol{I}$. Thus we are looking to minimize $\|\boldsymbol{G}\boldsymbol{P}^{-1}(\boldsymbol{I} - \boldsymbol{Q}) - \boldsymbol{I}\|^2$ over some free Q_{ij} and all P_{ii}. Since $\boldsymbol{P}$ is diagonal, its inverse $\boldsymbol{P}^{\boldsymbol{i}}$

is also diagonal. Multiplying $\boldsymbol{P}^i$ by $\boldsymbol{I} - \boldsymbol{Q}$ results in the ith row of $\boldsymbol{I} - \boldsymbol{Q}$ being multiplied by P^i_{ii}. That means that the diagonal of $\boldsymbol{P}^{-1}(\boldsymbol{I} - \boldsymbol{Q})$ is just the diagonal of $\boldsymbol{P}^i$, and all other terms are of the form $P^i_{ii}Q_{ij}$. Define a new variable matrix $\boldsymbol{X}$ with the same diagonal as $\boldsymbol{P}^i$ and the off-diagonal terms equal to $P^i_{ii}Q_{ij}$. Then the problem can be rewritten as

$$\inf_{\boldsymbol{X}} \|\boldsymbol{GX} - \boldsymbol{I}\|^2,$$

which is a convex problem. Note that some of the entries in $\boldsymbol{X}$ will be zero, corresponding to those $Q_{ij} = 0$.

This can be cast as a least squares problem. For the free terms in $\boldsymbol{X}$, write them in vector form $\boldsymbol{x}$. Rewrite $\boldsymbol{GX} - \boldsymbol{I}$ in a vector form instead of a matrix form, $\boldsymbol{Ax} - \boldsymbol{b}$, where the entries in $\boldsymbol{A}$ are transfer functions derived from $\boldsymbol{G}$, and $\boldsymbol{b}$ is the vector form of the identity matrix, obtained column by column. Note that there are at least three unknowns in $\boldsymbol{x}$ from the diagonal of $\boldsymbol{X}$, corresponding to the entries of the diagonal $\boldsymbol{P}$, which are not affected by the choice of zeros in $\boldsymbol{Q}$. If we use the norm given by $\|\boldsymbol{\Delta}\|^2 =$ sum of all $\|\Delta_{ij}\|^2_\infty$, where $\|\cdot\|_\infty$ denotes the L_∞ norm over $s = j\omega$, then the problem reduces to

$$\begin{aligned}\inf_{\boldsymbol{x}} \|\boldsymbol{Ax} - \boldsymbol{b}\|^2_\infty &= \inf_{\boldsymbol{x}} \sup_{\omega} \|\boldsymbol{Ax} - \boldsymbol{b}\|^2_2 \\ &= \sup_{\omega} \inf_{\boldsymbol{x}} \|\boldsymbol{Ax} - \boldsymbol{b}\|^2_2 \\ &= \sup_{\omega} \boldsymbol{b}^T(\boldsymbol{A}(\boldsymbol{A}^T\boldsymbol{A})^{-1}\boldsymbol{A}^T - \boldsymbol{I})^2\boldsymbol{b},\end{aligned}$$

where, for each ω, we use the least-squares solution. When experiments are repeated N times, $\boldsymbol{A}$ and $\boldsymbol{b}$ are matrices N times taller than with one experiment.

Penalizing Connections

The above methodology has a critical weakness: there are several Boolean networks with optimal distances smaller than or equal to the optimal distance to the true network. Take, for example, the fully connected network. The extra degrees of freedom allow the optimal distance to be the smallest of all, and zero if the experiments are performed only once. If the true network has l zero connections, then there are $2^l - 1$ networks that have a smaller or equal optimal distance. With no noise and perfect system identification, this is not a problem since we see from the optimization that the optimal $Q_{ij} = 0$ for all those l networks and thus recover the true network.

When noise is present, the true network will always have the worst optimal distance compared to the other l networks; given this limitation, how can we find it?

With repeated experiments, and small enough noise and nonlinearities, the optimal distances of the l networks become comparable and much smaller than that of the other networks. Thus, by penalizing connections, the true network should be revealed. This can be achieved, for example, with Akaike's information criterion (AIC; Akaike, 1974) or some of its variants such as AICc, which is AIC with a second-order correction for small sample sizes, or with the Bayesian information criterion (BIC; for a review and comparisons of these methods, see Burnham and Anderson, 2004, and the references therein).

In the example from section 13.4.4, because the network has three elements in $\boldsymbol{Q}$ equal to zero, we expect that there are $2^3 = 8$ networks with an optimal cost better than or equal to that of the true network. The important thing is that all other networks have an optimal cost over an order of magnitude larger than the true network. With AIC, the true network is selected; with AICc, it is further differentiated from the others.

14 Identification of Biochemical Reaction Networks Using a Parameter-Free Coordinate System

Dirk Fey, Rolf Findeisen, and Eric Bullinger

A fundamental step in systems biology is the estimation of kinetic parameters such as association and dissociation constants. Because estimating these from direct measurements of isolated reactions in vivo can be expensive, time consuming, or even infeasible, researchers often do so from indirect measurements such as time-series data. This chapter proposes an observer-based parameter estimation methodology particularly suited for biochemical reaction networks in which the reaction kinetics are described by polynomial or rational functions. Parameter estimation is performed in three steps: (1) the system is transformed into a new set of coordinates that makes it parameter free, which facilitates (2) the design of a standard observer; (3) parameter estimates are obtained in a straightforward way from the observer states, transforming them back to the original coordinates. The methodology is illustrated by application to a MAPK signaling pathway.

14.1 Biological Processes as Systems

In 1943, when Erwin Schrödinger gave three talks in Dublin, later published as *What Is Life?* (Schrödinger, 1944), one of his central, then revolutionary, assertions was that biological systems follow physical laws and thus can be described by mathematical models. A decade later, this was achieved for cell membrane potential by Alan Hodgkin and Andrew Huxley (1952), who explained their experimental data with a mathematical model, a breakthrough in understanding how neurons function. Denis Noble (1960) expanded upon their work to produce the first mathematical model of the heart. Modeling other intracellular systems proved to be quite difficult, in particular owing to the lack of sufficient suitable experimental data. Advances in biological experimental techniques of the last decades have led to a rapidly growing number of models (Le Novère et al., 2006). The main bottleneck in developing dynamical models of biological systems has been the difficulty of estimating biological parameters, even when structural information like stoichiometry is known.

Unknown parameters are commonly estimated from time-series data in technical applications. Several peculiarities of biological systems hinder a straightforward application of most existing identification methodologies as typically used for technical systems. Whereas most biological systems have a large number of parameters, often only a small number of experiments are possible, consisting of a few experimental steps and scarce time points. Moreover, the noise level is usually significant.

In recent years, control- and system-theoretic viewpoints and approaches have proven to be valuable tools for gaining a deeper understanding of biological systems (Gilles, 2002). As outlined above, however, biological systems have particular properties not often found in technical applications (Sontag, 2005; Sontag et al., 2004); this situation demands novel methodologies particularly suited for systems biology. For example, there is a need for approaches that guarantee the positivity of states and parameters or the monotonicity of their interactions.

To help meet such needs, we propose here an observer-based parameter estimation algorithm that considers structural information such as the stoichiometry and the type of flow models. In many practical applications, this information is known or fixed a priori. Our methodology makes explicit use of this knowledge in the design of the parameter estimator. Although it has its shortcomings, in particular when faced with noisy measurements, our approach provides us with insights into the possibilities and limitations of global parameter estimation.

Section 14.2 introduces the modeling of biochemical reaction networks and the model class considered. Section 14.3 gives a brief overview of existing parameter estimation techniques for these models. Section 14.4 describes the proposed observer-based parameter estimation methodology, as described above. Section 14.5 applies the proposed methodology to model the MAPK cascade, while section 14.6 presents our conclusions.

14.2 Mathematical Description of Biochemical Reaction Networks

A common framework for the modeling of biochemical reaction networks is provided by sets of reactions of the form

$$\alpha_1 \mathrm{S}_1 + \cdots + \alpha_{n_s} \mathrm{S}_{n_s} \rightarrow \beta_1 \mathrm{P}_1 + \cdots + \beta_{n_p} \mathrm{P}_{n_p}, \tag{14.1}$$

where S_i denotes substrates that are transformed into the products P_i. The factors α_i and β_i denote the stoichiometric coefficients of the reactants. Neglecting spatial and stochastic effects, one can model these reactions with the system of ordinary differential equations:

$$\frac{\mathrm{d}\boldsymbol{c}}{\mathrm{d}t} = \boldsymbol{N}\boldsymbol{v}(\boldsymbol{c}, \boldsymbol{p}), \tag{14.2}$$

where $c \in \mathbf{R}^n_{\geq 0}$ is the vector of concentrations, $p \in \mathbf{R}^m_{>0}$ is the vector of parameters and the vector of flows v is a function from $\mathbf{R}^n_{\geq 0} \times \mathbf{R}^m_{>0}$ into $\mathbf{R}^{\tilde{m}}_{\geq 0}$, where $\tilde{m}$ is the number of reactions.

The stoichiometric matrix $N \in \mathbf{R}^{n \times \tilde{m}}$ depends on the coefficients α_i, β_i, and may also depend on factors compensating for different units or volumes (for a more detailed introduction, see, for example, Klipp et al., 2005; Keener and Sneyd, 2001).

Although there is a large variety of possible reaction models (Cornish-Bowden, 2004), here we consider only those most commonly used to describe signaling networks:

- *mass-action*: The flow is proportional to each substrate concentration: $v = k \prod_{i \in I} c_i$, where I is a subset of $1, \ldots, n$ with possibly repeated entries;
- *power-law* (S-systems, generalized mass-action): The flow is polynomial in the substrate concentrations: $v = k \prod_{i \in [1,n]} c_i^{\alpha_i}$
- *Michaelis-Menten* or *Monod*: The flow depends linearly on the substrate concentration s for low concentrations and saturates for high substrate concentrations at $V_{\max}$: $v = V_{\max}(s/(K_M + s))$. At a substrate concentration of K_M, the flow is half the maximum rate.
- *Hill*: The flow is sublinear for low substrate concentrations and saturates for high substrate concentrations at $V_{\max}$: $v = V_{\max}(s^h/(K_M^h + s^h))$. The exponent h is larger than one, and at a substrate concentration of K_M the flow is half the maximum rate.

In biochemical reaction modeling, the stoichiometry is usually known, as is the type of reaction kinetics, as opposed to the often quite uncertain parameters.

14.3 Parameter Identification Approaches for Biochemical Reaction Networks

System identification, a classical area of systems theory, addresses the estimation of dynamical models based on input-output data. Estimation of the discrete-time models commonly used in engineering is often achieved by minimizing quadratic functionals (Gadkar et al., 2005; Ljung, 1987). In systems biology, however, the models are usually continuous-time, which allows for improved physical insight. As described in section 14.2, these models are the result of first-principle modeling and take simple forms. For example, in the case of mass-action kinetics, the models are linear in the parameters. Additionally, continuous-time models are less sensitive to the noisy measurements common in biology (Garnier et al., 2003). For these reasons, continuous-time models are preferable in systems biology, even though there are only few suitable continuous identification algorithms. Common in systems biology are global, optimization-based parameter estimation algorithms such as genetic or branch-and-bound algorithms (Feng et al., 2004; Moles et al., 2003; Singer et al.,

2006). By formulating parameter estimation as the minimization of the error between simulated and measured outputs, subject to the model dynamics as constraints, the resulting optimization problem is usually nonconvex and the complexity grows significantly with the size of the parameter space (Polisetty et al., 2006). Alternatives are local search strategies (Zak et al., 2003), whose main disadvantage is the need for good initial guesses if one is to obtain good (valid) parameter estimates.

As noted previously, there is a clear need for novel parameter estimation methodologies particularly suited for systems biology (Ljung, 2003). Examples of such methodologies are multiple shooting (Peifer and Timmer, 2007) or algebraic approaches, which take the model structure explicitly into account (Audoly et al., 2001; Ljung and Glad, 1994; Xia and Moog, 2003), although the latter can be applied only to small systems. For larger systems, numerical solutions that exploit structural information are preferable. This chapter proposes such an approach. The parameter estimation problem is transformed into a state estimation problem, using a special coordinate transformation, which renders the model parameter free. The method outlined here is an extension of previous results (Bullinger et al., 2008; Farina et al., 2006, 2007; Fey et al., 2008) that apply both to a more general class of systems and to systems with known parameters. The method is illustrated by applying it to a rather large system, the MAPK pathway.

14.4 An Observer-Based Approach for Parameter Estimation in Biochemical Reaction Networks

The proposed approach consists of three main steps, depicted in figure 14.1:

1. Transform coordinates into a parameter-free system;
2. Design an observer considering the parameter-free system;
3. Estimate parameters based on the estimated coordinates of the parameter-free system.

The observer itself may incorporate an additional coordinate transformation. This section describes the three steps in greater detail.

14.4.1 Parameter-Free System Representation

For a system (14.2), with constant parameter vector $\boldsymbol{p}$, one can construct a coordinate transformation under which the dynamics of the transformed system do not depend on the parameters. Rather, the parameters appear as functions of the transformed state variables and can therefore be identified through estimation of these new system states.

$$\dot{c} = Nv(c,p) \xrightarrow{} \dot{x} = f(x) \xrightarrow{} \dot{\xi} = A\xi + B\varphi(\xi)$$

$$x = \Psi(c,p) \qquad\qquad \xi = \Phi(x)$$

$$\Big\downarrow \text{design}$$

$$\dot{\hat{\xi}} = A\hat{\xi} + B\varphi(\hat{\xi})$$
$$+\Theta L(y - \hat{y})$$

$$\Big\downarrow \text{estimate}$$

$$p = \Psi^{\text{inv}}(\hat{x}) \qquad\qquad \hat{x} = \Phi^{\text{inv}}(\hat{\xi})$$
$$p(t) \longleftarrow \hat{x}(t) \longleftarrow \hat{\xi}(t)$$

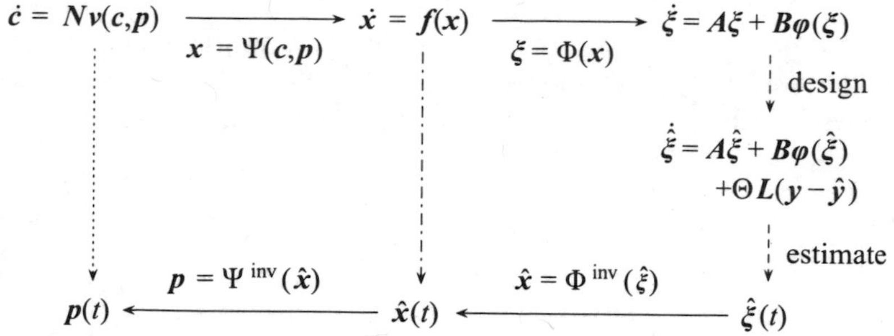

Figure 14.1
Overview of the proposed parameter estimation, depicting the direct (*dotted line*) and proposed observer's (*dashed line*) way of estimating the parameter, as well as the unavailable direct observation (*dashed-dotted line*). The proposed approach first transforms the system into parameter-free coordinates, using the mapping $\Psi(\cdot)$, then transforms it again to simplify the observer design, using $\Phi(\cdot)$. The proposed observer estimates these states, $\hat{\xi}(t)$, which are transformed back into the original coordinates, recovering first $\hat{x}(t)$ and then the original parameters $p(t)$. The primary difficulty herein is finding a nonlinear observer.

Mass-Action Flows

To simplify the presentation, we first consider only mass action flows; the general case of rational reaction kinetics is discussed below. To exploit the structure of biochemical reaction networks, it is helpful to split the stoichiometry into the difference of two matrices with nonnegative entries:

$$N = N_\text{p} - N_\text{s}, \tag{14.3}$$

where N_p denotes the product part of the stoichiometry and N_s the substrate part (Farina et al., 2006). Then

$$v = \text{diag}(p) \begin{bmatrix} \prod c_i^{N_{s_{i1}}} \\ \vdots \\ \prod c_i^{N_{s_{im}}} \end{bmatrix}, \tag{14.4}$$

where $N_{s_{ij}}$ is the i,jth element of the substrate stoichiometry matrix N_s. The transformation to a parameter-free system begins by treating the parameters as state variables. Assuming constant parameters, one can expand the state vector with the additional vector differential equation

$$\frac{\text{d}p}{\text{d}t} = 0,$$

resulting in an extended state vector containing the concentrations c and the parameters p. The dependence of the equations for c on the unknown parameters p

impedes the design of appropriate observers. For biochemical reaction networks, it is advantageous to transform the extended system by using concentrations $\boldsymbol{c}$ and fluxes $\boldsymbol{v}$ as states. The coordinate transformation requires the biologically reasonable assumption of positivity of the concentrations and is described in the following theorem.

Theorem 14.1 Let the concentrations c be strictly positive and all flows modeled according to mass action. Then the system

$$\frac{\mathrm{d}\boldsymbol{c}}{\mathrm{d}t} = \boldsymbol{N}\boldsymbol{v}(\boldsymbol{c}, \boldsymbol{p}), \tag{14.5a}$$

$$\frac{\mathrm{d}\boldsymbol{p}}{\mathrm{d}t} = \boldsymbol{0} \tag{14.5b}$$

is equivalent to

$$\frac{\mathrm{d}\boldsymbol{c}}{\mathrm{d}t} = \boldsymbol{N}\boldsymbol{v}, \tag{14.6a}$$

$$\frac{\mathrm{d}\boldsymbol{v}}{\mathrm{d}t} = \mathrm{diag}(\boldsymbol{v})\boldsymbol{N}_s^T(\mathrm{diag}(\boldsymbol{c}))^{-1}\boldsymbol{N}\boldsymbol{v}. \tag{14.6b}$$

■

The main advantage of (14.6) is that the right-hand side does not depend on parameters, but only on the stoichiometry and the states. The parameters are hidden in the initial conditions because each flux $\boldsymbol{v}_i$ is proportional to the parameter $\boldsymbol{p}_i$ (see equation (14.4)). A further advantage of the transformed system is that the measurements are often a subset of the coordinates $\boldsymbol{c}$ and $\boldsymbol{v}$ as fluxes can be measured, for example, using ^{13}C labeling (Costenoble et al., 2007). This simple output function also helps in the observer design.

Flows with Rational Terms

Although theorem 14.1 applies only to systems with mass-action kinetics, the proposed approach can be extended to cope with more complex flow models, allowing for fluxes of the form

$$\boldsymbol{v}_i = \boldsymbol{k}_i \prod_{j=1}^{n} \frac{\boldsymbol{c}_j^{\boldsymbol{v}_{ij}}}{\boldsymbol{K}_{ij}^{\boldsymbol{\eta}_{ij}} + \boldsymbol{c}_j^{\boldsymbol{\eta}_{ij}}}, \tag{14.7}$$

where $\boldsymbol{K}_{ij} > 0$, $\boldsymbol{v}_{ij} \geq 0$ and $\boldsymbol{\eta}_{ij} \geq 0$. (If $\boldsymbol{\eta}_{ij} = 0$, then the arbitrary parameter $\boldsymbol{K}_{ij}$ shall be set equal to 1.) The general form of (14.7) includes mass-action kinetics, generalized mass-action kinetics, Michaelis-Menten and Hill kinetics as well as products of these kinetic terms. For example, setting $\boldsymbol{\eta}_{ij} = 0$ leads to a mass-action model.

To simplify the presentation, define the following matrix-valued function $\boldsymbol{M}$

$$\boldsymbol{M}_{ij} = \boldsymbol{K}_{ij}^{\eta_{ij}} + \boldsymbol{c}_j^{\eta_{ij}}. \tag{14.8}$$

As in the mass-action case, one can find an extended system that is free of parameters.

Theorem 14.2 Let the concentrations be strictly positive and the flows be of the form (14.7), with known exponents $\boldsymbol{v}_{ij}$ and $\boldsymbol{\eta}_{ij}$. Then the following two systems are equivalent:

$$\frac{\mathrm{d}\boldsymbol{c}}{\mathrm{d}t} = \boldsymbol{N}\boldsymbol{v}(\boldsymbol{c}, \boldsymbol{p}), \tag{14.9a}$$

$$\frac{\mathrm{d}\boldsymbol{p}}{\mathrm{d}t} = 0 \tag{14.9b}$$

with

$$\boldsymbol{p} = [\boldsymbol{k}_1 \quad \cdots \quad \boldsymbol{k}_m \quad \boldsymbol{K}_{11} \quad \cdots \quad \boldsymbol{K}_{mn}]^T, \tag{14.9c}$$

and

$$\frac{\mathrm{d}\boldsymbol{c}}{\mathrm{d}t} = \boldsymbol{N}\boldsymbol{v}, \tag{14.10a}$$

$$\frac{\mathrm{d}\boldsymbol{M}_{ij}}{\mathrm{d}t} = \boldsymbol{\eta}_{ij}\boldsymbol{c}_j^{(\eta_{ij}-1)}\boldsymbol{e}_j^T\boldsymbol{N}\boldsymbol{v}, \tag{14.10b}$$

$$\frac{\mathrm{d}\boldsymbol{v}}{\mathrm{d}t} = \mathrm{diag}(\boldsymbol{v})(\boldsymbol{v}(\mathrm{diag}(\boldsymbol{c}))^{-1}\boldsymbol{N}\boldsymbol{v} - \tilde{\boldsymbol{m}}), \tag{14.10c}$$

where

$$\tilde{\boldsymbol{m}}_i = \sum_j \frac{\boldsymbol{\eta}_{ij}\boldsymbol{c}_j^{(\eta_{ij}-1)}\boldsymbol{e}_j^T\boldsymbol{N}\boldsymbol{v}}{\boldsymbol{M}_{ij}}.$$

■

To summarize, any biochemical reaction model consisting of flows modeled as in (14.7) can be transformed into a system that is free of parameters. This system has an extended-state vector and depends only on structural properties of the original system.

Remark 14.1 If some parameter values are already known, the proposed methodology can be adjusted in a straightforward way to avoid estimating them. There are two cases. First, if the parameter is a Hill or Michaelis-Menten constant $\boldsymbol{K}_{ij}$, then there exists a state $\boldsymbol{M}_{ij}$ that depends on $\boldsymbol{K}_{ij}$ and on some concentrations. Consequently, there is an algebraic equation which specifies this dependence and can replace the differential equation corresponding to this parameter. In the second case, if the parameter is proportional to a flow, that is, $\boldsymbol{k}_i$ in a flow $\boldsymbol{v}_j$, then this flow contains no unknowns, and can be written as a function of the other state variables. In either case, these dependent states can easily be eliminated, thus reducing the state-space dimension by the number of known parameters. ■

The extended state is denoted by

$$\boldsymbol{x} = \begin{bmatrix} \boldsymbol{c} \\ \boldsymbol{m} \\ \boldsymbol{v} \end{bmatrix}, \tag{14.11}$$

where $\boldsymbol{m}$ is the vector of all nonconstant entries of $[\boldsymbol{M}_{11} \ \cdots \ \boldsymbol{M}_{m1} \ \boldsymbol{M}_{12} \ \cdots \ \boldsymbol{M}_{mn}]^T$. The extended system can be written

$$\frac{\mathrm{d}\boldsymbol{x}}{\mathrm{d}t} = \boldsymbol{f}(\boldsymbol{x}), \tag{14.12a}$$

$$y = \boldsymbol{h}(\boldsymbol{x}), \tag{14.12b}$$

where $\boldsymbol{h}$ is the output map.

To simplify the observer design, we introduce the assumption that the output is a subset of the concentrations and flows. This is the case in many biological applications.

Assumption 14.1 The output $y(t) \in \mathbf{R}^{n_y}$ is a subset of the concentrations $\boldsymbol{c}$ and the flows $\boldsymbol{v}$:

$$y = \boldsymbol{h}\left(\begin{bmatrix} \boldsymbol{c} \\ \boldsymbol{m} \\ \boldsymbol{v} \end{bmatrix}\right) = \begin{bmatrix} \boldsymbol{H}_c & 0 & 0 \\ 0 & 0 & \boldsymbol{H}_v \end{bmatrix} \begin{bmatrix} \boldsymbol{c} \\ \boldsymbol{m} \\ \boldsymbol{v} \end{bmatrix}, \tag{14.13}$$

where the columns of $\boldsymbol{H}_c$ and of $\boldsymbol{H}_v$ are a subset of the columns of the corresponding identity matrices. ■

14.4.2 Observer Design

The parameter-free system (14.10) simplifies the design of an observer because the system does not contain any unknown parameters on the right-hand side. The estimation of the parameters of the original system requires the estimation of the extended-state vector of the transformed system. In systems theory, this estimation is carried out by the use of observers, which are mathematical systems often consist-

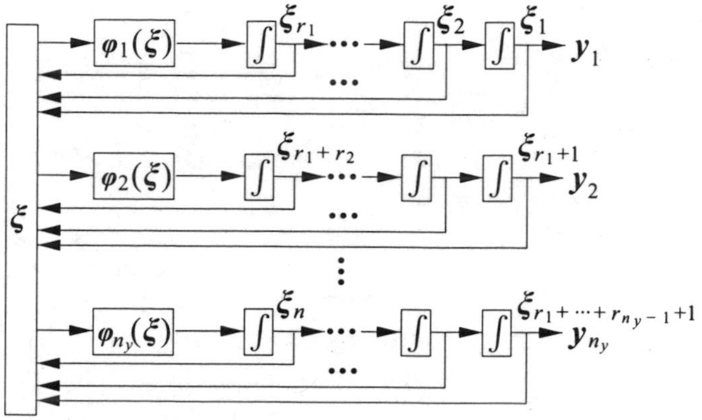

Figure 14.2
Sketch of the observer canonical form.

ing of an (approximated) copy of the true system and feedback injection of the predicted output error.

In contrast to the linear observers discussed in chapter 1, the complexity of the parameter estimation problem here requires more advanced nonlinear observer structures. The design uses a nonlinear analog to the observability matrix given by equation (1.20). The observability map $\Phi(\cdot)$ is used to transform the system into observability canonical coordinates (see figure 14.1). The transformation requires a suitable choice of outputs and a number of their derivatives such that Φ is invertible. In the transformed coordinates, the states correspond to the output y and its first r_i derivatives (see figure 14.2).

The corresponding state-space description is

$$\frac{\mathrm{d}\xi}{\mathrm{d}t} = \boldsymbol{J}\xi + \boldsymbol{B}\varphi(\xi), \tag{14.14a}$$

$$y = \boldsymbol{C}\xi, \tag{14.14b}$$

where

$$\boldsymbol{J} = \begin{bmatrix} \boldsymbol{J}_1 & & \\ & \ddots & \\ & & \boldsymbol{J}_{n_y} \end{bmatrix}, \qquad \boldsymbol{J}_i = \begin{bmatrix} 0 & 1 & 0 & \\ & & \ddots & \\ & & & 1 \\ 0 & \cdots & \cdots & 0 \end{bmatrix} \in \mathbf{R}^{r_i \times r_i},$$

$$\boldsymbol{B} = \begin{bmatrix} \boldsymbol{B}_1 & & \\ & \ddots & \\ & & \boldsymbol{B}_{n_y} \end{bmatrix}, \qquad \boldsymbol{B}_i = [0 \quad \cdots \quad 0 \quad 1]^T \in \mathbf{R}^{r_i \times 1},$$

$$C = \begin{bmatrix} C_1 & & \\ & \ddots & \\ & & C_{n_y} \end{bmatrix}, \qquad C_i = [1 \quad 0 \quad \cdots \quad 0] \in \mathbf{R}^{1 \times r_i}.$$

The nonlinearities are concentrated in the function $\varphi(\xi)$ (for a detailed discussion, see, for example, Xia and Zeitz, 1997; Schaffner and Zeitz, 1999).

The transformation into the observability canonical form (14.14) is only possible if $\Phi(\cdot)$ is invertible. Moreover, although $\varphi(\cdot)$ is continuous and bounded on the true trajectory, it might not be elsewhere (Bullinger et al., 2008). To resolve the problem, Vargas and Moreno (2005) proposed a continuous extension of $\Phi(\cdot)$, which may prove difficult to achieve in practice. Here we bound the nonlinearity $\varphi(\cdot)$ directly, as illustrated in figure 14.3, where the true trajectory corresponds to $\hat{\xi}_1 = \hat{\xi}_2$. This is possible because parameter estimation does not require convergence of the state estimate in physical coordinates x, but in observability coordinates ξ. Theorem 14.3 discusses the properties of a high-gain observer whose states converge with arbitrary precision to the states of the system in observability canonical form, based on the design proposed by Vargas and Moreno (2005).

Theorem 14.3 Choose coefficients $l_j^{(i)}$ in such a way that $s^{r_i} + \sum l_j^{(i)} s^j$ is a Hurwitz polynomial, or equivalently, that $J - LC$ with $L_i = [l_{r_i-1}^{(i)}, \ldots, l_0^{(i)}]$ is a Hurwitz matrix and a bounded approximation $\hat{\varphi}(\cdot)$ such that $\hat{\varphi}(\xi) = \varphi(\xi)$ on the true trajectory $\xi(t, \xi_0)$. Then, for any $\epsilon > 0$, there exists a $\theta > 0$ such that the observer

$$\frac{d\hat{\xi}}{dt} = J\hat{\xi} + B\hat{\varphi}(\hat{\xi}) + \Theta L(y - C\hat{\xi}) \tag{14.15a}$$

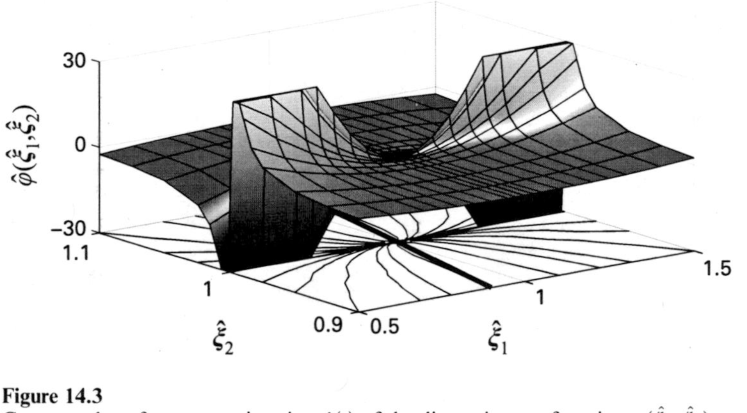

Figure 14.3
Contour plot of an approximation $\hat{\varphi}(\cdot)$ of the discontinuous function $\varphi(\hat{\xi}_1, \hat{\xi}_2) = -(\hat{\xi}_1 - 1)^2/(\hat{\xi}_2 - 1)$. The approximation $\hat{\varphi}(\cdot)$ is equal to $\varphi(\cdot)$ on $\hat{\xi}_1 = \hat{\xi}_2$ (*thick line*) and globally bounded, with magnitude 30.

$$\hat{\boldsymbol{x}} = \boldsymbol{\Phi}^{-1}(\hat{\xi}) \tag{14.15b}$$

$$\boldsymbol{\Theta} = \begin{bmatrix} \boldsymbol{\Theta}_1 & & \\ & \ddots & \\ & & \boldsymbol{\Theta}_{n_y} \end{bmatrix}, \qquad \boldsymbol{L} = \begin{bmatrix} \boldsymbol{L}_1 & & \\ & \ddots & \\ & & \boldsymbol{L}_{n_y} \end{bmatrix},$$

$$\boldsymbol{\Theta}_i = \operatorname{diag}(\theta, \ldots, \theta^{r_i}),$$

estimates the state ξ with ε precision in finite time. In other words, there exists a $T \geq 0$ such that

$$\|\hat{\xi}(t) - \xi(t)\| \leq \varepsilon, \qquad \text{for all } t \geq T. \qquad \blacksquare$$

The proof of theorem 14.3 is similar to the one of Vargas and Moreno (2005). Although the high-gain parameter θ can be used to tune the speed of convergence, θ also amplifies the noise in the data and should thus not be chosen too high.

One of the main disadvantages of the proposed methodology is the necessity of transforming the extended system into observability canonical form. Another disadvantage is the sensitivity to noise inherent to high-gain observers. Nevertheless, as shown in section 14.5, the methodology can be applied to relatively complex systems.

14.4.3 Parameter Estimation

The final step of the proposed methodology is the actual parameter estimation. The observer coordinates can be transformed back into the coordinates of the original system through the inverse of the observability map $\boldsymbol{\Phi}^{-1}$:

$$\hat{\boldsymbol{x}} = \boldsymbol{\Phi}^{-1}(\hat{\xi}). \tag{14.16a}$$

For each time point t, $\hat{\boldsymbol{x}}(t)$ can be calculated, leading directly to estimates of the concentrations $\hat{\boldsymbol{c}}$. Using (14.8), the parameters $\boldsymbol{K}_{ij}(t)$ can be estimated:

$$\hat{\boldsymbol{K}}_{ij}(t) = \begin{cases} (\hat{\boldsymbol{M}}_{ij}(t) - \hat{c}_j(t))^{1/\eta_{ij}}, & \text{for } \eta_{ij} > 0, \\ 1, & \text{for } \eta_{ij} = 0. \end{cases} \tag{14.16b}$$

Finally, the estimation of the parameters $\boldsymbol{k}_i(t)$ is possible using (14.7)

$$\hat{\boldsymbol{k}}_i(t) = \hat{v}_i(t) \prod_j^{n_c} \frac{\hat{\boldsymbol{M}}_{ij}(t)}{\hat{c}_j(t)^{v_{ij}}}. \tag{14.16c}$$

The presence of output noise or measurement errors demands that the estimates of the parameters be filtered, for example using a moving average, which can be achieved in a pre- or postprocessing step.

14.4.4 Summary of Proposed Methodology

The parameter estimation methodology we propose here consists of three main steps. First, the system is transformed into a parameter-free system as described in section 14.4.1. This requires only the knowledge of structural information such as the stoichiometry and the type of reaction kinetics. The second step is the design of an observer for the extended system in observability canonical form (see section 14.4.2). The third and final step is the back-transformation of the observer states into the state space of the extended system and is discussed in section 14.4.3. The parameters can be recovered in a straightforward manner. Contrary to other approaches, the proposed methodology guarantees the uniqueness of the parameter estimation. Its drawbacks are the necessity of transforming the extended system into observability canonical form, the need of continuous-time measurements or approximations thereof, and the possibility of noise sensitivity. The latter two points can be dealt with using appropriate measurement filtering. The first point, however, limits the applicability of the methodology to systems of not too large a dimension. Its main advantage is that it explicitly takes into account the structural information.

14.5 Application to the MAPK Cascade

The well-known mitogen-activated protein kinase (MAPK) cascade (Kholodenko, 2000) is an important signaling pathway connecting extracellular signals to the regulation of transcription. The proposed methodology is applied to a pathway model consisting of 8 states and 10 reactions, with a total of 21 parameters (see figure 14.4).

The reactions describe phosphorylation and dephosphorylation of the proteins Raf, MEK, and ERK. The latter two can be phosphorylated twice. The system acts as a three-step signal amplifier (Goldbeter and Koshland, 1981) and noise filter (Rai

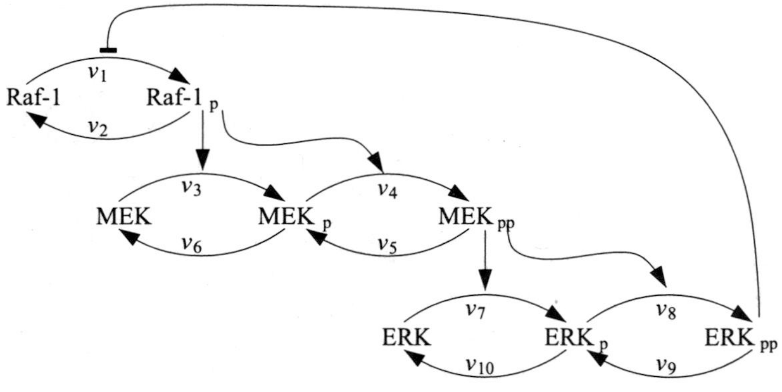

Figure 14.4
Reaction scheme of the mitogen-activated protein kinase (MAPK) cascade.

et al., 2008). The output ERK_{pp} inhibits the input flow v_1. This leads to oscillatory behavior. The substrate concentrations are

$$\boldsymbol{c} = [\text{Raf-1} \quad \text{Raf-1}_\text{p} \quad \text{MEK} \quad \text{MEK}_\text{p} \quad \text{MEK}_{\text{pp}} \quad \text{ERK} \quad \text{ERK}_\text{p} \quad \text{ERK}_{\text{pp}}]^T,$$

and the system dynamics can be described by

$$\frac{\mathrm{d}\boldsymbol{c}}{\mathrm{d}t} = \boldsymbol{N}\boldsymbol{v},$$

where

$$\boldsymbol{N} = \begin{bmatrix} -1 & 1 & 0 & 0 & 0 & 0 & 0 & 0 & 0 & 0 \\ 1 & -1 & 0 & 0 & 0 & 0 & 0 & 0 & 0 & 0 \\ 0 & 0 & -1 & 0 & 0 & 1 & 0 & 0 & 0 & 0 \\ 0 & 0 & 1 & -1 & 1 & -1 & 0 & 0 & 0 & 0 \\ 0 & 0 & 0 & 1 & -1 & 0 & 0 & 0 & 0 & 0 \\ 0 & 0 & 0 & 0 & 0 & 0 & -1 & 0 & 0 & 1 \\ 0 & 0 & 0 & 0 & 0 & 0 & 1 & -1 & 1 & -1 \\ 0 & 0 & 0 & 0 & 0 & 0 & 0 & 1 & -1 & 0 \end{bmatrix},$$

and the reaction kinetics are of the following form:

$$v_1 = \frac{k_1 K_I^{\nu_I}}{K_I^{\nu_I} + \text{ERK}_{\text{pp}}^{\nu_I}} \frac{\text{Raf-1}}{K_1 + \text{Raf-1}}, \qquad v_6 = k_6 \frac{\text{MEK}_\text{p}}{K_6 + \text{MEK}_\text{p}},$$

$$v_2 = k_2 \frac{\text{Raf-1}_\text{p}}{K_2 + \text{Raf-1}_\text{p}}, \qquad v_7 = k_7 \text{MEK}_{\text{pp}} \frac{\text{ERK}}{K_7 + \text{ERK}},$$

$$v_3 = k_3 \text{Raf-1}_\text{p} \frac{\text{MEK}}{K_3 + \text{MEK}}, \qquad v_8 = k_8 \text{MEK}_{\text{pp}} \frac{\text{ERK}_\text{p}}{K_8 + \text{ERK}_\text{p}},$$

$$v_4 = k_4 \text{Raf-1}_\text{p} \frac{\text{MEK}_\text{p}}{K_4 + \text{MEK}_\text{p}}, \qquad v_9 = k_9 \frac{\text{ERK}_{\text{pp}}}{K_9 + \text{ERK}_{\text{pp}}},$$

$$v_5 = k_5 \frac{\text{MEK}_{\text{pp}}}{K_5 + \text{MEK}_{\text{pp}}}, \qquad v_{10} = k_{10} \frac{\text{ERK}_\text{p}}{K_{10} + \text{ERK}_\text{p}}.$$

The extended-state vector consists of the concentrations $\boldsymbol{c}$, the flows $\boldsymbol{v}$ and the denominators of the flows

$$\boldsymbol{m} = [m_{11} \quad m_{12} \quad m_2 \quad m_3 \quad \cdots \quad m_{10}]^T,$$

with

Table 14.1
Parameters of the MAPK model (Kholodenko, 2000)

Parameter	Value	Parameter	Value
v_I	2.0	K_I	18
k_1	2.5	K_1	50
k_2	0.25	K_2	40
k_3	0.025	K_3	100
k_4	0.025	K_4	100
k_5	0.75	K_5	100
k_6	0.75	K_6	100
k_7	0.025	K_7	100
k_8	0.025	K_8	100
k_9	1.25	K_9	100
k_{10}	1.25	K_{10}	100

$$m_{11} = K_I^{v_I} + \mathrm{ERK}_{\mathrm{pp}}^{v_I}, \qquad m_6 = K_6 + \mathrm{MEK_p},$$

$$m_{12} = K_1 + \text{Raf-1}, \qquad m_7 = K_7 + \mathrm{ERK},$$

$$m_2 = K_2 + \text{Raf-1}_\mathrm{p}, \qquad m_8 = K_8 + \mathrm{ERK_p},$$

$$m_3 = K_3 + \mathrm{MEK}, \qquad m_9 = K_9 + \mathrm{ERK_{pp}},$$

$$m_4 = K_4 + \mathrm{MEK_p}, \qquad m_{10} = K_{10} + \mathrm{ERK_p},$$

$$m_5 = K_5 + \mathrm{MEK_{pp}}.$$

The parameters values are the same as in Kholodenko (2000), and are listed in table 14.1.

The parameters can be grouped into two vectors $\boldsymbol{k}$ and $\boldsymbol{K}$:

$$\boldsymbol{k} = \begin{bmatrix} k_1 K_I^{v_I} \\ k_2 \\ \vdots \\ k_{10} \end{bmatrix} \quad \text{and} \quad \boldsymbol{K} = \begin{bmatrix} K_I \\ K_1 \\ \vdots \\ K_{10} \end{bmatrix}.$$

The product $k_1 K_I^{v_I}$ is treated as a single parameter: because $K_I^{v_I}$ is estimated separately, k_1 can also be obtained. The exponent v_I is assumed to be known.

By construction, the time derivatives of $\boldsymbol{c}$ and $\boldsymbol{m}$ do not depend on the parameters $\boldsymbol{K}$ and $\boldsymbol{k}$. Simple calculations show that this also holds for the time derivatives of $\boldsymbol{v}$. For example,

$$\frac{\mathrm{d}v_1}{\mathrm{d}t} = -\frac{k_1 K_I^{v_I}}{(K_I^{v_I} + \mathrm{ERK}_{\mathrm{pp}}^{v_I})^2} v_I \mathrm{ERK}_{\mathrm{pp}}^{v_I - 1} \frac{\mathrm{dERK}_{\mathrm{pp}}}{\mathrm{d}t} \frac{\text{Raf-1}}{K_1 + \text{Raf-1}}$$

$$+ \frac{k_1 K_I^{v_I}}{K_I^{v_I} + \mathrm{ERK}_{\mathrm{pp}}^{v_I}} \frac{\frac{\mathrm{dRaf\text{-}1}}{\mathrm{d}t}(K_1 + \text{Raf-1}) - \text{Raf-1}\frac{\mathrm{dRaf\text{-}1}}{\mathrm{d}t}}{(K_1 + \text{Raf-1})^2}$$

$$= -v_1 \frac{v_I c_8^{v_I - 1} \frac{\mathrm{d}c_8}{\mathrm{d}t}}{m_{11}} + v_1 \frac{\frac{\mathrm{d}c_1}{\mathrm{d}t}(m_{12} - c_1)}{m_{12} c_1}$$

$$= -v_1 \frac{v_I c_8^{v_I - 1} e_8^T N v}{m_{11}} + v_1 \frac{e_1^T N v (m_{12} - c_1)}{m_{12} c_1}.$$

The extended state vector is given by

$$\boldsymbol{x} = \begin{bmatrix} \boldsymbol{c} \\ \boldsymbol{m} \\ \boldsymbol{v} \end{bmatrix}.$$

Using all concentrations $\boldsymbol{c}$ and flows $\boldsymbol{v}$ as outputs, one can construct the following observability map:

$$\boldsymbol{\Phi}(\boldsymbol{x}) = \begin{bmatrix} \boldsymbol{c} \\ \frac{\mathrm{d}\boldsymbol{c}}{\mathrm{d}t} \\ \boldsymbol{v} \\ \frac{\mathrm{d}\boldsymbol{v}}{\mathrm{d}t} \\ \frac{\mathrm{d}^2 \boldsymbol{v}_1}{\mathrm{d}t^2} \end{bmatrix}. \tag{14.17}$$

This observability mapping in not invertible at all points. At points of noninvertibility the nonlinearity $\varphi(\cdot)$ blows up, and so in the observer construction the mapping's values are truncated to

$$\hat{\varphi}(\xi) = \begin{cases} -\delta, & \text{if } \varphi(\xi) \le -\delta, \\ \varphi(\xi), & \text{if } -\delta < \varphi(\xi) < \delta, \\ \delta, & \text{if } \varphi(\xi) \ge \delta, \end{cases}$$

where $\delta = 30$.

Following Vargas and Moreno (2000), any time point for which Φ is nearly noninvertible is an *event*. Specifically, these are times when the observability map is ill conditioned in the following sense:

$$T_{\mathrm{Event}} = \left\{ t : \frac{\sigma_{\min}}{\sigma_{\max}} < 10^{-6} \right\},$$

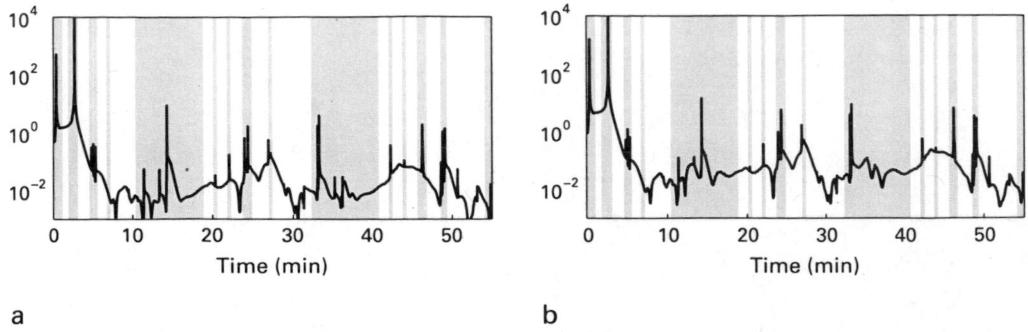

Figure 14.5
(a) Error in observer coordinates. (b) Relative parameter estimation error. The gray markings denote the events during which the extended system is not identifiable or only poorly so.

where $\sigma_{\min}$ and $\sigma_{\max}$ are the smallest and largest singular values of $\partial\Phi/\partial \boldsymbol{x}$. Events label times at which the procedure is not performing well, and so the estimates provided by (14.16) are likely not valid.

Applying the estimation methodology to simulated output from the MAPK model provides a test of its performance since the "true" values of the parameters are known. Figure 14.5 indicates the overall agreement of the parameter estimates. For this illustrative application the system outputs are presumed to be measured continuously.

A close analysis of the estimated parameters and, in particular their fluctuations, reveals that the parameters of $\boldsymbol{v}_1$, that is, k_1, K_1, and K_I are not well estimated. This is particularly visible whenever $\mathrm{ERK_{pp}}$ (c_8) is large. The inhibition term of $\boldsymbol{v}_1$

$$v_I(\mathrm{ERK_{pp}}) = -\frac{k_1 K_I^{\nu_I}}{K_I^{\nu_I} + \mathrm{ERK}_{\mathrm{pp}}^{\nu_I}} \tag{14.18}$$

is shown in figure 14.6. For $\mathrm{ERK_{pp}}$ much larger than K_I, its relative sensitivity

$$S_{v_I}^{K_I}(\mathrm{ERK_{pp}}) = \frac{K_I}{\mathrm{ERK_{pp}}} \frac{\partial v_I(\mathrm{ERK_{pp}}, K_I)}{\partial K_I} \tag{14.19}$$

approaches zero. Whenever $\mathrm{ERK_{pp}}$ is large, the flow v_1 and its sensitivity to K_I are significantly reduced, thus the effect of the parameters K_I is negligible and the parameters can hardly be identified, again as shown in figure 14.6.

To circumvent this identifiability problem, an iterative approach is taken. The three parameters of $\boldsymbol{v}_1$ are estimated when c_8 is low, for example, at $t = 26.5$ min ($V_1 = 150.01$, $K_1 = 50.01$, $K_I = 17.95$), and the differential equations corresponding

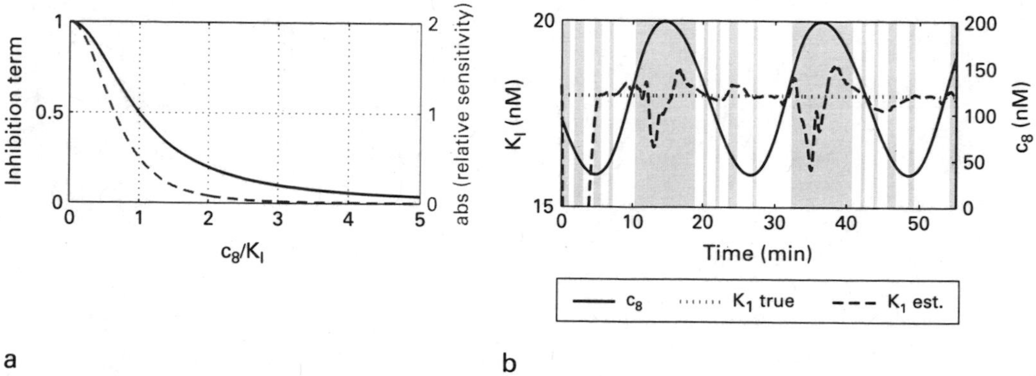

Figure 14.6
Poor identifiability of the parameters of v_1 for high concentrations of ERK_{pp} (c_8). (a) Inhibition term (*black, solid line*) and relative sensitivity (*gray dashed line*) as a function of the inhibitor concentration c_8. (b) Time course of ERK_{pp} (c_8, *solid line*), estimate of K_I (*dashed line*), and true value (*dotted line*). For large values of ERK_{pp}, the estimation deteriorates and the system is poorly observable (*gray markings*).

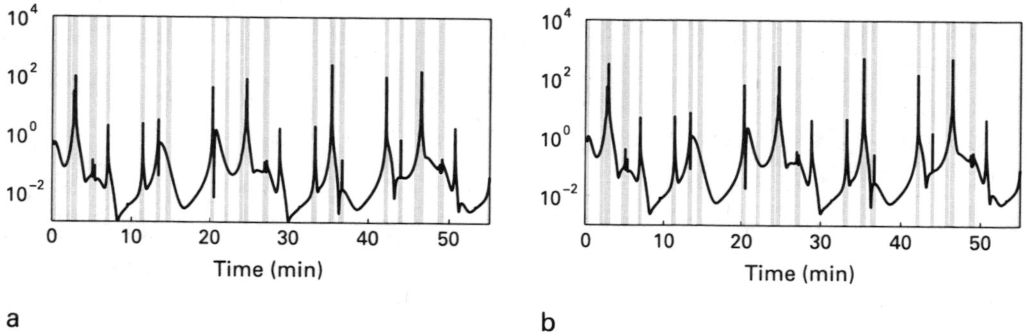

Figure 14.7
Error of the reduced estimation problem, in observer coordinates (a) and in parameter estimation (b).

to the extended states are replaced with algebraic equations, as described in remark 14.1. Then the reduced extended-state vector (without v_1 as output and without v_1, m_{11}, and m_{12} as states) is estimated.

Comparing the results with the estimate of the whole parameter vector reveals that the identifiability problem associated with v_1 not only causes large fluctuations in the estimation, but also increases the domains of nonobservability (gray marking; compare figure 14.7 with figure 14.5). The reduced estimation slightly degrades the estimation error, as shown for k_2 and K_7 (figure 14.8). This is in particular the case for the parameters in reactions close to v_1, while more downstream reaction parameters are well estimated (figure 14.8).

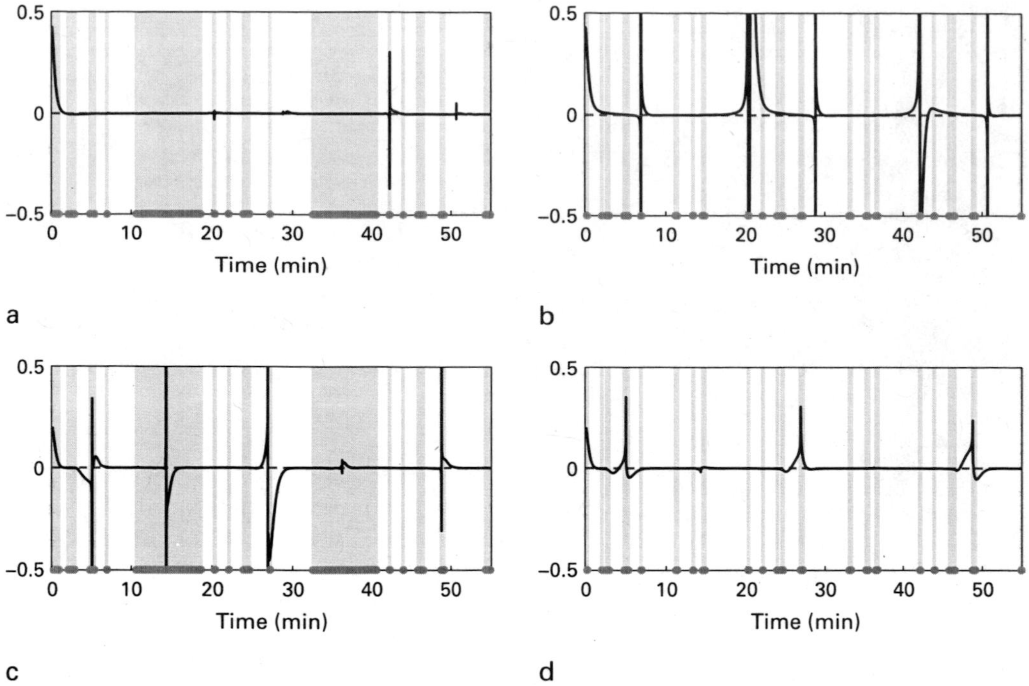

Figure 14.8
Relative estimation error for parameter k_2 (*top panels*) and K_7 (*bottom panels*), for full extended (*left*) and reduced (*right*) state. Gray marking indicate where the trajectory is poorly observable.

Our result clearly demonstrates the applicability of the proposed parameter estimation method. The parameters can be estimated with relatively small errors. The application highlights that practical identifiability is often time dependent. Here three parameters are identified in a first round, before all others are treated in a second simulation.

14.6 Conclusions

Mathematical models are necessary for a quantitative and dynamic understanding of biological systems. These models usually depend on many parameters that are unknown or difficult to measure. Major drawbacks of existing parameter estimation algorithms are that they do not explicitly use structural information, which is often available, and that they cannot guarantee convergence. The methodology presented here is tailored for biochemical reaction networks and allows the model structure to be taken into account directly. It is applicable to reaction networks where flows are described by polynomial or rational terms and concentrations are nonzero. A major advantage of the proposed parameter estimation approach is that it performs a global parameter search and ensures a guaranteed convergence. An important limitation is that it relies on estimates of derivatives of the output, which requires a large number of outputs or a high order of output derivatives. It is also limited with regard to time-varying practical identifiability, in which case, an interval-based approach may be preferable.

References

Akaike, H. (1974). A new look at the statistical model identification. *IEEE Transactions on Automatic Control* 19:716–723.

Alberts, B., Bray, D., Lewis, J., Raff, M., Roberts, K., and Watson, J. (1989). *Molecular Biology of the Cell.* 2nd ed. New York: Garland.

Alessi, D. R., Saito, Y., Campbell, D. G., Cohen, P., Sithanandam, G., Rapp, U., Ashworth, A., Marshall, C., and Cowley, S. (1994). Identification of the sites in MAP kinase kinase-1 phosphorylated by p74raf-1. *EMBO Journal* 13:1610–1619.

Alon, U. (2007). *An Introduction to Systems Biology: Design Principles of Biological Circuits.* Boca Raton, FL: Chapman & Hall/CRC.

Alon, U., Surette, M. G., Barkai, N., and Leibler, S. (1999). Robustness in bacterial chemotaxis. *Nature* 397:168–171.

Altschuler, S. J., Angenent, S. B., Wang, Y., and Wu, L. F. (2008). On the spontaneous emergence of cell polarity. *Nature* 454:886–889.

Alves, R., Antunes, F., and Salvador, A. (2006). Tools for kinetic modeling of biochemical networks. *Nature Biotechnology* 24:667–672.

Alves, R., and Savageau, M. A. (2000). Extending the method of mathematically controlled comparison to include numerical comparisons. *Bioinformatics* 16:786–798.

Anderson, N. G., Maller, J. L., Tonks, N. K., and Sturgill, T. (1990). Requirement for integration of signals from two distinct phosphorylation pathways for activation of MAP kinase. *Nature* 343:651–653.

Andrews, B. W., Yi, T. M., and Iglesias, P. A. (2006). Optimal noise filtering in the chemotactic response of *Escherichia coli. PLoS Computational Biology* 2:e154.

Andrews, S. S., and Bray, D. (2004). Stochastic simulation of chemical reactions with spatial resolution and single molecule detail. *Physical Biology* 1:137–151.

Andrianantoandro, E., Basu, S., Karig, D. K., and Weiss, R. (2006). Synthetic biology: New engineering rules for an emerging discipline. *Molecular Systems Biology* 2:1–14.

Angeli, D. (2006). Systems with counterclockwise input-output dynamics. *IEEE Transactions on Automatic Control* 51:1130–1143.

Angeli, D. (2007). Multistability in systems with counterclockwise input-output dynamics. *IEEE Transactions on Automatic Control* 52:596–609.

Angeli, D., de Leenheer, P., and Sontag, E. D. (2007). A Petri net approach to the study of persistence in chemical reaction networks. *Mathematical Biosciences* 210:598–618.

Angeli, D., and Sontag, E. D. (2003). Monotone control systems. *IEEE Transactions on Automatic Control* 48 (10): 1684–1698.

Angeli, D., and Sontag, E. D. (2004). Multi-stability in monotone input/output systems. *Systems & Control Letters* 51 (3–4): 185–202.

Arber, W., and Linn, S. (1969). DNA modification and restriction. *Annual Review of Biochemistry* 38:467–500.

Arcak, M., and Sontag, E. D. (2006). Diagonal stability for a class of cyclic systems and applications. *Automatica* 42:1531–1537.

Arcak, M., and Sontag, E. D. (2008). A passivity-based stability criterion for a class of interconnected systems and applications to biochemical reaction networks. *Mathematical Biosciences and Engineering* 5:1–19.

Arkin, A., Ross, J., and McAdams, H. H. (1998). Stochastic kinetic analysis of developmental pathway bifurcation in phage λ-infected *Escherichia coli* cells. *Genetics* 149:1633–1648.

Arkowitz, R. A., and Iglesias, P. A. (2008). Basic principles of polarity establishment and maintenance. *EMBO Reports* 9 (9): 847–852.

Asthagiri, A. R., and Lauffenburger, D. A. (2001). A computational study of feedback effects on signal dynamics in a mitogen-activated protein kinase (MAPK) pathway model. *Biotechnology Progress* 17:227–239.

Atkins, P. W., and de Paula, J. (2002). *Atkins Physical Chemistry*. 7th ed. Oxford: Oxford University Press.

Atkinson, M. R., Savageau, M. A., Meyers, J. T., and Ninfa, A. J. (2003). Development of genetic circuitry exhibiting toggle switch or oscillatory behavior in *Escherichia coli. Cell* 113:597–607.

Audoly, S., Bellu, G., D'Angiò, L., Saccomani, M. P., and Cobelli, C. (2001). Global identifiability of nonlinear models of biological systems. *IEEE Transactions on Biomedical Engineering* 48 (1): 55–65.

Avruch, J. (2007). MAP kinase pathways: The first twenty years. *Biochimica et Biophysica Acta* 1773 (8): 1150–1160.

Axelrod, D., Koppel, D. E., Schlessinger, J., Elson, E., and Webb, W. W. (1976). Mobility measurement by analysis of fluorescence photobleaching recovery kinetics. *Biophysical Journal* 16:1055–1069.

Bagheri, N., Stelling, J., and Doyle, F. J. (2007). Quantitative performance metrics for robustness in circadian rhythms. *Bioinformatics* 23:358–364.

Baker, D., Church, G., Collins, J., Endy, D., Jacobson, J., Keasling, J., Modrich, P., Smolke, C., and Weiss, R. (2006). Engineering life: Building a FAB for biology. *Scientific American* 294 (6): 44–51.

Balas, G. J., Chiang, R., Packard, A., and Safonov, M. (2008). *Robust Control Toolbox 3: User's Guide*. Natick, MA: MathWorks.

Balas, G. J., Doyle, J. C., Glover, K., Packard, A., and Smith, R. (2001). *μ-Analysis and Synthesis Toolbox*. Natick, MA: MathWorks.

Bardwell, L. (2005). A walk-through of the yeast mating pheromone response pathway. *Peptides* 26 (2): 339–350.

Barkai, N., and Leibler, S. (1997). Robustness in simple biochemical networks. *Nature* 387:913–917.

Barrio, R. A., Varea, C., Aragón, J. L., and Maini, P. K. (1999). A two-dimensional numerical study of spatial pattern formation in interacting Turing systems. *Bulletin of Mathematical Biology* 61:483–505.

Bayliss, L. E. (1966). *Living Control Systems*. San Francisco: Freeman.

Becker-Weimann, S., Wolf, J., Herzel, H., and Kramer, A. (2004). Modeling feedback loops of the mammalian circadian oscillator. *Biophysical Journal* 87:3023–3034.

Becskei, A., and Serrano, L. (2000). Engineering stability in gene networks by autoregulation. *Nature* 405:590–593.

Berg, H. C. (1993). *Random Walks in Biology*. Expanded ed. Princeton, NJ: Princeton University Press.

Bergmann, S., Sandler, O., Sberro, H., Shnider, S., Schejter, E., Shilo, B. Z., and Barkai, N. (2007). Pre-steady-state decoding of the bicoid morphogen gradient. *PLoS Biology* 5:e46.

Bhalla, U. S., and Iyengar, R. (1999). Emergent properties of networks of biological signaling pathways. *Science* 283:381–387.

Bhalla, U. S., Ram, T., and Iyengar, R. (2002). MAP kinase phosphatase as a locus of flexibility in a mitogen activated protein kinase signalling network. *Science* 297:1018–1023.

Blüthgen, N., Bruggemann, F. J., Legewie, S., Herzel, H., Westerhoff, H. V., and Kholodenko, B. N. (2006). Effects of sequestration on signal transduction cascades. *FEBS Journal* 273 (5): 895–906.

Blüthgen, N., and Herzel, H. (2003). How robust are switches in intracellular signaling cascades? *Journal of Theoretical Biology* 225 (3): 293–300.

Bode, H. W. (1945). *Network Analysis and Feedback Amplifier Design.* New York: Van Nostrand.

Boveri, T. (1901). Die Polaritöt von Ovocyte, Ei, und Larve des *Strongylocentrus lividus. Zoologische Jahrbücher* 14:630–653.

Braatz, R. P., Young, P. M., Doyle, J. C., and Morari, M. (1994). Computational complexity of μ calculation. *IEEE Transactions on Automatic Control* 39 (5): 1000–1002.

Bray, D. (2000). *Cell Movements: From Molecules to Motility.* 2nd ed. New York: Garland.

Breiman, L. (1992). *Probability.* Vol. 7 of *Classics in Applied Mathematics.* Philadelphia: SIAM.

Brown, E., Moehlis, J., and Holmes, P. (2004). On the phase reduction and response dynamics of neural oscillator populations. *Neural Computation* 16:673–715.

Bullinger, E., Fey, D., Farina, M., and Findeisen, R. (2008). Identifikation biochemischer Reaktionsnetzwerke: Ein beobachterbasierter Ansatz. *AT-Automatisierungstechnik* 56 (5): 269–279.

Burnham, K. P., and Anderson, D. R. (2004). Multimodel inference: Understanding AIC and BIC in model selection. *Sociological Methods & Research* 33 (2): 261–304.

Butler, G., and Waltman, P. (1986). Persistence in dynamical systems. *Journal of Differential Equations* 63:255–263.

Cabeen, M. T., and Jacobs-Wagner, C. (2007). Skin and bones: The bacterial cytoskeleton, cell wall, and cell morphogenesis. *Journal of Cell Biology* 179:381–387.

Cagampang, F. R., Sheward, W. J., Harmar, A. J., Piggins, H. D., and Coen, C. W. (1998). Circadian changes in the expression of vasoactive intestinal peptide-2 receptor mRNA in the rat suprachiasmatic nuclei. *Molecular Brain Research* 54:108–112.

Cannon, W. B. (1932). *The Wisdom of the Body.* New York: Norton.

Cascante, M., Boros, L. G., Comin-Anduix, B., de Atauri, P., Centelles, J. J., and Lee, P. W.-N. (2002). Metabolic control analysis in drug discovery and design. *Nature Biotechnology* 20:243–249.

Cascante, M., Franco, R., and Canela, E. I. (1989a). Use of implicit methods from general sensitivity theory to develop a systematic approach to metabolic control: 1. Unbranched pathways. *Mathematical Biosciences* 94:271–288.

Cascante, M., Franco, R., and Canela, E. I. (1989b). Use of implicit methods from general sensitivity theory to develop a systematic approach to metabolic control: 2. Complex systems. *Mathematical Biosciences* 94:289–309.

Caudron, M., Bunt, G., Bastiaens, P., and Karsenti, E. (2005). Spatial coordination of spindle assembly by chromosome-mediated signaling gradients. *Science* 309:1373–1376.

Cedersund, G. (2004). System identification of nonlinear thermochemical systems with dynamical instabilities. Ph.D. diss., Linköping University, Linköping, Sweden.

Chen, B. S., Wang, Y. C., Wu, W. S., and Li, W. H. (2005). A new measure of the robustness of biochemical networks. *Bioinformatics* 21 (11): 2698–2705.

Chen, C.-T. (1999). *Linear System Theory and Design.* 3rd ed. New York: Oxford University Press.

Chen, T., and Francis, B. (1996). *Optimal Sampled-Data Control Systems.* Communications and Control Engineering Series. London: Springer.

Coates, J. C., and Harwood, A. J. (2001). Cell-cell adhesion and signal transduction during *Dictyostelium* development. *Journal of Cell Science* 114 (24): 4349–4358.

CompBio Group, Institute for Systems Biology (2006). Dizzy. Version 1.11.4 (software package). http://magnet.systemsbiology.net/software/Dizzy.

Conrad, E., Mayo, A. E., Ninfa, A. J., and Forger, D. B. (2008). Rate constants rather than biochemical mechanism determine behaviour of genetic clocks. *Journal of the Royal Society Interface* 5:S9–S15.

Cornish-Bowden, A. (2004). *Fundamentals of Enzyme Kinetics.* 3rd ed. London: Portland Press.

Cornish-Bowden, A., and Cárdenas, M. L., eds. (1999). *Technological and Medical Implications of Metabolic Control Analysis.* Dordrecht: Kluwer Academic.

Costenoble, R., Müller, D., Barl, T., van Gulik, W. M., van Winden, W. A., Reuss, M., and Heijnen, J. J. (2007). ^{13}C-labeled metabolic flux analysis of a fed-batch culture of elutriated *Saccharomyces cerevisiae. FEMS Yeast Research* 7 (4): 511–526.

Cox, R. S., III, Surette, M. G., and Elowitz, M. B. (2007). Programming gene expression with combinatorial promoters. *Molecular Systems Biology* 3:145.

Craciun, G., and Feinberg, M. (2005). Multiple equilibria in complex chemical reaction networks: 1. The injectivity property. *SIAM Journal of Applied Mathematics* 65:1526–1546.

Craciun, G., and Feinberg, M. (2006). Multiple equilibria in complex chemical reaction networks: 2. The species-reactions graph. *SIAM Journal of Applied Mathematics* 66:1321–1338.

Cross, F. R. (2003). Two redundant oscillatory mechanisms in the yeast cell cycle. *Developmental Cell* 4 (5): 741–752.

Davis, L. (1991). *Handbook of Genetic Algorithms*. VNR Computer Library. New York: Van Nostrand Reinhold.

Day, S. J., and Lawrence, P. A. (2000). Measuring dimensions: The regulation of size and shape. *Development* 127:2977–2987.

Dean, J. P., and Dervakos, G. A. (1998). Redesigning metabolic networks using mathematical programming. *Biotechnology and Bioengineering* 58:267–271.

de Atauri, P., Orrell, D., Ramsey, S., and Bolouri, H. (2004). Evolution of "design" principles in biochemical networks. *IEE Systems Biology* 1 (1): 28–40.

Del Vecchio, D. (2007). Design of an activator-repressor clock in *E. coli.* In *Proceedings of the American Control Conference*, 1589–1594. New York: IEEE.

Del Vecchio, D., Ninfa, A. J., and Sontag, E. D. (2008a). A systems theory with retroactivity: Application to transcriptional modules. In *Proceedings of the American Control Conference*, 1368–1373. Seattle: IEEE.

Del Vecchio, D., Ninfa, A. J., and Sontag, E. D. (2008b). Modular cell biology: Retroactivity and insulation. *Molecular Systems Biology* 4:161.

Desharnais, J., Gupta, V., Jagadeesan, R., and Panangaden, P. (2004). Metrics for labelled Markov processes. *Journal of Theoretical Computer Science* 318:323–354.

Doyle, J. C., Francis, B. A., and Tannenbaum, A. R. (1992). *Feedback Control Theory*. New York: Macmillan.

Doyle, J. C., and Stein, G. (1979). Robustness and observers. *IEEE Transactions on Automatic Control* 24 (4): 607–611.

Doyle, F. J., III, and Stelling, J. (2005). Robust performance in biophysical networks. In *Proceedings of the IFAC World Congress* 16:31–36. Prague: IFAC.

Driever, W., and Nüsslein-Volhard, C. (1988). The bicoid protein determines position in the *Drosophila* embryo in a concentration-dependent manner. *Cell* 54:95–104.

Dudley, R. M. (2002). *Real Analysis and Probability*. Cambridge: Cambridge University Press.

Einstein, A. (1905). Über die von der molekularkinetischen Theorie der Wärme geforderte Bewegung von in ruhenden Flüssig kieten suspendierten Teilchen. *Annalen der Physik* 17:549–560.

Eissing, T., Allgöwer, F., and Bullinger, E. (2005). Robustness properties of apoptosis models with respect to parameter variations and intrinsic noise. *IEE Proceedings: Systems Biology* 152 (4): 221–228.

Elf, J., and Ehrenberg, M. (2003). Fast evaluation of fluctuations in biochemical networks with the linear noise approximation. *Genome Research* 13:2475–2484.

Elf, J., and Ehrenberg, M. (2004). Spontaneous separation of bi-stable biochemical systems into spatial domains of opposite phases. *Systems Biology* 1:230–236.

Elowitz, M., and Leibler, S. (2000). A synthetic oscillatory network of transcriptional regulators. *Nature* 403:335–338.

Elowitz, M., Levine, A., Siggia, E., and Swain, P. (2002). Stochastic gene expression in a single cell. *Nature* 297:1183–1186.

El-Samad, H., Del Vecchio, D., and Khammash, M. (2005). Repressilators and promotilators: Loop dynamics in synthetic gene networks. In *Proceedings of the American Control Conference*, 4405–4410. Portland, OR: IEEE.

El-Samad, H., and Khammash, M. (2006a). Coherence resonance: A mechanism for noise induced stable oscillations in gene regulatory networks. In *Conference on Decision and Control*, 2382–2387. San Diego, CA: IEEE.

El-Samad, H., and Khammash, M. (2006b). Regulated degradation is a mechanism for suppressing stochastic fluctuations in gene regulatory networks. *Biophysical Journal* 90:3749–3761.

El-Samad, H., Kurata, H., Doyle, J. C., Gross, C. A., and Khammash, M. (2005). Surviving heat shock: Control strategies for robustness and performance. *Proceedings of the National Academy of Sciences USA* 102:2736–2741.

Enciso, G. A., and Sontag, E. D. (2005). Monotone systems under positive feedback: Multistability and a reduction theorem. *Systems & Control Letters* 54 (2): 159–168.

Enciso, G. A., and Sontag, E. D. (2006). Global attractivity, I/O monotone small-gain theorems, and biological delay systems. *Discrete and Continuous Dynamical Systems* 14 (3): 549–578.

Enciso, G. A., and Sontag, E. D. (2008). Monotone bifurcation graphs. *Journal of Biological Dynamics* 2 (2): 121–139.

Endy, D. (2005). Foundations for engineering biology. *Nature* 438:449–453.

Ephrussi, A., and St Johnston, D. (2004). Seeing is believing: The bicoid morphogen gradient matures. *Cell* 116:143–152.

Ethier, S. N., and Kurtz, T. G. (1986). *Markov Processes: Characterization and Convergence*. Wiley Series in Probability and Mathematical Statistics. New York: Wiley.

Farina, M., Bullinger, E., Findeisen, R., and Bittanti, S. (2007). An observer-based strategy for parameter identification in systems biology. In *Proceedings of the Foundations of Systems Biology in Engineering*, 521–526. Stuttgart: FOSBE.

Farina, M., Findeisen, R., Bullinger, E., Bittanti, S., Allgöwer, F., and Wellstead, P. (2006). Results towards identifiability properties of biochemical reaction networks. In *Conference on Decision and Control*, 2104–2109. San Diego, CA: IEEE.

Feinberg, M. (1987). Chemical reaction network structure and the stability of complex isothermal reactors: 1. The deficiency zero and deficiency one theorems. *Chemical Engineering Science* 42:2229–2268.

Feinberg, M., and Horn, F. J. M. (1974). Dynamics of open chemical systems and algebraic structure of underlying reaction network. *Chemical Engineering Science* 29:775–787.

Fell, D. A. (1992). Metabolic control analysis: A survey of its theoretical and experimental development. *Biochemical Journal* 286:313–330.

Fell, D. A. (1997). *Understanding the Control of Metabolism*. London: Portland Press.

Fell, D. A., and Sauro, H. M. (1985). Metabolic control and its analysis: Additional relationships between elasticities and control coefficients. *European Journal of Biochemistry* 148:555–561.

Feng, X.-J., Hooshangi, S., Chen, D., Li, G., Weiss, R., and Rabitz, H. (2004). Optimizing genetic circuits by global sensitivity analysis. *Biophysical Journal* 87 (4): 2195–2202.

Ferrell, J. E., Jr. (1996). Tripping the switch fantastic: How a protein kinase cascade can convert graded inputs into switch-like outputs. *Trends in Biochemical Sciences* 21 (12): 460–466.

Ferrell, J. E., Jr., and Machleder, E. M. (1998). The biochemical basis of an all-or-none cell fate switch in *Xenopus* oocytes. *Science* 280:895–898.

Fey, D., Findeisen, R., and Bullinger, E. (2008). Parameter estimation in kinetic reaction models using nonlinear observers facilitated by model extensions. In *Proceedings of the IFAC World Congress*, 313–318. Seoul: IFAC.

Flach, E., and Schnell, S. (2006). Use and abuse of the quasi-steady-state approximation. *IEE Proceedings: Systems Biology* 153 (4): 187–191.

Forger, D. B., and Peskin, C. S. (2003). A detailed predictive model of the mammalian circadian clock. *Proceedings of the National Academy of Sciences USA* 100 (25): 14806–14811.

Frank, P. M. (1978). *Introduction to System Sensitivity Theory*. New York: Academic Press.

Franklin, G. F., Powell, J. D., and Emami-Naeini, A. (2006). *Feedback Control of Dynamic Systems*. 5th ed. Upper Saddle River, NJ: Prentice Hall.

Frey, S., Millat, T., Hohmann, S., and Wolkenhauer, O. (2008). How quantitative measures unravel design principles in multi-stage phosphorylation cascades. *Journal of Theoretical Biology* 254:27–36.

Funahashi, A., Morohashi, M., and Kitano, H. (2003). CellDesigner: A process diagram editor for gene-regulatory and biochemical networks. *BIOSILICO* 1 (5): 159–162.

Gadkar, K. G., Gunawan, R., and Doyle, F. J., III (2005). Iterative approach to model identification of biological networks. *BMC Bioinformatics* 6:155.

Gard, T. C. (1980). Persistence in food webs with general interactions. *Mathematical Biosciences* 51:165–174.

Gardner, T. S., Cantor, C. R., and Collins, J. J. (2000). Construction of a genetic toggle switch in *Escherichia coli. Nature* 403:339–342.

Garnier, H., Mensler, M., and Richard, A. (2003). Continuous-time model identification from sampled data: Implementation issues and performance evaluation. *International Journal of Control* 76 (13): 1337–1357.

Geier, F., Becker-Weimann, S., Kramer, A., and Herzel, H. (2005). Entrainment in a model of the mammalian circadian oscillator. *Journal of Biological Rhythms* 20:83–93.

Gibbs, A. L., and Su, F. E. (2002). On choosing and bounding probability metrics. *International Statistical Review* 70 (3): 419–435.

Gierer, A., and Meinhardt, H. (1972). A theory of biological pattern formation. *Kybernetik* 12:30–39.

Giersch, C. (1988a). Control analysis of metabolic networks: 1. Homogeneous functions and the summation theorems for control coefficients. *European Journal of Biochemistry* 174:509–513.

Giersch, C. (1988b). Control analysis of metabolic networks: 2. Total differentials and general formulation of the connectivity relations. *European Journal of Biochemistry* 174:515–519.

Gilles, E. D. (2002). Control: Key to better understanding biological systems. *At: Automatisierungstechnik* 50:7–17.

Gillespie, D. T. (1976). A general method for numerically simulating the stochastic time evolution of coupled chemical reactions. *Journal of Computational Physics* 22 (4): 403–434.

Gillespie, D. T. (1977). Exact stochastic simulation of coupled chemical reactions. *Journal of Physical Chemistry* 81 (25): 2340–2361.

Gillespie, D. T. (2000). The chemical Langevin equation. *Journal of Chemical Physics* 113 (1): 297–306.

Gillespie, D. T. (2001). Approximate accelerated stochastic simulation of chemically reacting systems. *Journal of Chemical Physics* 115 (4): 1716–1733.

Gillespie, D. T. (2002). The chemical Langevin and Fokker-Planck equations for the reversible isomerization reaction. *Journal of Physical Chemistry* 106:5063–5071.

Gillespie, D. T. (2007). Stochastic simulation of chemical kinetics. *Annual Review of Physical Chemistry* 58 (1): 35–55.

Gillespie, D. T., and Petzold, L. (2003). Improved leap-size selection for accelerated stochastic simulation. *Journal of Chemical Physics* 119:8229–8234.

Gingle, A. R., and Robertson, A. (1976). The development of the relaying competence in *Dictyostelium discoideum. Journal of Cell Science* 20:21–27.

Goldberg, D. E. (1989). *Genetic Algorithms for Search, Optimization, and Machine Learning*. Reading, MA: Addison-Wesley.

Goldbeter, A. (1996). *Biochemical Oscillations and Cellular Rhythms: The Molecular Bases of Periodic and Chaotic Behaviour*. Cambridge: Cambridge University Press.

Goldbeter, A., and Koshland, D. E. (1981). An amplified sensitivity arising from covalent modification in biological systems. *Proceedings of the National Academy of Sciences USA* 78:6840–6844.

Gomez-Uribe, C. A., and Verghese, G. C. (2007). Mass fluctuation kinetics: Capturing stochastic effects in systems of chemical reactions through coupled mean-variance computations. *Journal of Chemical Physics* 126 (2): 024109–024112.

Gonçalves, J., and Warnick, S. (2008). Necessary and sufficient conditions for dynamical structure reconstruction of LTI networks. *IEEE Transactions on Automatic Control* 53 (7): 1670–1674.

Gonze, D., Bernard, S., Waltermann, C., Kramer, A., and Herzel, H. (2005). Spontaneous synchronization of coupled circadian oscillators. *Biophysical Journal* 89 (1): 120–129.

Goodwin, B. C. (1965). Oscillatory behavior in enzymatic control processes. *Advances in Enzyme Regulation* 3:425–438.

Görlich, D., Seewald, M. J., and Ribbeck, K. (2003). Characterization of Ran-driven cargo transport and the RanGTPase system by kinetic measurements and computer simulation. *EMBO Journal* 22:1088–1100.

Grigorova, I. L., Chaba, R., Zhong, H. J., Alba, B. M., Rhodius, V., Herman, C., and Gross, C. A. (2004). Fine-tuning of the *Escherichia coli* σ^E envelope stress response relies on multiple mechanisms to inhibit signal-independent proteolysis of the transmembrane anti-sigma factor, RseA. *Genes & Development* 18:2686–2697.

Grodins, F. S. (1963). *Control Theory and Biological Systems.* New York: Columbia University Press.

Grodins, F. S., Gray, J. S., Schroeder, K. R., Norins, A. L., and Jones, R. W. (1954). Respiratory responses to CO_2 inhalation. A theoretical study of a nonlinear biological regulator. *Journal of Applied Physiology* 7:283–308.

Gross, C. (1996). Function and regulation of the heat shock proteins. In F. C. Neidhardt et al., eds., Escherichia coli *and* Salmonella: *Cellular and Molecular Biology* 1:1382–1399. 2nd ed. Washington, DC: ASM Press.

Guckenheimer, J., and Holmes, P. (1986). *Nonlinear Oscillations, Dynamical Systems, and Bifurcations of Vector Fields.* 2nd ed. New York: Springer.

Haberman, R. (1983). *Elementary Applied Partial Differential Equations: with Fourier Series and Boundary Value Problems.* Englewood Cliffs, NJ: Prentice-Hall.

Härkegård, O., and Glad, S. T. (2005). Resolving actuator redundancy: Optimal control vs. control allocation. *Automatica* 41:137–144.

Hartman, P. (1963). On the local linearization of differential equations. *Proceedings of the American Mathematical Society* 14:568–573.

Hartwell, L., Hopfield, J., Leibler, S., and Murray, A. (1999). From molecular to modular cell biology. *Nature* 402:47–52.

Hastings, S., Tyson, J., and Webster, D. (1977). Existence of periodic solutions for negative feedback cellular systems. *Journal of Differential Equations* 25:39–64.

Hasty, J., McMillen, D., and Collins, J. J. (2002). Engineered gene circuits. *Nature* 420:224–230.

Hatzimanikatis, V., Floudas, C. A., and Bailey, J. E. (1996). Analysis and design of metabolic reaction networks via mixed-integer linear optimization. *AIChE Journal* 42:1277–1292.

Haustein, E., and Schwille, P. (2007). Fluorescence correlation spectroscopy: Novel variations of an established technique. *Annual Review of Biophysics & Biomolecular Structure* 36:151–169.

Heinrich, R., Neel, B. G., and Rapoport, T. A. (2002). Mathematical models of protein kinase signal transduction. *Molecular Cell* 9 (5): 957–970.

Heinrich, R., and Rapoport, T. A. (1974a). A linear steady-state treatment of enzymatic chains: General properties, control and effector strength. *European Journal of Biochemistry* 42:89–95.

Heinrich, R., and Rapoport, T. A. (1974b). A linear steady-state treatment of enzymatic chains: Critique of the crossover theorem and a general procedure to identify interaction sites with an effector. *European Journal of Biochemistry* 42:97–105.

Heinrich, R., and Schuster, S. (1996). *The Regulation of Cellular Systems.* New York: Chapman & Hall.

Hespanha, J. P. (2005). Polynomial stochastic hybrid systems. In M. Morari and L. Thiele, eds., *Hybrid Systems: Computation and Control*, 322–338. Berlin: Springer.

Hespanha, J. P. (2006). `StochDynTools`—a MATLAB toolbox to compute moment dynamics for stochastic networks of bio-chemical reactions. http://www.ece.ucsb.edu/~hespanha/.

Hirsch, M. (1983). Differential equations and convergence almost everywhere in strongly monotone flows. *Contemporary Mathematics* 17:267–285.

Hirsch, M., and Smith, H. L. (2005). Monotone dynamical systems. In *Handbook of Differential Equations, Ordinary Differential Equations.* Vol. 2. Amsterdam: Elsevier.

Hodgkin, A. L., and Huxley, A. F. (1952). A quantitative description of membrane current and its application to conduction and excitation in nerve. *Journal of Physiology* 117 (4): 500–544.

Hofmeyr, J.-H. S. (2001). Metabolic control analysis in a nutshell. In *Proceedings of the International Conference on Systems Biology*, 291–300. Pasadena, CA: ICSB.

Hofmeyr, J.-H. S., and Cornish-Bowden, A. (1996). Co-response analysis: A new experimental strategy for metabolic control analysis. *Journal of Theoretical Biology* 182:371–380.

Horn, F. J. M. (1974). The dynamics of open reaction systems. In *Proceedings of the SIAM-AMS Symposium on Applied Mathematics* 8:125–137. Providence, RI: AMS.

Horn, R. A., and Johnson, C. R. (1985). *Matrix Analysis*. Cambridge: Cambridge University Press.

Hornberg, J. J., Bruggeman, F. J., Binder, B., Geest, C. R., De Vaate, A. J. M. B., Lankelma, J., Heinrich, R., and Westerhoff, H. V. (2005). Principles behind the multifarious control of signal transduction: ERK phosphorylation and kinase/phosphatase control. *FEBS Journal* 272 (1): 244–258.

Huang, C. Y., and Ferrell, J. E. (1996). Ultrasensitivity in the mitogen-activated protein kinase cascade. *Proceedings of the National Academy of Sciences USA* 93:10078–10083.

Hufnagel, L., Teleman, A. A., Rouault, H., Cohen, S. M., and Shraiman, B. I. (2007). On the mechanism of wing size determination in fly development. *Proceedings of the National Academy of Sciences USA* 104:3835–3840.

Ideker, T., Galitski, T., and Hood, L. (2001). A new approach to decoding life: Systems biology. *Annual Review Genomics and Human Genetics* 2:343–372.

Isaacs, F. J., Hasty, J., Cantor, C. R., and Collins, J. J. (2003). Prediction and measurement of an autoregulatory genetic module. *Proceedings of the National Academy of Sciences USA* 100:7714–7719.

Jacob, F., and Monod, J. (1961). Genetic regulatory mechanisms in the synthesis of proteins. *Journal of Molecular Biology* 3:318–56.

Jacobsen, E. W., and Cedersund, G. (2008). Structural robustness of biochemical network models-with application to the oscillatory metabolism of activated neutrophils. *IET Systems Biology* 2 (1): 39–47.

Kacser, H., and Acerenza, L. (1993). A universal method for achieving increases in metabolite production. *European Journal of Biochemistry* 216:361–367.

Kacser, H., and Burns, J. A. (1973). The control of flux. *Symposia Society for Experimental Biology* 27:65–104.

Kacser, H., and Small, J. R. (1994). A method for increasing the concentration of a specific internal metabolite in steady-state systems. *European Journal of Biochemistry* 226:649–656.

Kalman, R. E. (1960). A new approach to linear filtering and prediction problems. *Transactions of ASME, Journal of Basic Engineering* D 82:35–45.

Kalmus, H., ed. (1966). *Regulation and Control in Living Systems*. London: Wiley.

Kanemori, M., Nishihara, K., Yanagi, H., and Yura, T. (1997). Synergistic roles of HslVU and other ATP-dependent proteases in controlling in vivo turnover of σ^{32} and abnormal proteins in *Escherichia coli*. *Journal of Bacteriology* 179:7219–7225.

Keeling, M. J. (2000). Multiplicative moments and measures of persistence in ecology. *Journal of Theoretical Biology* 205:269–281.

Keener, J., and Sneyd, J. (2001). *Mathematical Physiology*. Vol. 8 of *Interdisciplinary Applied Mathematics*. 2nd ed. New York: Springer.

Khalil, H. K. (2002). *Nonlinear Systems*. 3rd ed. Upper Saddle River, NJ: Prentice-Hall.

Khammash, M., and El-Samad, H. (2005). Stochastic modeling and analysis of genetic networks. In *Conference on Decision and Control*, 2320–2325, Seville: IEEE.

Kholodenko, B. N. (2000). Negative feedback and ultrasensitivity can bring about oscillations in the mitogen-activated protein kinase cascades. *European Journal of Biochemistry* 267 (6): 1583–1588.

Kholodenko, B. N., Cascante, M., Hoek, J. B., Westerhoff, H. V., and Schwaber, J. (1998). Metabolic design: How to engineer a living cell to desired metabolite concentrations and fluxes. *Biotechnology and Bioengineering* 59:239–247.

Kholodenko, B. N., Demin, O. V., Moehren, G., and Hoek, J. B. (1999). Quantification of short-term signaling by the epidermal growth factor receptor. *Journal of Biological Chemistry* 274 (42): 30169–30181.

Kholodenko, B. N., Demin, O. V., and Westerhoff, H. V. (1997). Control analysis of periodic phenomena in biological systems. *Journal of Physical Chemistry* B 101 (11): 2070–2081.

Kholodenko, B. N., and Westerhoff, H. V., eds. (2004). *Metabolic Engineering in the Post Genomic Era.* Norfolk, UK: Horizon Biosciences.

Kholodenko, B. N., Westerhoff, H. V., Schwaber, J., and Cascante, M. (2000). Engineering a living cell to desired metabolite concentrations and fluxes: Pathways with multifunctional enzymes. *Metabolic Engineering* 2:1–13.

Kim, J., Bates, D. G., and Postlethwaite, I. (2006). Robustness analysis of linear periodic time-varying systems subject to structured uncertainty. *Systems & Control Letters* 55 (9): 719–725.

Kim, J., Bates, D. G., and Postlethwaite, I. (2008). Evaluation of stochastic effects on biomolecular networks using the generalized Nyquist stability criterion. *IEEE Transactions on Automatic Control* 53 (8): 1937–1941.

Kim, J., Bates, D. G., Postlethwaite, I., Ma, L., and Iglesias, P. A. (2006). Robustness analysis of biochemical network models. *Systems Biology* 153:96–104.

Kim, J., Heslop-Harrison, P., Postlethwaite, I., and Bates, D. G. (2007). Stochastic noise and synchronisation during *Dictyostelium* aggregation make cAMP oscillations robust. *PLoS Computational Biology* 3:e218.

Kindzelskii, A. L., and Petty, H. R. (2002). Apparent role of travelling metabolic waves in oxidant release by living neutrophils. *Proceedings of the National Academy Sciences USA* 99:9207–9212.

Kitano, H., ed. (2001). *Foundations of Systems Biology*. Cambridge, MA: MIT Press.

Kitano, H. (2002). Systems biology: A brief overview. *Science* 295:1662–1664.

Kitano, H. (2004). Biological robustness. *Nature Reviews Genetics* 5 (11): 826–837.

Kitano, H. (2007). Towards a theory of biological robustness. *Molecular Systems Biology* 3:137.

Klipp, E., Herwig, R., Kowald, A., Wierling, C., and Lehrach, H. (2005). *Systems Biology in Practice: Concepts, Implementation and Application*. Weinheim: Wiley.

Kokotović, P., Khalil, H. K., and O'Reilly, J. (1999). *Singular Perturbation Methods in Control: Analysis and Design*. Vol. 25 of *Classics in Applied Mathematics*. Philadelphia: SIAM.

Körner, T. W. (1988). *Fourier Analysis*. Cambridge: Cambridge University Press.

Kramer, M. A., Rabitz, H., and Calo, J. M. (1984). Sensitivity analysis of oscillatory systems. *Applied Mathematical Modelling* 8:328–340.

Kuramoto, Y. (1984). *Chemical Oscillations, Waves, and Turbulence*. Berlin: Springer.

Kurtz, T. G. (1978). Strong approximation theorems for density dependent Markov chains. *Stochastic Processes and Their Applications* 6 (3): 223–240.

Kwei, E. C., Sanft, K. R., Petzold, L. R., and Doyle, F. J., III (2008). Systems analysis of the insulin signaling pathway. In *Proceedings of the IFAC World Congress*, 17:15,891–15,896. Seoul: IFAC.

Kyriakis, J. M., and Avruch, J. (2001). Mammalian mitogen-activated protein kinase signal transduction pathways activated by stress and inflammation. *Physiological Reviews* 81 (2): 807–869.

Lamkanfi, M., Festjens, N., Declercq, W., Vanden Berghe, T., and Vandenabeele, P. (2007). Caspases in cell survival, proliferation and differentiation. *Cell Death and Differentiation* 14 (1): 44–55.

Lander, A. D. (2007). Morpheus unbound: Reimagining the morphogen gradient. *Cell* 128:245–256.

Laub, M. T., and Loomis, W. F. (1998). A molecular network that produces spontaneous oscillations in excitable cells of *Dictyostelium. Molecular Biology of the Cell* 9:3521–3532.

Lawrence, M. C., Jivan, A., Shao, C., Duan, L., Goad, D., Zaganjor, E., Osborne, J., McGlynn, K., Stippec, S., Earnest, S., Chen, W., and Cobb, M. H. (2008). The roles of MAPKs in disease. *Cell Research* 18 (4): 436–442.

Leloup, J. C., and Goldbeter, A. (2001). A molecular explanation for the long-term suppression of circadian rhythms by a single light pulse. *American Journal of Physiology: Regulatory, Integrative and Comparative Physiology* 280:1206–1212.

Leloup, J. C., and Goldbeter, A. (2003). Toward a detailed computational model for the mammalian circadian clock. *Proceedings of the National Academy of Sciences USA* 100:7051–7056.

Leloup, J. C., and Goldbeter, A. (2004). Modeling the mammalian circadian clock: sensitivity analysis and multiplicity of oscillatory mechanisms. *Journal of Theoretical Biology* 230 (4): 541–562.

Le Novère, N., Bornstein, B., Broicher, A., Courtot, M., Donizelli, M., Dharuri, H., Li, L., Sauro, H., Schilstra, M., Shapiro, B., Snoep, J., and Hucka, M. (2006). BioModels Database: A free, centralized database of curated, published, quantitative kinetic models of biochemical and cellular systems. *Nucleic Acids Research* 34:D689–D691.

Li, H. Y., Ng, W. P., Wong, C. H., Iglesias, P. A., and Zheng, Y. (2007). Coordination of chromosome alignment and mitotic progression by the chromosome-based Ran signal. *Cell Cycle* 6:1886–1895.

Li, S., and Petzold, L. (2000). Software and algorithms for sensitivity analysis of large-scale differential algebraic systems. *Journal of Computational and Applied Mathematics* 125 (1–2): 131–146.

Linkens, D. A., ed. (1979). *Biological Systems: Modelling and Control.* IEE Control Engineering Series, vol. 11. Stevenage, UK: Peregrinus.

Liu, R. T., Liaw, S. S., and Maini, P. K. (2006). Two-stage Turing model for generating pigment patterns on the leopard and the jaguar. *Physical Review E: Statistical, Nonlinear, and Soft Matter Physics* 74:011914.

Ljung, L. (1987). *System Identification: Theory for the User*. Englewood Cliffs, NJ: Prentice-Hall.

Ljung, L. (2003). Challenges of non-linear identification. Bode Lecture, 42nd Conference on Decision and Control, Maui, HI.

Ljung, L., and Glad, T. (1994). On global identifiability for arbitrary model parametrization. *Automatica* 30 (2): 265–276.

Llorens, M., Nuño, J. C., Rodríguez, Y., Meléndez-Hevia, E., and Montero, F. (1999). Generalization of the theory of transition times in metabolic pathways: A geometrical approach. *Biophysical Journal* 77 (1): 23–36.

Lobo, F., and Goldberg, D. (1996). Decision making in a hybrid genetic algorithm. Technical report. IlliGAL Report no. 96009.

Locke, J. C., Westermark, P. O., Kramer, A., and Herzel, H. (2008). Global parameter search reveals design principles of the mammalian circadian clock. *BMC Systems Biology* 2:22.

Ma, L., and Iglesias, P. A. (2002). Quantifying robustness of biochemical network models. *BMC Bioinformatics* 3:38.

Maertens, G., Vercammen, J., Debyser, Z., and Engelborghs, Y. (2005). Measuring protein-protein interactions inside living cells using single color fluorescence correlation spectroscopy. Application to human immunodeficiency virus type 1 integrase and LEDGF/p75. *FASEB Journal* 19:1039–1041.

Maini, P. K., Baker, R. E., and Chuong, C. M. (2006). Developmental biology: The Turing model comes of molecular age. *Science* 314:1397–1398.

Mallet-Paret, J., and Smith, H. L. (1990). The Poincaré-Bendixson theorem for monotone cyclic feedback systems. *Journal of Dynamics and Differential Equations* 2:367–421.

Markevich, N. I., Hoek, J. B., and Kholodenko, B. N. (2004). Signaling switches and bistability arising from multisite phosphorylation in protein kinase cascades. *Journal of Cell Biology* 164 (3): 353–359.

Marshall, C. J. (1995). Specificity of receptor tyrosine kinase signaling: Transient versus sustained extracellular signal-regulated kinase activation. *Cell* 80 (2): 179–185.

MathWorks (2003). *MATLAB User's Guide*. Natick, MA.

MathWorks (2006). *MATLAB Optimization Toolbox*. 3rd ed. Natick, MA.

Maywood, E. S., Reddy, A. B., Wong, G. K., O'Neill, J. S., O'Brien, J. A., McMahon, D. G., Harmar, A. J., Okamura, H., and Hastings, M. H. (2006). Synchronization and maintenance of timekeeping in suprachiasmatic circadian clock cells by neuropeptidergic signaling. *Current Biology* 16:599–605.

McAdams, H., and Arkin, A. (1997). Stochastic mechanisms in gene expression. *Proceedings of the National Academy of Sciences USA* 94:814–819.

McAdams, H. H., and Arkin, A. (1999). It's a noisy business! Genetic regulation at the nanomolar scale. *Trends in Genetics* 15 (2): 65–69.

McClean, M. N., Mody, A., Broach, J. R., and Ramanathan, S. (2007). Cross-talk and decision making in MAP kinase pathways. *Nature Genetics* 39 (3): 409–414.

McQuarrie, D. (1967). Stochastic approach to chemical kinetics. *Journal of Applied Probability* 4:413–478.

Meinhardt, H. (1982). *Models of Biological Pattern Formation.* London: Academic Press.

Menon, P. P., Kim, J., Bates, D. G., and Postlethwaite, I. (2006). Clearance of nonlinear flight control laws using hybrid evolutionary optimisation. *IEEE Transactions on Evolutionary Computation* 10 (6): 689–699.

Mettetal, J. T., and van Oudenaarden, A. (2007). Necessary noise. *Science* 317:463–464.

Meyers, J., Craig, J., and Odde, D. J. (2006). Potential for control of signaling pathways via cell size and shape. *Current Biology* 16:1685–1693.

Milhorn, H. T. (1966). *The Application of Control Theory to Physiological Systems.* Philadelphia: Saunders.

Millat, T., Bullinger, E., Rohwer, J., and Wolkenhauer, O. (2007). Approximations and their consequences for dynamic modelling of signal transduction pathways. *Mathematical Biosciences* 207 (1): 40–57.

Moles, C. G., Mendes, P., and Banga, J. R. (2003). Parameter estimation in biochemical pathways: A comparison of global optimization methods. *Genome Research* 13 (11): 2467–2474.

Morgan, T. H. (1901). *Regeneration.* New York: Macmillan.

Morita, M., Kanemori, M., Yanagi, H., and Yura, T. (1999). Heat-induced synthesis of σ^{32} in *Escherichia coli*: Structural and functional dissection of rpoH mRNA secondary structure. *Journal of Bacteriology* 181:401–410.

Morita, M. T., Kanemori, M., Yanagi, H., and Yura, T. (2000). Dynamic interplay between antagonistic pathways controlling the σ^{32} level in *Escherichia coli. Proceedings of the National Academy of Sciences USA* 97:5860–5865.

Morohashi, M., Winn, A. E., Borisuk, M. T., Bolouri, H., Doyle, J., and Kitano, H. (2002). Robustness as a measure of plausibility in models of biochemical networks. *Journal of Theoretical Biology* 216 (1): 19–30.

Munsky, B., Hernday, A. Low, D., and Khammash, M. (2005). Stochastic modeling of the Pap pili epigenetic switch. In *Proceedings of the Foundations of Systems Biology in Engineering*, 145–148. Santa Barbara, CA: FOSBE.

Munsky, B., and Khammash, M. (2006). The finite state projection algorithm for the solution of the chemical master equation. *Journal of Chemical Physics* 124 (4): 044104.

Munsky, B., and Khammash, M. (2008). The finite state projection approach for the analysis of stochastic noise in gene networks. *IEEE Transactions on Automatic Control* 53:201–214.

Nagrath, I. J., and Gopal, M. (1986). *Control Systems Engineering.* 2nd ed. New Delhi: Wiley Eastern.

Nasell, I. (2003a). An extension of the moment closure method. *Theoretical Population Biology* 64:233–239.

Nasell, I. (2003b). Moment closure and the stochastic logistic model. *Theoretical Population Biology* 63:159–168.

Newman, S. A., Christley, S., Glimm, T., Hentschel, H. G., Kazmierczak, B., Zhang, Y. T., Zhu, J., and Alber, M. (2008). Multiscale models for vertebrate limb development. *Current Topics in Developmental Biology* 81:311–340.

Noble, D. (1960). Cardiac action and pacemaker potentials based on the Hodgkin-Huxley equations. *Nature* 188:495–497.

Olsen, L. F., Kummer, U., Kindzelskii, A. L., and Petty, H. R. (2003). A model of the oscillatory metabolism of activated neutrophils. *Biophysical Journal* 84 (4): 69–81.

Orton, R. J., Strum, O. E., Vysheiresky, V., Calder, M., Gilbert, D. R., and Kolch, W. (2005). Computational modelling of the receptor-tyrosine-kinase-activated MAPK pathway. *Biochemical Journal* 392 (2): 249–261.

Othmer, H. G., and Schaap, P. (1998). Oscillatory cAMP signalling in the development of *Dictyostelium discoideum*. *Comments on Theoretical Biology* 5:175–282.

Painter, K. J., Maini, P. K., and Othmer, H. G. (1999). Stripe formation in juvenile *Pomacanthus* explained by a generalized Turing mechanism with chemotaxis. *Proceedings of the National Academy of Sciences USA* 96:5549–5554.

Pardee, A. B., and Reddy, G. P.-V. (2003). Beginnings of feedback inhibition, allostery, and multi-protein complexes. *Gene* 321:17–23.

Parent, C. A., and Devreotes, P. N. (1996). Molecular genetics of signal transduction in *Dictyostelium*. *Annual Review Biochemistry* 65:411–440.

Paulsson, J. (2004). Summing up the noise in gene networks. *Nature* 427:415–418.

Paulsson, J., Berg, O. G., and Ehrenberg, M. (2000). Stochastic focusing: Fluctuation-enhanced sensitivity of intracellular regulation. *Proceedings of the National Academy of Sciences USA* 97:7148–7153.

Pearson, G., Robinson, F., Gibson, T. B., Xu, B., Karandikar, M., Berman, K., and Cobb, M. H. (2001). Mitogen-activated protein (MAP) kinase pathways: Regulation and physiological functions. *Endocrine Reviews* 22 (2): 153–183.

Peifer, M., and Timmer, J. (2007). Parameter estimation in ordinary differential equations for biochemical processes using the method of multiple shooting. *IET Systems Biology* 1 (2): 78–88.

Peles, S., Munsky, B., and Khammash, M. (2006). Reduction and solution of the chemical master equation using time scale separation and finite state projection. *Journal of Chemical Physics* 20:204104.

Petty, H. R. (2000). Neutrophil oscillations: Temporal and spatiotemporal aspects of cell behavior. *Immunologic Research* 23:85–94.

Pircher, T. J., Petersen, H., Gustafsson, J.-Å., and Haldosén, L.-A. (1999). Extracellular signal-regulated kinase (ERK) interacts with signal transducer and activator of transcription (STAT) 5a. *Molecular Endocrinology* 13 (4): 555–565.

Polisetty, P. K., Voit, E. O., and Gatzke, E. P. (2006). Identification of metabolic system parameters using global optimization methods. *Theoretical Biology and Medical Modelling* 3:4.

Potma, E. O., de Boeij, W. P., Bosgraaf, L., Roelofs, J., van Haastert, P. J., and Wiersma, D. A. (2001). Reduced protein diffusion rate by cytoskeleton in vegetative and polarized *Dictyostelium* cells. *Biophysical Journal* 81:2010–2019.

Price, N. D., and Shmulevich, I. (2007). Biochemical and statistical network models for systems biology. *Current Opinion in Biotechnology* 18 (4): 365–370.

Qiao, L., Nachbar, R. B., Kevrekidis, I. G., and Shvartsman, S. Y. (2007). Bistability and oscillations in the Huang-Ferrell model of MAPK signaling. *PLoS Computational Biology* 3 (9): e184.

Rai, N., Ramkumar, K., Venkatesh, K. V., Dhabolkar, S., Thattai, M., and Sreenivasan, V. (2008). Inferring closed-loop responses from open-loop characteristics for a family of synthetic transcriptional feedback systems. Paper presented at IET BioSysBio 2008, London, April 20–22.

Raman, M., Chen, W., and Cobb, M. H. (2007). Differential regulation and properties of MAPKs. *Oncogene* 26 (22): 3100–3112.

Raman, R. K., Hashimoto, Y., Cohen, M. H., and Robertson, A. (1976). Differentiation for aggregation in the cellular slime moulds: The emergence of autonomously signalling cells in *Dictyostelium discoideum*. *Journal of Cell Science* 21:243–259.

Rapp, P. E. (1975). A theoretical investigation of a large class of biochemical oscillators. *Mathematical Biosciences* 25:165–188.

Rathinam, M., Petzold, L. R., Cao, Y., and Gillespie, D. T. (2005). Consistency and stability of tau-leaping schemes for chemical reaction systems. *Multiscale Modeling and Simulation* 4:867–895.

Rathinam, M. M., Petzold, L. R., Cao, Y., and Gillespie, D. T. (2003). Stiffness in stochastic chemically reacting systems: The implicit tau-leaping method. *Journal of Chemical Physics* 119:12784–12794.

Reddy, V. N., Mavrovouniotis, M. L., and Liebman, M. N. (1993). Petri net representations in metabolic pathways. In *Proceedings of the 1st International Conference on Intelligent Systems for Molecular Biology*, 328–336. Bethesda, MD: AAAI Press.

Reder, C. (1988). Metabolic control theory: A structural approach. *Journal of Theoretical Biology* 135:175–201.

Reed, J. L., Vo, T. D., Schilling, C. H., and Palsson, B. O. (2003). An expanded genome-scale model of *Escherichia coli* K-12 (iJR904 GSM/GPR). *Genome Biology* 4:R54.1–12.

Reppert, S. M., and Weaver, D. R. (2002). Coordination of circadian timing in mammals. *Nature* 418:935–941.

Rosenfeld, N., Elowitz, M., and Alon, U. (2002). Negative autoregulation speeds the response times of transcription networks. *Journal of Molecular Biology* 323:785–793.

Rosenwasser, E., and Yusupov, R. (2000). *Sensitivity of Automatic Control Systems*. New York: CDC Press.

Rubertis, G. D., and Davies, S. W. (2003). A genetic circuit amplifier: Design and simulation. *IEEE Transactions on Nanobioscience* 2 (4): 239–246.

Rugh, W. J. (1996). *Linear System Theory*. Prentice Hall Information and System Sciences Series. 2nd ed. Upper Saddle River, NJ: Prentice Hall.

Sabbagh, W., Flatauer, L. J., Bardwell, A. J., and Bardwell, L. (2001). Specificity of MAP kinase signaling in yeast differentiation involves transient versus sustained MAPK activation. *Molecular Cell* 8:683–691.

Saez-Rodriguez, J., Kremling, A., Conzelmann, H., Bettenbrock, K., and Gilles, E. D. (2004). Modular analysis of signal transduction networks. *IEEE Control Systems Magazine* 24:35–52.

Saez-Rodriguez, J., Kremling, A., and Gilles, E. D. (2005). Dissecting the puzzle of life: Modularization of signal transduction networks. *Computers and Chemical Engineering* 29:619–629.

Saha, K., and Schaffer, D. V. (2006). Signal dynamics in Sonic hedgehog tissue patterning. *Development* 133:889–900.

Samoilov, M., Plyasunov, S., and Arkin, A. P. (2005). Stochastic amplification and signaling in enzymatic futile cycles through noise-induced bistability with oscillations. *Proceedings of the National Academy of Sciences USA* 102:2310–2315.

Sasagawa, S., Ozaki, Y.-I., Fujita, K., and Kuroda, S. (2005). Prediction and validation of the distinct dynamics of transient and sustained ERK activation. *Nature Cell Biology* 7 (4): 365–373.

Sauro, H. (2004). The computational versatility of proteomic signaling networks. *Current Proteomics* 1 (1): 67–81.

Sauro, H. M., and Ingalls, B. (2007). MAPK cascades as feedback amplifiers. Technical report http://arxiv.org/abs/0710.5195.

Sauro, H. M., and Kholodenko, B. N. (2004). Quantitative analysis of signaling networks. *Progress in Biophysics & Molecular Biology* 86:5–43.

Savageau, M. A. (1972). The behavior of intact biochemical control systems. *Current Topics Cellular Regulation* 6:63–130.

Savageau, M. A. (1976). *Biochemical Systems Analysis: A Study of Function and Design in Molecular Biology*. Reading, MA: Addison-Wesley.

Schaffner, J., and Zeitz, M. (1999). Variants of nonlinear normal form observer design. In H. Nijmeijer and T. I. Fossen, eds., *New Directions in Nonlinear Observer Design*, 161–180. London: Springer.

Schmidt, H., and Jacobsen, E. W. (2004). Linear systems approach to analysis of complex dynamic behaviours in biochemical networks. *Systems Biology* 1:149–158.

Schnell, S., and Mendoza, C. (1997). Closed form solution for time-dependent enzyme kinetics. *Journal of Theoretical Biology* 187 (2): 207–212.

Schoeberl, B., Eichler-Jonsson, C., Gilles, E. D., and Müller, G. (2002). Computational modeling of the dynamics of the MAP kinase cascade activated by surface and internalized EGF receptors. *Nature Biotechnology* 20 (4): 370–375.

Schrödinger, E. (1944). *What Is Life? The Physical Aspect of the Living Cell*. Cambridge: Cambridge University Press.

Schwarz, G. (1968). Kinetic analysis by chemical relaxation methods. *Review of Modern Physics* 40 (1): 206–218.

Scott, M., Hwa, T., and Ingalls, B. (2007). Deterministic characterization of stochastic genetic circuits. *Proceedings of the National Academy of Sciences USA* 104:7402–7407.

Seborg, D., Edgar, T., and Mellichamp, D. (1989). *Process Dynamics and Control*. 2nd ed. New York: Wiley.

Sedaghat, A. R., Sherman, A., and Quon, M. J. (2002). A mathematical model of metabolic insulin signaling pathways. *American Journal of Physiology: Endocrinology and Metabolism* 283:E1084–1101.

Segel, I. H. (1993). *Enzyme Kinetics: Behavior and Analysis of Rapid Equilibrium and Steady State Enzyme Systems*. New York: Wiley.

Seger, R., and Krebs, E. G. (1995). The MAPK signaling cascade. *FASEB Journal* 9 (9): 726–735.

Seydel, R. (1994). *Practical Bifurcation and Stability Analysis: From Equilibrium to Chaos*. Vol. 5 of *Interdisciplinary Applied Mathematics*. 2nd ed. New York: Springer.

Shao, J., and Tu, D. (1995). *The Jackknife and Bootstrap*. New York: Springer.

Shoemaker, J. E., and Doyle, F. J., III (2008). Identifying fragilities in biochemical networks: Robust performance analysis of Fas signaling-induced apoptosis. *Biophysical Journal* 95:2610–2623.

Sick, S., Reinker, S., Timmer, J., and Schlake, T. (2006). WNT and DKK determine hair follicle spacing through a reaction-diffusion mechanism. *Science* 314:1447–1450.

Singer, A. B., Taylor, J. W., Barton, P. I., and Green, W. H. (2006). Global dynamic optimization for parameter estimation in chemical kinetics. *Journal of Physical Chemistry* A 110 (3): 971–976.

Singh, A., and Hespanha, J. P. (2006). Moment closure techniques for stochastic models in population biology. In *Proceedings of the American Control Conference*, 4730–4735. Minneapolis: IEEE.

Singh, A., and Hespanha, J. P. (2007a). A derivative matching approach to moment closure for the stochastic logistic model. *Bulletin of Mathematical Biology* 69:1909–1925.

Singh, A., and Hespanha, J. P. (2007b). Stochastic analysis of gene regulatory networks using moment closure. In *Proceedings of the American Control Conference*, 1299–1304. New York: IEEE.

Singh, A., and Hespanha, J. P. (2008). Scaling of stochasticity in gene cascades. In *Proceedings of the American Control Conference*, 2780–2785. Seattle: IEEE.

Skogestad, S., and Postlethwaite, I. (2005). *Multivariable Feedback Control: Analysis and Design*. 2nd ed. Chichester: Wiley.

Smith, H. (1995). *Monotone Dynamical Systems: An Introduction to the Theory of Competitive and Cooperative Systems*. Vol. 41 of *Mathematical Surveys and Monographs*. Providence, RI: AMS.

Sontag, E. D. (2001). Structure and stability of certain chemical networks and applications to the kinetic proofreading model of T-cell receptor signal transduction. *IEEE Transactions on Automatic Control* 46 (7): 1028–1047.

Sontag, E. D. (2005). Molecular systems biology and control. *European Journal of Control* 11 (4–5): 396–435.

Sontag, E. D. (2006). Passivity gains and the "secant condition" for stability. *Systems & Control Letters* 55 (3): 177–183.

Sontag, E. D. (2007). Monotone and near-monotone biochemical networks. *Systems and Synthetic Biology* 1:59–87.

Sontag, E. D., Kiyatkin, A., and Kholodenko, B. N. (2004). Inferring dynamic architecture of cellular networks using time series of gene expression, protein and metabolite data. *Bioinformatics* 20 (12): 1877–1886.

Soumpasis, D. M. (1983). Theoretical analysis of fluorescence photobleaching recovery experiments. *Biophysical Journal* 41:95–97.

Spall, J. C. (2003). *Introduction to Stochastic Search and Optimization: Estimation, Simulation, and Control*. Hoboken, NJ: Wiley-Interscience.

Stelling, J., Gilles, E. D., and Doyle, F. J. (2004a). Robustness properties of circadian clock architectures. *Proceedings of the National Academy of Sciences USA* 101:13210–13215.

Stelling, J., Sauer, U., Szallasi, Z., Doyle, F. J., and Doyle, J. (2004b). Robustness of cellular functions. *Cell* 118:675–685.

Stephanopoulos, G., Aristidou, A., and Nielsen, J. (1998). *Metabolic Engineering: Principles and Applications.* San Diego, CA: Academic Press.

Straus, D., Walter, W., and Gross, C. A. (1990). DnaK, DnaJ, and GrpE heat shock proteins negatively regulate heat shock gene expression by controlling the synthesis and stability of σ^{32}. *Genes & Development* 4:2202–2209.

Swain, P. (2004). Efficient attenuation of stochasticity in gene expression through post-transcriptional control. *Journal of Molecular Biology* 344:965–976.

Swain, P., Elowitz, M., and Siggia, E. (2002). Intrinsic and extrinsic contributions to stochasticity in gene expression. *Proceedings of the National Academy of Sciences USA* 99:12795–12800.

Tatsuta, T., Joob, D. M., Calendar, R., Akiyama, Y., and Ogura, T. (2000). Evidence for an active role of the DnaK chaperone system in the degradation of σ^{32}. *FEBS Letters* 478:271–275.

Taylor, S., Gunawan, R., Petzold, L., and Doyle, F. J., III (2008a). Sensitivity measures for oscillating systems: Application to mammalian circadian gene network. *IEEE Transactions on Automatic Control* 53:177–188.

Taylor, S. R., Doyle, F. J., and Petzold, L. R. (2008b). Oscillator model reduction preserving the phase response: application to the circadian clock. *Biophysical Journal* 95 (4): 1658–1673.

Taylor, S. R., Gadkar, K., Gunawan, R., and Doyle, F. J., III (2008c). BioSens: A sensitivity analysis toolkit for Bio-SPICE. University of California, Santa Barbara.

Teusink, B., Walsh, M. C., van Dam, K., and Westerhoff, H. V. (1998). The danger of metabolic pathways with turbo design. *Trends in Biochemical Sciences* 23:162–169.

Thattai, M. (2007). From open-loop characteristics to closed-loop responses: Experimental validation of a powerful bottom-up design principle. Best Modeling Prize project, International Genetically Engineered Machine competition, MIT. November.

Thattai, M., and van Oudenaarden, A. (2001). Intrinsic noise in gene regulatory networks. *Proceedings of the National Academy of Sciences USA* 98:8614–8619.

Thorsley, D., and Klavins, E. (2008). Model reduction of stochastic processes using Wasserstein pseudometrics. In *Proceedings of the American Control Conference*, 1374–1381, Seattle: IEEE.

To, T. L., Henson, M. A., Herzog, E. D., and Doyle, F. J. (2007). A molecular model for intercellular synchronization in the mammalian circadian clock. *Biophysical Journal* 92:3792–3803.

Tomioka, R., Kimura, H., Koboyashi, T. J., and Aihara, K. (2004). Multivariate analysis of noise in genetic regulatory networks. *Journal of Theoretical Biology* 229 (3): 501–521.

Tomović, R. (1963). *Sensitivity Analysis of Dynamic Systems.* New York: McGraw-Hill.

Tong, A. H., Evangelista, M., Parsons, A. B., Xu, H., Bader, G. D., Pagé, N., Robinson, M., Raghibizadeh, S., Hogue, C. W., Bussey, H., Andrews, B., Tyers, M., and Boone, C. (2001). Systematic genetic analysis with ordered arrays of yeast deletion mutants. *Science* 294:2364–2368.

Torres, N. V., and Voit, E. O. (2002). *Pathway Analysis and Optimization in Metabolic Engineering.* Cambridge: Cambridge University Press.

Turing, A. M. (1952). The chemical basis of morphogenesis. *Philosophical Transactions of the Royal Society* B 237:37–72.

Vallender, S. S. (1974). Calculation of the Wasserstein distance between probability distributions on the line. *Theory of Probability and Its Applications* 18 (4): 784–786.

van Breugel, F., and Worrell, J. (2006). Approximating and computing behavioural distances in probabilistic transition systems. *Journal of Theoretical Computer Science* 360:373–385.

van Kampen, N. G. (1981). *Stochastic Processes in Physics and Chemistry.* Amsterdam: North-Holland.

Vargas, A., and Moreno, J. A. (2000). Approximate high-gain observers for uniformly observable nonlinear systems. In *Conference on Decision and Control*, 784–789. Sydney: IEEE.

Vargas, A., and Moreno, J. A. (2005). Approximate high-gain observers for non-Lipschitz observability forms. *International Journal of Control* 78 (4): 247–253.

Varma, A., Morbidelli, M., and Wu, H. (1999). *Parametric Sensitivity in Chemical Systems.* Cambridge: Cambridge University Press.

Vaudry, D., Stork, P. J. S., Lazarovici, P., and Eiden, L. E. (2002). Signaling pathways for PC12 cell differentiation: Making the right connections. *Science* 296:1648–1649.

Vayttaden, S. J., Ajay, S. M., and Bhalla, U. S. (2004). A spectrum of models of signaling pathways. *ChemBioChem* 5 (10): 1365–1374.

Vershik, A. M. (2006). Kantorovich metric: Initial history and little-known applications. *Journal of Mathematical Science* 133 (4): 1410–1417.

Vilar, J. M. G., Kueh, H. Y., Barkai, N., and Leibler, S. (2002). Mechanisms of noise-resistance in genetic oscillators. *Proceedings of the National Academy of Sciences USA* 99:5988–5992.

Villa-Komaroff, L., Efstratiadis, A., Broome, S., Lomedico, P., Tizard, R., Naber, S. P., Chick, W. L., and Gilbert, W. (1978). A bacterial clone synthesizing proinsulin. *Proceedings of the National Academy of Sciences USA* 75:3727–3731.

Wagner, A. (2005). Circuit topology and the evolution of robustness in two-gene circadian oscillators. *Proceedings of the National Academy of Sciences USA* 102:11775–11780.

Wang, L., and Sontag, E. D. (2008). Singularly perturbed monotone systems and an application to double phosphorylation cycles. *Journal of Nonlinear Science* 18 (5): 527–550.

Wang, X., Hao, N., Dohlman, H. G., and Elston, T. C. (2006). Bistability, stochasticity, and oscillations in the mitogen-activated protein kinase cascade. *Biophysical Journal* 90 (6): 1961–1978.

Westerhoff, H. V., and Chen, Y.-D. (1984). How do enzyme activities control metabolite concentrations? An additional theorem in the theory of metabolic control. *European Journal of Biochemistry* 142:425–430.

Westerhoff, H. V., Hofmeyr, J.-H., and Kholodenko, B. N. (1994). Getting to the inside of cells using metabolic control analysis. *Biophysical Chemistry* 50:273–283.

Westerhoff, H. V., and Kell, D. B. (1996). What biotechnologists knew all along . . . ? *Journal of Theoretical Biology* 182:411–420.

Whittle, P. (1957). On the use of the normal approximation in the treatment of stochastic processes. *Journal of the Royal Statistical Society* B 19:268–281.

Widmann, C., Gibson, S., Jarpe, M. B., and Johnson, G. L. (1999). Mitogen-activated protein kinase: Conservation of a three-kinase module from yeast to human. *Physiological Reviews* 79 (1): 143–180.

Wiener, N. (1948). *Cybernetics; or, Control and Communication in the Animal and the Machine*. New York: Wiley.

Wilhelm, T., Behre, J., and Schuster, S. (2004). Analysis of structural robustness of metabolic networks. *Systems Biology* 1 (1): 114–120.

Willems, J. C. (1999). Behaviors, latent variables, and interconnections. *Systems, Control and Information* 43 (9): 453–464.

Williams, R. S. B., Boeckeler, K., Graf, R., Muller-Taubenberger, A., Li, Z., Isberg, R. R., Wessels, D., Soll, D. R., Alexander, H., and Alexander, S. (2006). Towards a molecular understanding of human diseases using *Dictyostelium discoideum*. *Trends in Molecular Medicine* 12 (9): 415–424.

Winfree, A. T. (2001). *The Geometry of Biological Time*. 2nd ed. New York: Springer.

Wolpert, L. (1969). Positional information and the spatial pattern of cellular differentiation. *Journal of Theoretical Biology* 25:1–47.

Wong, W. W., Tsai, T. Y., and Liao, J. C. (2007). Single-cell zeroth-order protein degradation enhances the robustness of synthetic oscillator. *Molecular Systems Biology* 3:130.

Xia, X., and Moog, C. H. (2003). Identifiability of nonlinear systems with application to HIV/AIDS models. *IEEE Transactions on Automatic Control* 48 (2): 330–336.

Xia, X., and Zeitz, M. (1997). On nonlinear continuous observers. *International Journal of Control* 66:943–954.

Xiong, W., and Ferrell, J. E., Jr. (2003). A positive-feedback-based bistable memory module that governs a cell fate decision. *Nature* 426:460–465.

Yen, J., Randolph, D., Liao, J. C., and Lee, B. (1995). A hybrid approach to modeling metabolic systems using genetic algorithm and simplex method. In *Proceedings of the 11th Conference on Artificial Intelligence for Applications*, 277–283. Benalmadena, Spain: IEEE Computer Society Press.

Yoda, M., Ushikubo, T., Inoue, W., and Sasai, M. (2007). Roles of noise in single and coupled multiple genetic oscillators. *Journal of Chemical Physics* 126:115101.

Zak, D. E., Gonye, G. E., Schwaber, J. S., and Doyle, F. J., III (2003). Importance of input perturbations and stochastic gene expression in the reverse engineering of genetic regulatory networks: Insights from an identifiability analysis of an in silico network. *Genome Research* 13 (11): 2396–2405.

Zevedei-Oancea, I., and Schuster, S. (2003). Topological analysis of metabolic networks based on Petri net theory. *In Silico Biology* 3:0029.

Contributors

David Angeli
Department of Systems and Information
University of Florence
angeli@dsi.unifi.it.

Declan G. Bates
Control and Instrumentation Research Group
Department of Engineering
University of Leicester
dgb3@leicester.ac.uk.

Eric Bullinger
Industrial Control Centre
Department of Electronic and Electrical Engineering
University of Strathclyde, Glasgow
eric.bullinger@eee.strath.ac.uk.

Peter S. Chang
Department of Chemical Engineering
University of California, Santa Barbara
psochang@engr.ucsb.edu.

Domitilla Del Vecchio
Department of Electrical Engineering and Computer Science
University of Michigan, Ann Arbor
ddv@umich.edu.

Francis J. Doyle III
Department of Chemical Engineering
University of California, Santa Barbara
doyle@engineering.ucsb.edu.

Hana El-Samad
Department of Biochemistry and Biophysics
University of California, San Francisco
helsamad@biochem.ucsf.edu.

Dirk Fey
Industrial Control Centre
Department of Electronic and Electrical Engineering
University of Strathclyde, Glasgow
dirk.fey@eee.strath.ac.uk.

Rolf Findeisen
Institute for Automation Engineering
Otto-von-Guericke University, Magdeburg
rolf.findeisen@ovgu.de.

Simone Frey
Systems Biology and Bioinformatics Group
Department of Computer Science
University of Rostock
frey@informatik.uni-rostock.de.

Jorge Gonçalves
Control Group
Department of Engineering
University of Cambridge
jmg77@cam.ac.uk.

Pablo A. Iglesias
Department of Electrical and Computer Engineering
Johns Hopkins University, Baltimore
pi@jhu.edu.

Brian P. Ingalls
Department of Mathematics
University of Waterloo
bingalls@math.uwaterloo.ca.

Elling W. Jacobsen
Automatic Control Lab
School of Electrical Engineering, KTH
Royal Institute of Technology,
Stockholm
jacobsen@s3.kth.se.

Mustafa Khammash
Department of Mechanical Engineering
University of California, Santa Barbara
khammash@engineering.ucsb.edu.

Jongrae Kim
Department of Aerospace Engineering
University of Glasgow
jkim@aero.gla.ac.uk.

Eric Klavins
Department of Electrical Engineering
University of Washington, Seattle
klavins@ee.washington.edu.

Eric C. Kwei
Department of Chemical Engineering
University of California, Santa Barbara
kwei@engineering.ucsb.edu.

Thomas Millat
Systems Biology and Bioinformatics Group
Department of Computer Science
University of Rostock
thomas.millat@informatik.uni-rostock.de.

Jason E. Shoemaker
Department of Chemical Engineering
University of California, Santa Barbara
jshoe@engineering.ucsb.edu.

Eduardo D. Sontag
Department of Mathematics
Rutgers, State University of New Jersey, Piscataway
sontag@math.rutgers.edu.

Stephanie R. Taylor
Department of Chemical Engineering
University of California, Santa Barbara
staylor@cs.ucsb.edu.

David Thorsley
Department of Electrical Engineering
University of Washington, Seattle
thorsley@u.washington.edu.

Camilla Trané
Automatic Control Lab
School of Electrical Engineering, KTH
Royal Institute of Technology,
Stockholm
camilla.trane@ee.kth.se.

Sean Warnick
Department of Computer Science
Brigham Young University, Provo, Utah
sean.warnick@gmail.com.

Olaf Wolkenhauer
Systems Biology and Bioinformatics Group
Department of Computer Science
University of Rostock
ow@informatik.uni-rostock.de.

Index